35th International Symposium on Superconductivity (ISS 2022)

Journal of Physics: Conference Series Volume 2545

Nagoya, Japan
29 November - 1 December 2022

Editors:

Atsutaka Maeda
Hiraku Ogino
Akiyoshi Matsumoto
Yasuhiro Iijima
Yasunori Mawatari
Hirotaka Terai
Hirotake Yamamori
Naoyuki Amemiya
Tsuyoshi Wakuda
Hiroshi Miyazaki

ISBN: 978-1-7138-7751-6
ISSN: 1742-6588

Printed from e-media with permission by:

Curran Associates, Inc.
57 Morehouse Lane
Red Hook, NY 12571

CURRAN ASSOCIATES INC. proceedings.com

Some format issues inherent in the e-media version may also appear in this print version.

Institute of Physics
Dirac House, Temple Back
Bristol BS1 6BE UK

Phone: 44 1 17 929 7481
Fax: 44 1 17 920 0979

techtracking@iop.org

Additional copies of this publication are available from:

Curran Associates, Inc.
57 Morehouse Lane
Red Hook, NY 12571 USA
Phone: 845-758-0400
Fax: 845-758-2633
Email: curran@proceedings.com
Web: www.proceedings.com

TABLE OF CONTENTS

Journal of Physics: Conference Series **2545** (2023) 011001 doi:10.1088/1742-6596/2545/1/011001

35th International Symposium on Superconductivity

I 超 S 電 S 導 2022

WEB-BASED SYMPOSIUM
AIST and VUW
November 29 – December 1, 2022

35th International Symposium on Superconductivity (ISS2022)

Preface

It is our pleasure to report that the Proceedings of the 35th International Symposium on Superconductivity (ISS2022) held on November 29–December 1, 2022, are now published in *Journal of Physics: Conference Series* (JPCS). ISS2022 was co-organized by the Victoria University of Wellington (VUW) and the National Institute of Advanced Industrial Science and Technology (AIST). ISS2022 was held online owing to the continued travel restrictions by COVID-19 pandemic, but it had also in-person option at Winc-Aichi (Aichi Industry & Labour Center), Nagoya, Japan. Invited speakers, session chairs, program committee members and speakers were welcome to join in at Winc-Aichi.

Total of 315 scientists, engineers, students, corporate executives and so on (132 from abroad) made advanced registrations for ISS2022, and fruitful discussions were made to promote superconductor sciences and technologies. Total of 190 papers were presented, which included 6 plenary lectures (45 min), 71 invited talks (25–30 min) and 113 contributed oral presentations (15–20 min). Among the papers presented in ISS2022, 34 manuscripts were submitted for ISS2022 Proceedings, and 33 papers were accepted for publication. The papers published in JPCS were categorized into relevant topics in the following four fields; (a) Physics and Chemistry, (b) Wires and Bulk, (c) Electronic Devices, (d) Large Scale System Applications.

To ensure the high publication standard mandated by JPCS, every paper was peer reviewed by a reviewer with expertise before it was accepted for publication. As editors of the Proceedings, we would like to express our sincere appreciation to all the reviewers involved in the evaluation of the papers for their invaluable contribution.

Journal of Physics: Conference Series **2545** (2023) 011001 doi:10.1088/1742-6596/2545/1/011001

The Editors

Physics and Chemistry: Atsutaka Maeda (The University of Tokyo), Hiraku Ogino (AIST)

Wires and Bulk: Akiyoshi Matsumoto (National Institute for Materials Science), Yasuhiro Iijima (Fujikura Ltd.), Yasunori Mawatari (AIST)

Electronic Devices: Hirotaka Terai (National Institute of Information and Communications Technology), Hirotake Yamamori (AIST)

Large Scale System Applications: Naoyuki Amemiya (Kyoto University), Tsuyoshi Wakuda (Hitachi Ltd.), Hiroshi Miyazaki (Kyushu University)

Journal of Physics: Conference Series **2545** (2023) 011001 doi:10.1088/1742-6596/2545/1/011001

Organization of ISS2022

General Chair:
Dr. Nicholas Long
Associate Professor, Robinson Research Institute, Victoria University of Wellington (VUW)

Co-Chair:
Dr. Haruhiko Obara
Senior Executive Officer, Director-General, National Institute of Advanced Industrial Science and Technology (AIST)

Special Advisor
Dr. Toshihiko Kanayama
Special Emeritus Advisor, AIST

Secretary General
Dr. Michiya Okada
Deputy Director, TIA Central Office, AIST
michiya.okada@aist.go.jp

Local Organizing Committee:
Members:
Michiya Okada (AIST), Nicholas Long (VUW), Zhenan Jiang (VUW), Jeannie Redman (VUW), Mutsuo Hidaka (AIST), Yoshiyuki Yoshida (AIST), Mitsuho Furuse (AIST), Nao Takeshita (AIST), Shigeyuki Ishida (AIST), Susumu Ogawa (AIST), Kim Hyunjeong (AIST), Tsuyoshi Yanagibayashi (AIST)

Supported by:
New Energy and Industrial Technology Development Organization (NEDO)
Japan Science and Technology Agency (JST)
Cryogenics and Superconductivity Society of Japan (CSSJ)
The Institute of Electrical Engineers of Japan (IEEJ)
The Japan Society of Applied Physics (JSAP)
Applied Superconductivity Constellations of Tsukuba (ASCOT)
TIA—a platform for open innovation

Journal of Physics: Conference Series **2545** (2023) 011001 doi:10.1088/1742-6596/2545/1/011001

Committees of ISS2022

Program committee

Physics and Chemistry

Chair: Atsutaka Maeda (The Univ. of Tokyo), Deputy-Chair: Hiraku Ogino (AIST)

Members: Tetsuo Hanaguri (RIKEN), Tsutomu Nojima (Tohoku Univ.), Satoshi Okuma (Tokyo Inst. of Technology), Tsuyoshi Tamegai (The Univ. of Tokyo), Hiroyuki Yamase (National Inst. for Materials Science), Teppei Yoshida (Kyoto Univ.)

Advisors: Akira Iyo (AIST), Hiroshi Eisaki (AIST)

Wires and Bulk

Chair: Akiyoshi Matsumoto (National Inst. for Materials Science), Deputy-Chair: Yasuhiro Iijima (Fujikura Ltd.), Deputy-Chair: Yasunori Mawatari (AIST)

Members: Satoshi Awaji (Tohoku Univ.), Hiroshi Ikuta (Nagoya Univ.), Takanobu Kiss (Kyushu Univ.), Kaname Matsumoto (Kyushu Inst. of Technology), Hisaki Sakamoto (Furukawa Electric Co., Ltd.), Hideki Tanaka (Hitachi Ltd.), Akiyasu Yamamoto (Tokyo Univ. of Agriculture and Technology)

Advisors: Yoshiyuki Yoshida (AIST)

Electronic Devices

Chair: Hirotaka Terai (National Institute of Information and Communications Technology), Deputy-Chair: Hirotake Yamamori (AIST)

Members: Seiichiro Ariyoshi (Toyohashi Univ. of Technology), Kenji Sakai (Okayama Univ.), Naoto Sekiya (Univ. of Yamanashi), Yutaka Tabuchi (RIKEN), Masamitsu Tanaka (Nagoya Univ.),

Advisors: Masataka Ohkubo (AIST), Mutsuo Hidaka (AIST)

Large Scale System Application

Chair: Naoyuki Amemiya (Kyoto Univ.), Deputy-Chair: Hiroshi Miyazaki (Kyushu Univ.), Deputy-Chair: Tsuyoshi Wakuda (Hitachi Ltd.)

Members: Kohei Higashikawa (Kyushu Univ.), Kazuhiro Kajikawa (Kyushu Univ.), Kei Koyanagi (Toshiba Energy Systems & Solutions Corporation), Shinji Matsumoto (National Inst. for Materials Science), Naoko Nakamura (MAYEKAWA MFG. Co., Ltd.), Taketsune Nakamura (Kyoto Univ.), Tetsuo Oka (Shibaura Inst. of Technology), Shun Tonooka (Mitsubishi Electric Co.), Makoto Tsuda (Tohoku Univ.), Tomonori Watanabe (Chubu Electric Power Co., Inc.)

Advisors: Mitsuho Furuse (AIST)

International Advisory Committee

Physics and Chemistry

Jun Akimitsu (Okayama Univ.), Hideo Aoki (The Univ. of Tokyo/AIST), Xianhui Chen (Univ. of Science and Technology of China), Shen Chong (VUW), Atsushi Fujimori (The Univ. of Tokyo), Zenji Hiroi (The Univ. of Tokyo), Takekazu Ishida (Osaka Prefecture Univ.), Maarit Karppinen (Aalto Univ.), Wai-Kwong Kwok (Argonne National Laboratory), Victor Moshchalkov (KU Leuven), Satoshi Okuma (Tokyo Inst. of Technology), James Storey (VUW), Hermann Suderow (Autonomous Univ. of Madrid), Setsuko Tajima (Osaka Univ.), Jeffrey L. Tallon (VUW), Takami Tohyama (Tokyo Univ. of Science), Dirk van der Marel (Univ. of Geneva), Hai-Hu Wen (Nanjing Univ.), Grant William (VUW), Eli Zeldov (Weizmann Inst. of Science)

Wires and Bulk

Amalia Ballarino (CERN), David Cardwell (Univ. of Cambridge), Michael Eisterer (Vienna Univ. of Technology), Hiroyuki Fujishiro (Iwate Univ.), Fedor Gömöry (Slovak Academy of Sciences), Timothy J. Haugan (Air Force Research Laboratory), Kazuhiko Hayashi (Sumitomo Electric Industries), Hiroaki Kumakura (National Inst. for Materials Science), David Larbalestier (Florida State Univ.), Qiang Li (Brookhaven National Laboratory), Nicholas Long (VUW), Yanwei Ma (Chinese Academy of Sciences), Xavier Obradors (Institut Ciencia de Materials de Barcelona), Marina Putti (Univ. of Genoa), Venkat Selvamanickam (Univ. of Houston), Nick Strickland (VUW), Michael Sumption (The Ohio State Univ.), Masaru Tomita (Railway Technical Research Institute), Stuart Wimbush (VUW), Yutaka Yoshida (Nagoya Univ.)

Electronic Devices

Pascal Febvre (Univ. of Savoie Mont Blanc), Simon Granville (VUW), Shigetoshi Ohshima (Yamagata Univ.), Horst Rogalla (NIST/Univ. of Colorado at Boulder), Zhen Wang (Chinese Academy of Sciences)

Large Scale System Application

Rodney Badcock (VUW), Robert G. Buckley (VUW), Chris Bumby (VUW), Toru Fukushima (SuperPower Inc.), Jianzhao Geng (Huazhong Univ. of Science & Technology), Zhenan Jiang (VUW), Mathias Noe (Karlsruhe Inst. of Technology), Hiroyuki Ohsaki (The Univ. of Tokyo), Minwon Park (Changwon National Univ.), Christopher M. Rey (Energy-to-Power Solutions), Pascal Tixador (Grenoble-INP), Hubertus Weijers (VUW), Liye Xiao (Chinese Academy of Sciences)

Peer Review Statement

All papers published in this volume have been reviewed through processes administered by the Editors. Reviews were conducted by expert referees to the professional and scientific standards expected of a proceedings journal published by IOP Publishing.

1. Type of peer review: Single anonymous
2. Conference submission management system: Morressier
3. Number of submissions received: 34
4. Number of submissions sent for review: 34
5. Number of submissions accepted: 33
6. Acceptance Rate (Submissions Accepted / Submissions Received × 100): 97.1 %
7. Average number of reviews per paper: 2.15
8. Total number of reviewers involved: 31
9. Contact person for queries:
   Name: Hirofumi Yamasaki
   Affiliation: Secretariat of International Symposium on Superconductivity
   Email: hirofumi.yamasaki@iss-secretariat.org

* means value has been edited

High temperature spin-triplet topological superconductivity in $K_2Cr_3As_3$

Guo-qing Zheng

Department of Physics, Okayama University, Okayama 700-8530, Japan

E-mail: zheng@psun.phys.okayama-u.ac.jp

Abstract. Spin-triplet superconductors are novel materials capable of harboring Majorana bound states that can be used in topological quantum computing. However, such bulk materials are still rare. Here we review the results that established $K_2Cr_3As_3$ as a spin-triplet superconductor with transition temperature T_c as high as 6.5 K. We focus on the multiple-phases feature, and its exquisite distance to a ferromagnetic quantum critical point which is likely responsible for the high T_c. We touch on the topological aspects of the superconducting state, and suggest that it is a new route to the technical implementation using a topological spin-triplet superconductor at the highest temperature ever.

1. Introduction

In contrast to a spin-singlet state in materials including copper-oxide and iron-pnictide high temperature superconductors, where the electron pairs (Cooper pairs) are in the state with total spin S=0 [1, 2], spin-triplet superconductivity or fluidity with S=1 possesses internal structure, which gives rise to rich physics such as additional symmetry breaking, order parameter collective modes, and multiple phases of the condensate. Therefore, spin-triplet superconductors are an excellent playgound for testing and creating new concepts of condensed matter physics. In addition, a spin-triplet state with odd-parity can be topological and Majorana bound states or chiral Majorana fermions can emerge in the vortex cores or on boundary, which are robust against perturbation. Thus, spin-triplet superconductivity are of great interests and importance not only in fundamental physics, but also in applications as their edge or bound states can be used to implement topological quantum computing based on non-Abelian statistics [3]. However, such materials are still rare. Historically, efforts of searching for spin-triplet superconductivity was devoted to strongly correlated electron systems. In recent years, doped topological insulators have been proposed to be a new playground for spin-triplet superconductivity, and indeed unambiguous evidence has been obtained for $Cu_xBi_2Se_3$ through the discovery of spin-rotation symmetry breaking by nuclear magnetic resonance (NMR) [4].

More recently, a new superconducting family containing 3d transition-metal element Cr, $A_2Cr_3As_3$ (A = Na, K, Rb, Cs) was discovered [5, 6, 7, 8], with T_c as high as 8 K. Early NMR measurements in a poly-crystal sample found ferromagnetic spin fluctuation in the normal state and point nodes in the gap function in the superconducting state [9], indicating that this family is a strongly-correlated 3d electron system. Recent Knight shift measurements in a single crystal of $K_2Cr_3As_3$ with T_c = 6.5 K revealed that the spin susceptibility is unchanged below T_c when the magnetic field H_0 is applied in the ab plane, but vanishes toward zero when H_0 is along the c

axis, which unambiguously establishes this compound as a spin-triplet superconductor described by a vector order-parameter **d** parallel to the c-axis at low fields [10]. In this presentation, we focus on the multiple phases of $K_2Cr_3As_3$ and its equisite clossness to a ferromagnetic quantum critical point (QCP), which may be responsible for the high T_c. We will also touch on the topological aspects of the superconducting state.

2. Multiple superconducting phases and rotation of the vector order parameter at high fields

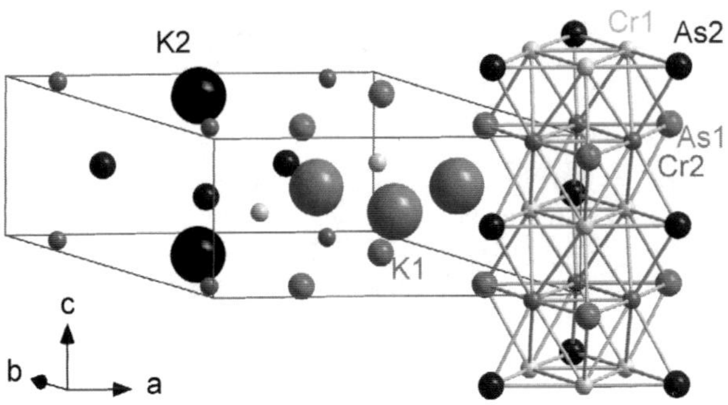

Figure 1. Crystal structure of $K_2Cr_3As_3$. The Cr_3 tube runs along the c-axis, surrounding which runs the As_3 tube.

Figure 1 shows the crystal structure of $K_2Cr_3As_3$. There are two Cr sites, both of which form a triangle shape. The Cr_3 tube runs along the c-axis, which is the reason why the compound is often called "quasi-one dimensional". However, band calculation shows that there are three pieces of Fermi surfaces, among which the three-dimensional one makes the main contribution to the density of states (DOS) [11]. Also, very recent neutron scattering measurements found that spin fluctuations due to nesting of the two one-dimensional Fermi surfaces do not contribute to superconductivity that must exist in the three-dimensional Fermi surface [12]

Figure 2 shows the ^{75}As-NMR Knight shift in a single crystal above and below T_c for different field orientations. The Knight shift K consists of two contributions, namely, K_s from spin susceptibility and $K_{\rm orb}$ from orbital susceptibility. The diamagnetic contribution due to a vortex lattice formation in the superconducting state is negligibe as the penetration depth is long [10]. The position of $K_{\rm orb}$ is shown by the horizontal arrows in Fig. 2. It was estimated from a novel analysis using the relation between K and the spin-lattice relaxation rate devided by T, $1/T_1T$. Recently, by using a more conventional method, the so-called $Kvs\chi$ plot, very close values have been obtained [13]. Thus, K_s has a substantial contribution for both $H \parallel c$ and $H \perp c$ directions.

As can be seen in Fig. 2, K (and thus K_s) shows a very anisotropic temperature dependence below T_c between $H_0 \parallel c$ and $H_0 \perp c$. Namely, For $H_0 \perp c$, K_s does not change across T_c. In contrast, for $H_0 \parallel c$, K_s is reduced in the superconducting state. In particular, For H_0=9.6955 T, K_s decreases toward zero in the T=0 limit. These results are firm evidence of spin-rotation symmetry breaking or an emergence of spin nematicity, and establish $K_2Cr_3As_3$ as a spin triplet superconductor described by the vector order parameter $\mathbf{d} \parallel c$-axis.

A novel aspect of Fig. 2 is that K_s does not decrease right at T_c but at a lower temperature T^*. It is emphasized that both the NMR intensity and $1/T_1$ drop sharply at T_c. The disparity

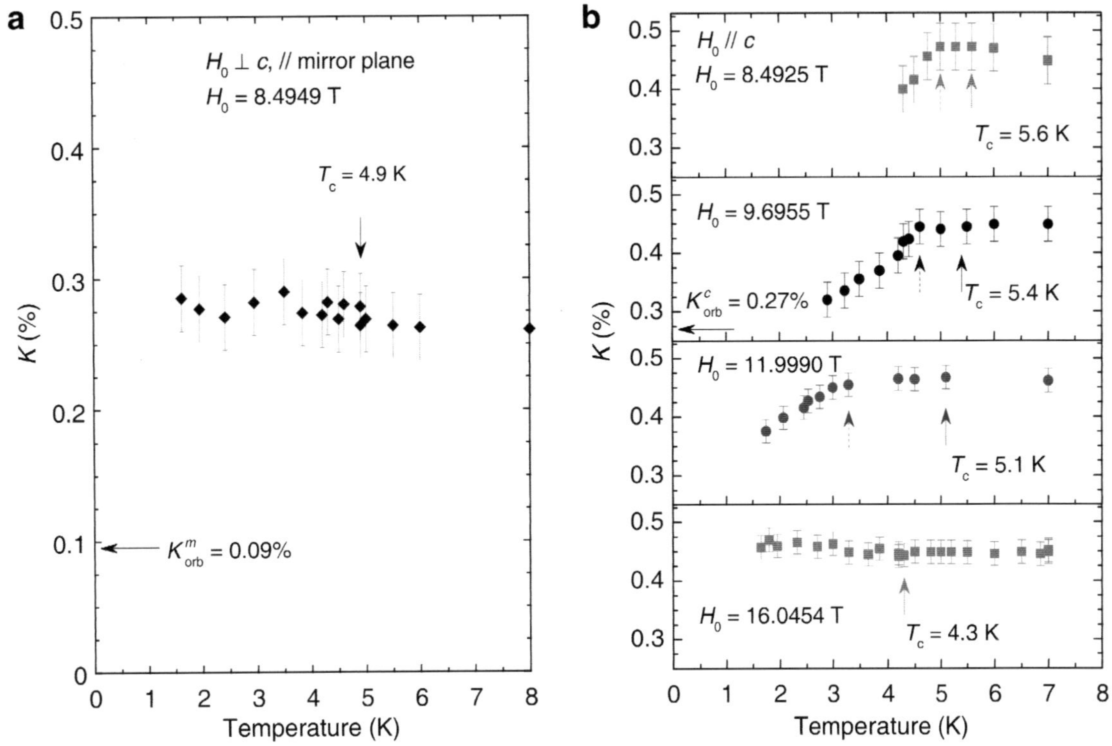

Figure 2. (a) The temperature dependence of the Knight shift with the magnetic field applied in the *ab* plane (and along the mirror plane of the lattice). (b) The Knight shift with $H_0 \parallel c$ axis. The solid arrows indicate T_c determined by ac susceptibility, spin-lattice relaxation rate $1/T_1$ and the NMR intensity. The dotted arrows indicate the temperature T^* below which the Knight shift starts to drop.

between T_c and T^* becomes larger as H_0 increases. Most intriguingly, under H_0=16.0454 T at which $T_c(H)$=4.3 K, K shows no reduction at all. These results are summarized in Fig. 3.

The boundary marked by T^* in Fig.3 is consistent with the Pauli-limited field H_p which is estimated to be about 13 T at T=0. It is thus consistent with **d** $\parallel$ *c*-axis for the low fields as discussed above. Above the T^* boundary, as the Knight shift shows no reduction, the **d**-vector must be flipped by 90 degrees. The mechanism for the **d**-vector rotation is unclear at the moment. There are two possibilities. One is the SO(5)-like transition to gain Zeeman energy. The other is that the T^* boundary marks a phase transition from one symmetry to another. Measurements at high fields are now in progress to resolve this issue. In any case, the multiple-phases feature is a direct consequence of a spin-triplet condensate due to the degrees of internal freedom of the pairing and provides good opportunity to investigate novel phases of matter.

3. Why T_c is high?

The Alkali element A in the formula $A_2Cr_3As_3$ can be changed and superconductivity with a maximal T_c =8 K was obtained. A natural question is what controls T_c within the family and why T_c is much higher than other systems with ferromagntic interactions. To answer the question, the temperature dependence of $1/T_1T$ was measured across the family and analyzed

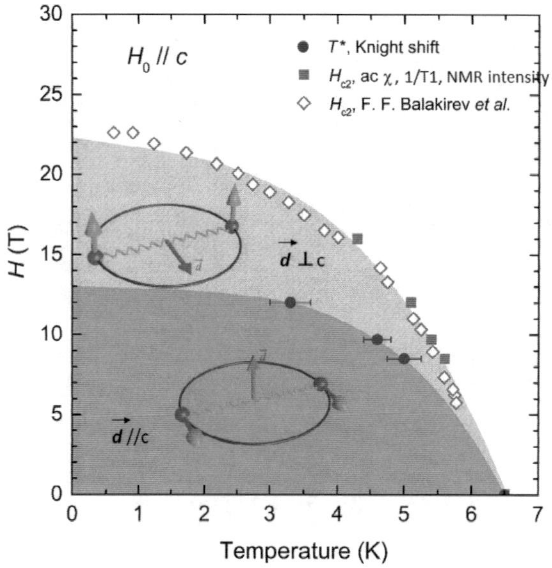

Figure 3. H-T phase diagram for $K_2Cr_3As_3$ with the field applied along the c-axis. The shaded regions are the superconducting state. In the illustration, the purple balls represent electrons that form a Cooper pair, the wavy lines depict the attractive force between two electrons. The green trigonal (pyramid) heads indicate the spin direction and the red arrorws show the direction of the vector order parameter **d**. The Pauli-limit field is estimated to be 13 T at T=0, being consistent with the boundary between the green and yellow regions. The data points marked by open diamonds are from Ref. [14].

[15]. The quantity $1/T_1T$ measures spin correlations through probing the imaginary part of transverse dynamic susceptibility $\chi"(\omega,\mathbf{q})$ summed over all **q**. In the case of strong electron correlations, we may write $1/T_1T$ as:

$$1/T_1T = (1/T_1T)_{\rm SF} + (1/T_1T)_0. \tag{1}$$

where the first part originates from spin fluctuations and the second part is due to the DOS at the Fermi level. According to self-consistent renormalization theory for three-dimensional ferromagnetic fluctuation [16], $(1/T_1T)_{\rm SF}$ follows a Curie-Weiss temperature dependence

$$(1/T_1T)_{\rm SF} = C/(T+\theta) \tag{2}$$

Here, θ describes the distance to a ferromagnetic QCP. The obtained parameter θ from the fittings is shown in Fig. 4(a). The θ gradually increases from 4.1 K for $Rb_2Cr_3As_3$ to 56.8 K for $Na_2Cr_3As_3$.

Figure 4 compares our phase diagram of the $A_2Cr_3As_3$ family (a) with that of UGe_2 (b) [20] and with theories for ferromagnetic superconductors (c) [21, 22, 23]. In this figure, the superconducting transition temperature is represented by $T_{\rm sc}$, in order to distinguish between the Curie temperature $T_{\rm C}$. In Fig.4(a) is plotted the parameter θ and $T_{\rm sc}$ as a function of the ionic radius of the alkali element A. There is a strong correlation between the ionic radius and the Cr2-As2-Cr2 bond angle α (see Fig. 1) which is shown as the upper horizontal axis. The angle α may be related to double exchange interaction through the Cr-As-Cr path. At α=90°, the As-4p_x and As-4p_y orbitals become degenerated with respect to Cr-3d orbitals, which will maximize the double exchange between the two Cr2 along the c-axis via the As-4p_x and As-4p_y orbitals [17]. Therefore, on going from A = Na to $Na_{0.75}K_{0.25}$, K, and Rb, an increase in the ferromagnetic interaction can be expected, which drives the system towards a ferromagnetic QCP. Our finding that $Na_2Cr_3As_3$ has a weaker spin correlation was subsequently supported by a theoretical evaluation based on the first-principle calculation [18]

In Fig. 4(a), one sees that T_{sc} increases upon moving away from the putative ferromagnetic QCP. This is in sharp contrast to the antiferromagnetic case where T_{sc} usually forms a peak around a QCP, but is consistent with theoretical expectations. Superconductivity in both the ferromagnetic ordered state and near a ferromagnetic QCP in paramagnetic side was first discussed by Fay and Appel [21]. Later on, Monthoux and Lonzarich [22], and Wang *et al* [23] improved the calculation for the paramagnetic side. In the antiferromagnetic case, the pairing

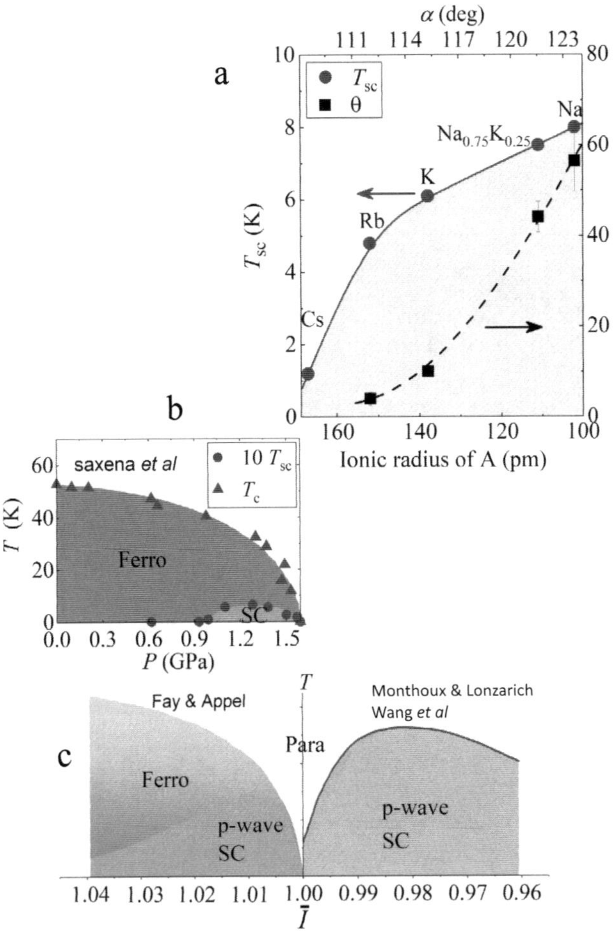

Figure 4. (a) Phase diagram of the $A_2Cr_3As_3$ (A = Na, $Na_{0.75}K_{0.25}$, K, Rb, Cs) family [15, 19]. The α is the angle of the Cr2-As2-Cr2 bonds (see Fig. 1). In this figure, T_C is reserved to represent the Curie temperature and the superconducting transition temperature is marked by T_{sc}. (b) Phase diagram of UGe_2 under pressure. [20]. (c) Theoretically proposed phase diagram for superconductivity in ferromagnetic ("Ferro") ordered state [21] and paramagnetic ("Para") state with ferromagnetic correlation. [22, 23]. The horizontal axis is the exchange interaction I with $I = 1$ at the QCP.

interaction is enhanced due to increased quantum fluctuations [22]. In the ferromagnetic case, when it is approached from the paramagnetic side, increased quantum fluctuations also enhances pairing strength [21, 22], but mass enhancement and a finite quasiparticle life time act as pair breaking and suppress T_{sc} [21, 22, 23]. As can be seen in Fig. 4(a), $K_2Cr_3As_3$ is away from the ferromagnetic QCP yet with spin correlation, which probably explains why T_{sc} is high.

4. Topological aspects

Having established the spin symmetry, the next issue is the orbital wave function of the Cooper pairs. Group theory analysis [10] shows that only E' states (p_x+ip_y and p_x-ip_y) are consistent with the Knight shift result and the $1/T_1$ result revealing a nodal gap [9, 15]. This state is analogous to the A phase (or Anderson-Brinkman-Morel state) in superfluid ^{3}He, and is topological. Therefore Majorana zero modes can be expected in vortex cores [24]. In particular, if a superconducting thin film with its thickness smaller than the superconducting coherence length is available, a single Majorana zero mode will be expected in the core of a half-quantum vortex. Table I lists the topological classification together with $Cu_xBi_2Se_3$ ($x < 0.46$) [4], $Cu_xBi_2Se_3$ ($x \geq 0.46$) [25] and Li_2Pt_3B [26].

Table 1. Topological aspects of $K_2Cr_3As_3$ and other superconductors and superfluid ^{3}He.

Class	Time Reversal Symmetry	Two Dimension	Three Dimension
D	Broken	^{3}He-A film $K_2Cr_3As_3$ film $Cu_xBi_2Se_3$ ($x \geq 0.46$) film	
DIII	Conserved		^{3}He-B Li_2Pt_3B $Cu_xBi_2Se_3$ ($x < 0.46$)

5. Summary

By NMR measurements in a single crystal of $K_2Cr_3As_3$, we discover a nematic response of the spin susceptibility between $H_0 \parallel c$-axis and $H_0 \perp c$-axis in the superconducting state, establishing this compound as a spin-triplet superconductor described by the vector order parameter **d** parallel to the c-axis. However, at high fields above 13 T, the **d**-vector is flipped to lie in the ab-plane. The multiple phases at different magnetic fields is another manifestation of the spin-triplet pairing due to its internal degrees of freedom and merits further investigation. Our results demonstrate that $K_2Cr_3As_3$ is a new platform for basic research of topological materials and technical applications of topological superconductivity.

Acknowledgments

This work was performed in collaboration with J. Yang, J. Luo, Y. Zhou, Y.G. Shi, Z.A. Ren, G.H. Cao, S. Ogawa, K. Matano, and Y. Inada. We thank T. Hanaguri and K. Machida for stimulating discussions. The work was partially supported by JSPS Grants (JP19H00657 and JP22H04482).

References

[1] Tsuei C C, Kirtley J R 2000 Pairing symmetry in cuprate superconductors *Rev. Mod. Phys.* **72** 969
[2] Matano K, Ren Z A, Dong X L, Sun, L L, Zhao Z X, Zheng G Q 2008 Spin-singlet superconductivity with multiple gaps in $PrFeAsO_{0.89}F_{0.11}$ *Europhys. Lett.* **83** 57001
[3] Kitaev A Y 2001 Unpaired Majorana fermions in quantum wires. *Physics-Uspekhi* **44** 131

[4] Matano K, Kriener M, Segawa K, Ando Y, Zheng GQ 2016 Spin-rotation symmetry breaking in the superconducting state of $Cu_xBi_2Se_3$. *Nat. Phys.* **12** 852

[5] Bao J K, Liu J Y, Ma C W, Meng Z H, Tang Z T, Sun Y L, Zhai H F, Jiang H, Bai H, Feng C M, Xu Z A, Cao G H 2015 Superconductivity in Quasi-One-Dimensional K2Cr3As3 with Significant Electron Correlations *Phys. Rev. X* **5** 011013

[6] Tang Z T, Bao J K, Liu J Y, Sun Y L, Ablimit A, Zhai H F, Jiang H, Feng C M, Xu Z A, Cao G H 2015 Unconventional superconductivity in quasi-one-dimensional $Rb_2Cr_3As_3$. *Phys. Rev. B* **91** 020506

[7] Tang Z T, Bao J K, Wang Z, Bai H, Jiang H, Liu J Y, Zhai H F, Feng C M, Xu Z A, Cao G H 2015 Superconductivity in quasi-one-dimensional $Cs_2Cr_3As_3$ with large interchain distance *Sci. China Mater.* **58** 16

[8] Mu Q G, Ruan B B, Pan B J, Liu T, Yu J, Zhao K, Chen G F, Ren Z A 2018 Ion-exchange synthesis and superconductivity at 8.6 K of $Na_2Cr_3As_3$ with quasi-one-dimensional crystal structure *Phys. Rev. Mater.* **2** 034803

[9] Yang J, Tang Z T, Cao G H, Zheng G Q 2015 Ferromagnetic Spin Fluctuation and Unconventional Superconductivity in $Rb_2Cr_3As_3$ Revealed by ^{75}As NMR and NQR *Phys. Rev. Lett.* **114** 147002

[10] Yang J, Luo J, Yi C J, Shi Y G, Zhou Y, Zheng G Q 2021, Spin-triplet superconductivity in $K_2Cr_3As_3$ *Sci. Adv.* **7** eabl4432

[11] Jiang H, Cao G H, Cao C 2015 Electronic structure of quasione-dimensional superconductor $K_2Cr_3As_3$ from first principles calculations *Sci. Rep.* **5** 16054

[12] Taddei K M, Lei B H, Susner M A, Zhai H F, Bullard T J, Sanjeewa L D, Zheng Q, Sefat A S, Chi S, dela Cruz C, Singh D J, Lv B 2022, Gapless spin-excitations in the superconducting state of a quasi-one-dimensional spin-triplet superconductor *Preprint* arXiv:2206.11757v1.

[13] Ogawa S, Miyoshi T, Matano K, Inada Y, Zheng GQ 2023 Single crystal growth of and hyperfine couplings in the spin-triplet superconductor $K_2Cr_3As_3$ *Preprint* arXiv:2302.03402v2.

[14] Balakirev F F, Kong T, Jaime M, McDonald R D, Mielke C H, Gurevich A., Canfield P C, Bud'ko S L 2015, Anisotropy reversal of the upper critical field at low temperatures and spin-locked superconductivity in $K_2Cr_3As_3$. *Phys. Rev. B* **91** 220505(R)

[15] Luo J, Yang J, Zhou R, Mu Q G, Liu T, Ren Z A, Yi C J, Shi Y G, Zheng G Q 2019 Tuning the Distance to a Possible Ferromagnetic Quantum Critical Point in $A_2Cr_3As_3$ *Phys. Rev. Lett.* **123** 047001

[16] Moriya T 1985 *Spin Fluctuations in Itinerant Electron Magnetism* Springer-Verlag, Berlin

[17] Wu X X, Le C C, Yuan J, Fan H, Hu J P 2015 Magnetism in Quasi-One-Dimensional $A_2Cr_3As_3$ (A = K,Rb) Superconductors *Chin. Phys. Lett.* **5** 057401

[18] Xu C, Wu N, Zhi G X, Lei B H, Duan B X, Ning F, Cao C, Chen Q 2020 Coexistence of nontrivial topological properties and strong ferromagnetic uctuations in quasi-one-dimensional $A_2Cr_3As_3$ *Npj Comput. Mater.* **6** 1

[19] Sumiyama A, Nagaeki T, Yamaguchi A, Jia K, Yang J, Shi Y G, Zheng GQ 2023 Superconducting Transition in Quasi-One-Dimensional Superconductor $Cs_2Cr_3As_3$ Probed by DC Magnetic Susceptibility *J. Phys. Soc. Jpn.* **92** 015001

[20] Saxena S S, Agarwal P, Ahilan K, Grosche F M, Haselwimmer R K W, Steiner M J, Pugh E, Walker I R, Julian S R, Monthoux P, Lonzarich G G, Huxley A, Sheikin I, Braithwaite D, Flouquet J 2000 Superconductivity on the border of itinerant-electron ferromagnetism in UGe_2 *Nature* **406** 587

[21] Fay D, Appel J 1980 Coexistence of p-state superconductivity and itinerant ferromagnetism *Phys. Rev. B* **22** 3173

[22] Monthoux P, Lonzarich G G 2001 Magnetically mediated superconductivity in quasi-two and three dimensions *Phys. Rev. B* **63** 054529

[23] Wang Z, Mao W, Bedell K 2001 Superconductivity near Itinerant Ferromagnetic Quantum Criticality *Phys. Rev. Lett.* **87** 257001

[24] Ivanov D A 2001 Non-Abelian Statistics of Half-Quantum Vortices in *p*-Wave Superconductors *Phys. Rev. Lett.* **86** 268

[25] Kawai T, Wang C G, Kandori Y, Honoki Y, Matano K, Kambe T, Zheng G Q 2020 Direction and symmetry transition of the vector order parameter in topological superconductors $Cu_xBi_2Se_3$ *Nat. Commun.* **11** 235

[26] Nishiyama M, Inada Y, Zheng G Q 2007 Spin Triplet Superconducting State due to Broken Inversion Symmetry in Li_2Pt_3B *Phys. Rev. Lett.* **98** 047002

Two-dimensional Superconductivity in Misfit Layered Compound $(BiSe)_{1.10}NbSe_2$

S Matsuzawa[1], H Kitano[2], S Pyon[1] and T Tamegai[1]

[1]Department of Applied Physics, The University of Tokyo, 7-3-1 Hongo, Bunkyo-ku, Tokyo 113-8656, Japan

[2]Department of Physics, Aoyama Gakuin University, 5-10-1 Fuchinobe, Chuo-ku, Sagamihara 252-5258, Japan

E-mail: `creator-m318@g.ecc.u-tokyo.ac.jp`

Abstract. Novel two-dimensional superconductivity in a misfit layered compound $(BiSe)_{1.10}NbSe_2$ with a critical temperature T_c of 2.7 K is reported. The temperature dependence of H_{c2} is linear close to T_c for both $H \parallel$ c-axis and $H \parallel$ ab-plane, suggesting that the superconductivity is three-dimensional with some anisotropy. However, the angular dependence of H_{c2} clearly reveals a cusp-like behavior, which can be explained by Tinkham model, indicating that the system is two-dimensional. In order to clarify the relationship between stripe structures on the surface and the two-dimensional superconductivity, we also performed the same measurement on a uniform microbridge sample. We found similar two-dimensional behavior even in a microbridge without linear defects, indicating that the origin of two-dimensional superconductivity in $(BiSe)_{1.10}NbSe_2$ is not related to the stripe structure.

1. Introduction

Misfit layered compounds, denoted by $(MX)_{1+\delta}(TX_2)_n$ (M=Sn, Pb, Sb, Bi, and rare earth metals, T=Ti, V, Nb, Ta, and Cr, X=S and Se, n=1, 2, 3), consists of alternating double layers of MX with quasi NaCl-type structure and TX_2 with slightly distorted trigonal prism [1]. These two sublattices have different spatial symmetry and periodicity from each other, resulting in incommensurability in the plane with a misalignment of lattice constants along the a-axis. Misfit layered compounds have attracted attention not only for their physical properties such as superconductivity and charge-density wave, but also as materials that can naturally realize van der Waals heterostructures.

$(BiSe)_{1.10}NbSe_2$ is one of superconducting misfit layer compounds with T_c of 2.6-3.0 K, composed of a semiconductor BiSe and a superconducting transition metal dichalcogenide $NbSe_2$ [2, 3]. In our previous report, the multi-phase character depending on the nominal composition, the anisotropic superconductivity, and its relevance to another ternary system consisting of the same elements, Nb doped Bi_2Se_3, were discussed. In that study, we suggested a possibility of two-dimensionality in this system. In the present study, we mainly focused on the out-of-plane anisotropy of $(BiSe)_{1.10}NbSe_2$ and performed some experiments in order to confirm whether the two-dimensional behavior is intrinsic or not. For that purpose, we prepared a microbridge sample to probe superconducting properties in homogeneous part of the single crystal.

2. Experimental Methods

Single crystals of $(BiSe)_{1.10}NbSe_2$ were synthesized by the flux method using CsCl/KCl as flux [4]. A photo of a piece of the crystal is shown in Fig 1(a). Presence of stripe structures on the surface is clear. A magnified image using scanning ion microscopy (SIM) is shown in Fig. 1(b), which reveals that these stripe structures are not grooves but rather protrudes outward with widths of 5-10 µm. The direction of the principal axes of the crystal plane was determined by X-ray diffraction. Stripe structures on the surface run along the *b*-axis, indicating that crystal growth in misfit layered compounds tends to proceed along the commensulate direction rather than incommensulate direction. In the transport measurements described below, the direction of the current was set parallel either to the *a*- or *b*-axis.

Transport measurements were performed by the four-probe method using a physical property measurement system (PPMS, Quantum Design) or a superconducting quantum interference device (SQUID) magnetometer (MPMS-XL5, Quantum Design). The upper critical field (H_{c2}) was determined from resistive transitions under magnetic fields. The magnetic field dependence of resistivity was measured at various temperatures and angles to determine temperature and angular dependences of H_{c2}, respectively.

In order to investigate transport properties in a uniform region without stripe structures in a single crystal, a microbridge device shown in Figs. 1(c) and (d) was fabricated by using the focused ion beam (FIB, Hitachi High-Tech). We selected the region for the current flow so that it will not contain any stripe structures, ensuring that transport properties are not affected by the presence of stripe structures.

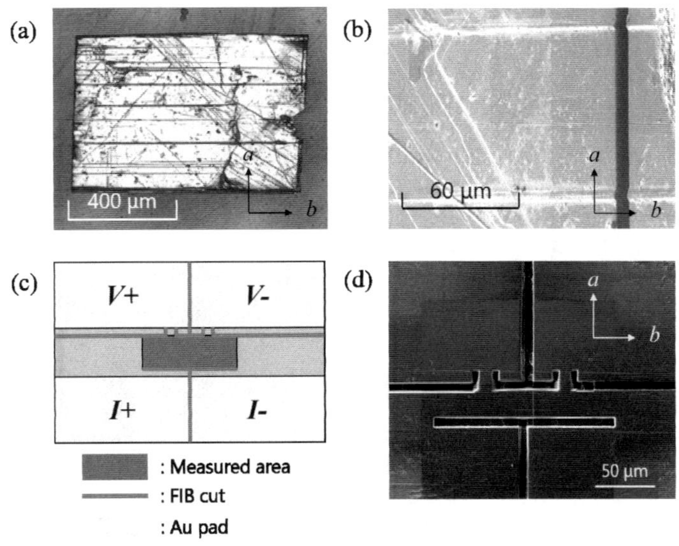

Figure 1. (a) Optical image of $(BiSe)_{1.10}NbSe_2$ single crystal. (b) SIM image of the crystal surface. (c) Schematic image of microbridge by FIB processing. (d) SIM image of the microbridge fabricated by FIB processing.

3. Results and Discussion

Figures 2(a)-(c) show the temperature dependence of in-plane resistivity under various magnetic fields. The field is applied along three principal axes. The critical temperature (T_c) is estimated as 2.7 K, somewhat higher than the reported value [2], but within a range of our former study [3].

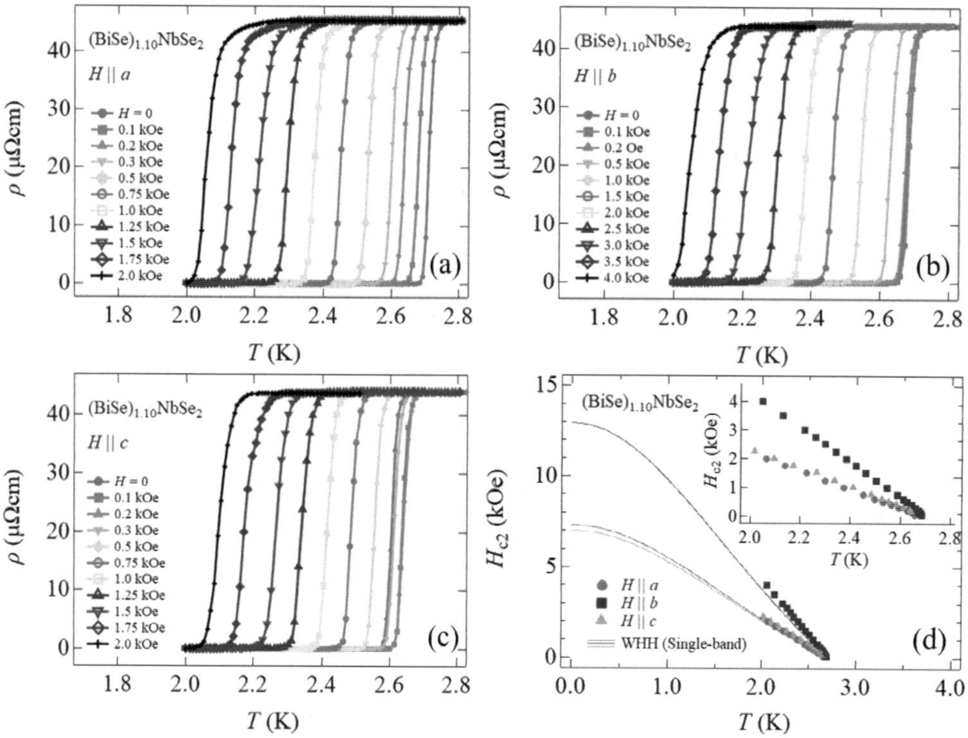

Figure 2. (a)-(c) Temperature dependence of in-plane resistivity of $(BiSe)_{1.10}NbSe_2$ single crystal for each axis with different magnetic fields. (d) Temperature dependence of H_{c2} of $(BiSe)_{1.10}NbSe_2$ single crystal.

Figure 2(d) shows the temperature dependence of H_{c2} ($H_{c2}(T)$) for the magnetic fields $H \parallel a$-, b-, and c-axes. Here, H_{c2} are determined by the midpoints of normal state resistivity. The superconducting anisotropy is evaluated to be about 1.0 in ac-plane and about 2.0 in bc-plane. This suggeseted that $(BiSe)_{1.10}NbSe_2$ is also anisotropic in the in-plane direction, which can be attributed to an averaged orthorhombic structure due to the mismatch of two subsystems.

The Ginzburg-Landau theory describes that $H_{c2}(T)$ has different temperature dependence depending on the dimensionality of the system as shown by the following formulae:

$$H_{c2}^{\parallel(3D)} = \frac{\Phi_0}{2\pi\xi_{ab}\xi_c} \propto (T_c - T) \tag{1}$$

$$H_{c2}^{\parallel(2D)} = \frac{\sqrt{12}\Phi_0}{2\pi\xi_{ab}d} \propto \sqrt{T_c - T}. \tag{2}$$

The observed temperature dependences of H_{c2} are linear in all directions, indicating that there is no evident sign of dimensional crossover at least in the measured temperature range, in contrast to another misfit layer compound $(LaSe)_{1.14}NbSe_2$ with much larger anisotropy [5]. In conventional type-II superconductors, the upper critical field due to the orbital effect is explained by the Werthamer-Helfand-Hohenberg (WHH) theory [6]. The WHH theory gives the expression for the zero-temperature value of upper critical field

$$H_{c2}(0) = -0.693\, T_c \left(\frac{dH_{c2}}{dT}\right)_{T=T_c} \tag{3}$$

for dirty superconductors. $H_{c2}(0)$ for each axis is calculated as $H_{c2}^{\|a}(0)$ =7.3 kOe, $H_{c2}^{\|b}(0)$ =13.0 kOe and $H_{c2}^{\|c}(0)$ =7.0 kOe, much lower than the Pauli paramagnetic limit field H_P [kOe] = $18.6 T_c$ [K] = 50 kOe.

The magnetic field dependences of in-plane resistivity for the single crystal and the microbridge sample measured at T = 2.0 K with $I \parallel b$ are shown in Figs. 3(a), (b) and (c), (d), respectively. The magnetic field was rotated in the *ac*- and the *bc*-planes, for Figs. 3(a), (c) and (b), (d), respectively, with θ being the angle between the field and the *ab*-plane.

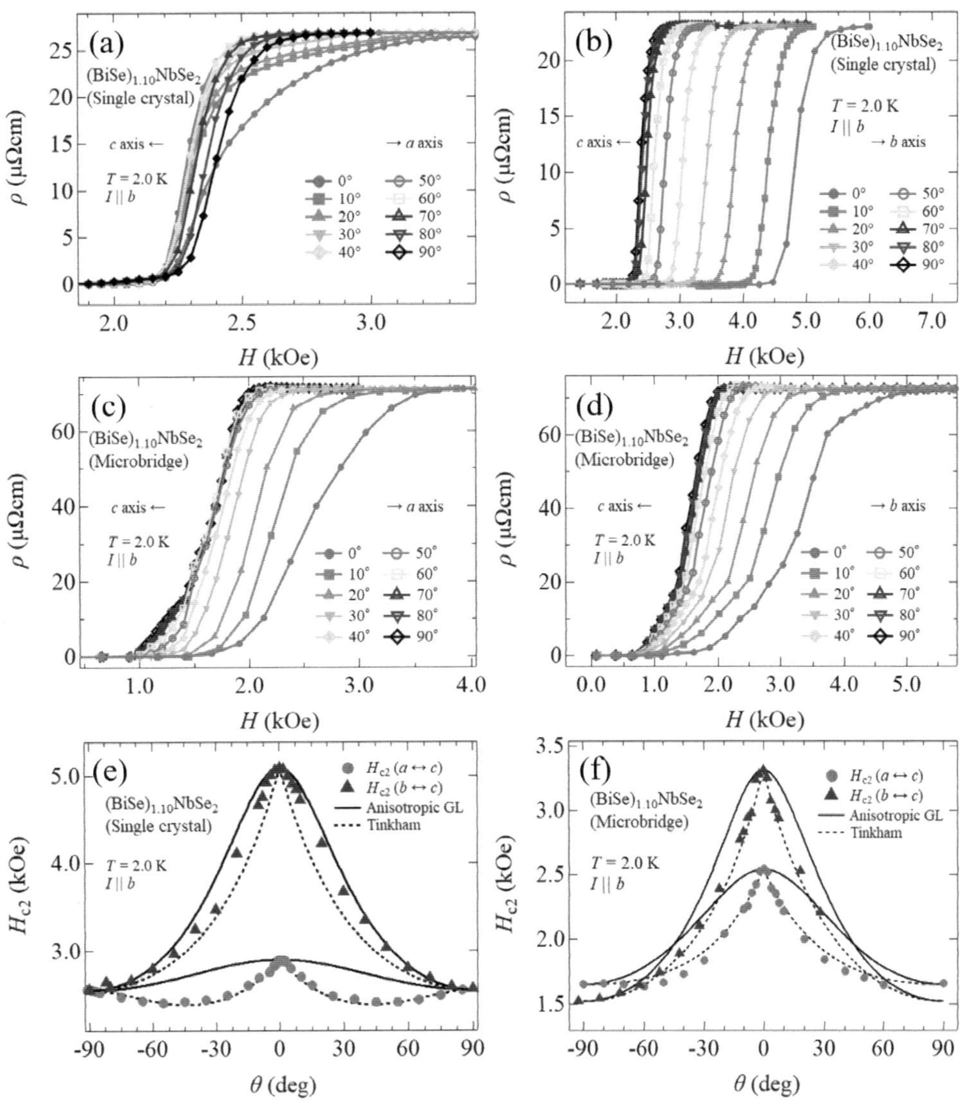

Figure 3. (a)-(d) Magnetic field dependence of in-plane resistivity of $(BiSe)_{1.10}NbSe_2$ with different field angles rotated in *ac*- and *bc*-planes measured at T = 2.0 K with $I \parallel b$. ((a), (b) Single crystal with the field rotated in *ac*- and *bc*-planes. (c), (d) Microbridge sample with the field rotated in *ac*- and *bc*-planes.) (e), (f) Angular dependence of H_{c2} of $(BiSe)_{1.10}NbSe_2$ for the single crystal and microbridge sample.

For both samples, the general trend is that the curves shift toward higher magnetic fields and the transition becomes broader as the magnetic field approaches 0°, which are typical behavior of layered superconductors with aisotropy. In addition, the curve shifts toward higher fields for the *bc*-plane than for the *ac*-plane, reflecting the in-plane anisotropy due to misfit nature. Figures 3(e) and (f) show the angular dependence of H_{c2} ($H_{c2}(\theta)$) for the single crystal and the microbridge sample, respectively. $H_{c2}(\theta)$ for the single crystal exhibits a characteristic cusp-like behavior close to 0° in the *ac*-plane, but slightly rounds off in the *bc*-plane. On the other hand, for the microbridge sample, clear cusp like behavior is observed in both field directions.

The $H_{c2}(\theta)$ expression for anisotropic 3D superconductors from the Ginzburg-Landau theory can be described by Eq.(4):

$$\left(\frac{H_{c2}(\theta)\sin\theta}{H_{c2}^{\perp}}\right)^2+\left(\frac{H_{c2}(\theta)\cos\theta}{H_{c2}^{\parallel}}\right)^2=1, \tag{4}$$

where $H_{c2}^{\parallel}$ amd $H_{c2}^{\perp}$ are the upper critical fields for directions parallel and perpendicular to the plane, respectively. On the other hand, the formula of the $H_{c2}(\theta)$ for 2D superconductors is given by Eq.(5) (Tinkham model):

$$\left|\frac{H_{c2}(\theta)\sin\theta}{H_{c2}^{\perp}}\right|+\left(\frac{H_{c2}(\theta)\cos\theta}{H_{c2}^{\parallel}}\right)^2=1, \tag{5}$$

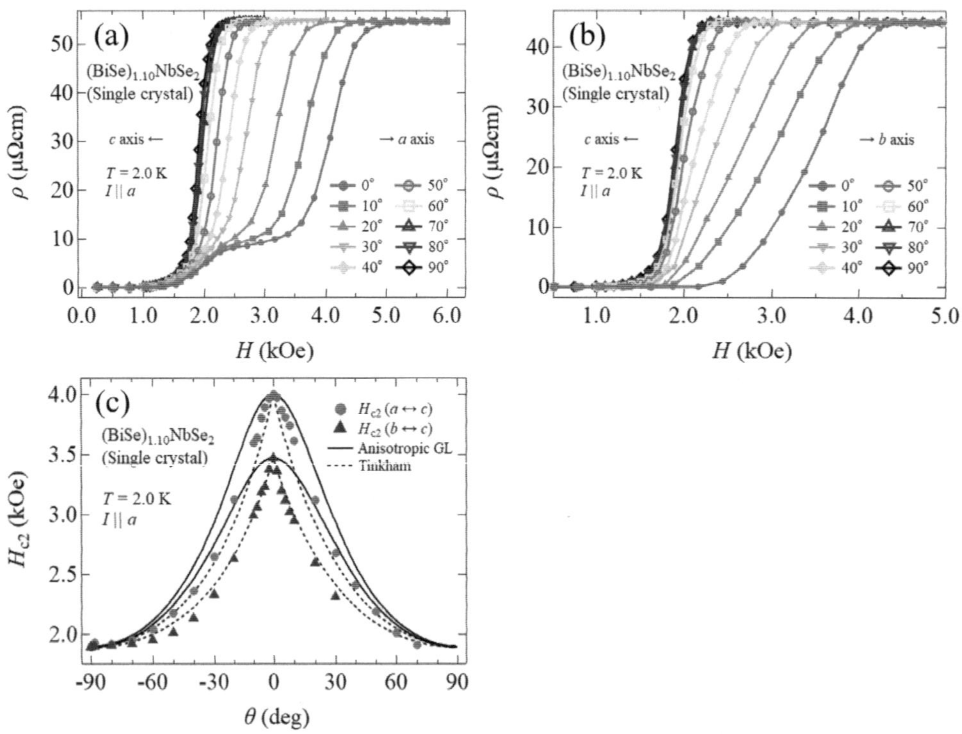

Figure 4. Magnetic field dependence of in-plane resistivity of $(BiSe)_{1.10}NbSe_2$ single crystal with different field angles rotated in (a) *ac*- and (b) *bc*-planes measured at $T = 2.0$ K with $I \parallel a$. (c) Angular dependence of H_{c2} of $(BiSe)_{1.10}NbSe_2$ both in the *ac*- and *bc*-planes.

with a finite slope at $\theta = 0°$ making a cusp. Tinkham model is usually applied to thin-film superconductors and anisotropic layered superconductors with large anisotropy, where the c-axis coherence length $\xi_c(0)$ is much smaller than the interlayer spacing. By contrast, experimentally estimated $\xi_c(0)$ in $(BiSe)_{1.10}NbSe_2$ is much larger than the c-axis lattice constant, and the anisotropy is at most ~ 3. Neverthelss, our data is fitted much better by Tinkham model than the anisotropic GL model, indicating the system contains two-dimensional region. This kind of anomalous two-dimensional behavior in $H_{c2}(\theta)$ is also reported in a siliside superconductor CaAlSi [7], and GICs such as CaC_6 [8] and YbC_6 [9]. A common feature of these superconductors are the presence of irregular stacking. In CaAlSi, the presence of superlattice with five-fold or six-fold modulations along the c-axis is confirmed [10]. On the other hand, it is well established that intercalation of foreign elements into graphite do not occur homogeneously, and local variations of staging is possible. Moreover, the reason why $H_{c2}(\theta)$ in the bc-plane appears to be different between the single crystal and the microbridge sample is as following. Although $(BiSe)_{1.10}NbSe_2$ is an essencially two-dimensional superconductor, sample inhomogeneities and/or presence of mosaic microstructures may broaden the cusp peak.

Figures 4(a) and (b) show the field dependence of in-plane resistivity for the single crystal with different field angles in the ac- and the bc-planes, which were measured at $T = 2.0$ K with $I \parallel a$. In the ac-rotation, two-step transition is observed at several angles close to 0°, and each transition is sharp. These two-step structures may be related to the presence of stripe structures in the crystal. By contrast, the bc-rotation shows single but broader transitions. Figure 4(c) shows $H_{c2}(\theta)$ determined from Figs. 4(a) and (b). $H_{c2}(\theta)$ exhibits a steep behavior with a cusp and is fitted by Tinkham model, showing that two-dimensional behavior is not related to the direction of the current. Surprisingly, however, the anisotropy is inverted between a- and b-axes, by the reason that is not clear for now. We need to investigate sample dependence of H_{c2} for different directions of current to clarify this point.

4. Summary

We present the measurements of the upper critical field in a misfit layered compound $(BiSe)_{1.10}NbSe_2$ to demonstrate the dimensionality of this system. To confirm the effect originating from the stripe structures on the crystal surface, a microbridge sample fabricated by FIB technique is measured as well as single crystals. The temperature dependence of H_{c2} is linear dependence close to T_c and does not show any sign of dimensional crossover. Moreover, it is revealed that the anisotropy in $(BiSe)_{1.10}NbSe_2$ is small for layered superconductors. Cusp-like behavior in the angular dependence of H_{c2} explained by Tinkham model rather than the anisotropic GL model is observed in both samples. This means that $(BiSe)_{1.10}NbSe_2$ has a two-dimensional property which is not related to the stripe structures on the surface of the crystal.

References

[1] Wiegers G A 1996 *Prog. Solid State Chem.* **24** 1
[2] Nader A, Briggs A and Gotoh Y 1997 *Solid State Commun.* **101** 149
[3] Matsuzawa S, Pyon S and Tamegai T 2022 *J. Phys. Conf. Ser.* **2323** 012008
[4] Nagao M, Miura A, Horibe Y, Maruyama Y, Watauchi S, Takano Y and Tanaka I 2020 *Solid State Commun.* **321** 114051
[5] Samuely P, Szabó P, Kačmarčík J, Meerschaut A, Cario L, Jansen A G M, Cren T, Kuzmiak M, Śofranko O and Samuely T 2021 *Phys. Rev. B* **104** 224507
[6] Werthamer N, Helfand E and Hohenberg P 1966 *Phys. Rev.* **147** 295
[7] Ghosh A K, Tokunaga M and Tamegai T 2003 *Phys. Rev. B* **68** 054507
[8] Jobiliong E, Zhou H D, Janik J A, Balicas Y -J Jo, Brooks J S and Wiebe C R 2007 *Phys. Rev. B* **76** 052511
[9] Kawai N F and Fukuyama H 2008 *Physica C* **468** 2403
[10] Sagayama H, Wakabayashi Y, Sawa H, Kamiyama T, Hoshikawa A, Harjo S, Uozato K, Ghosh A K, Tokunaka M and Tamegai T 2006 *J. Phys. Soc. Jpn.* **75** 043713

Unusual Transport Properties with In-plane Magnetic Field in $Sr_xBi_2Se_3$ Single Crystals

Y. Zu, T. Matsumae, S. Pyon and T. Tamegai

Department of Applied Physics, The University of Tokyo, 7-3-1 Hongo, Bunkyo-ku, Tokyo 113-8656, Japan

E-mail: zuyuhang@g.ecc.u-tokyo.ac.jp

Abstract. A topological insulator Bi_2Se_3 becomes a topological superconductor by doping Cu or Sr. Due to the nematicity of these $A_xBi_2Se_3$ (A = Cu, Sr) materials, physical quantities show two-fold rotational symmetry as a function of in-plane magnetic field angle φ despite their three-fold structural symmetry in the plane. In this study, single crystals of $Sr_xBi_2Se_3$ ($0.0 \leqslant x \leqslant 0.17$) were grown by the self-flux method. Using these crystals, extensive measurements of the magnetoresistivity as a function of φ ($\rho(\varphi)$) were performed. Angular dependence of magnetoresistance shows an unusual $\cos(\varphi)$ term, which can be related to the c-axis current as well as higher harmonics components. Origins of higher harmonics components of $\rho(\varphi)$ are discussed.

1. Introduction

As a mathematical concept, topology is invented to classify shapes. Topology only concerns about the number of 'holes' in the shape. When two objects have the same number of holes (two shapes can change continuously to each other), they are classified into the same topological class. Inspired by this mathematical concept, physicists define 'topologically identical' to describe two different quantum mechanical wavefunctions when they can be adiabatically connected to each other. The theoretical discovery of the quantum spin Hall insulator phase in a two-dimensional (2D) time-reversal-invariant insulator in 2005 raises the worldwide attention to topological quantum systems [1]. Later, some three-dimensional (3D) systems were also discovered [2,3] and were named as topological insulators.

Bi_2Se_3 is one of the representative topological insulators that show unique metallic surface states despite their insulating nature in the bulk [4]. One reason for the large-scale research on Bi_2Se_3 is that it can easily be turned into a superconductor by simply doping ions into it. Such a class of superconductors are called topological superconductors. In the previous study, several superconductors based on Bi_2Se_3 were discovered such as $A_xBi_2Se_3$ (A = Cu, Sr). When new ions are doped into Bi_2Se_3, they are inserted into the van der Waals (vdW) gap between two Bi_2Se_3 quintuple layers (QL) in the crystal structure, and do not change the crystal structure of Bi_2Se_3 too much, keeping a three-fold rotational symmetry in the ab-plane. However, some physical quantities of $A_xBi_2Se_3$ show two-fold rotational symmetry as a function of in-plane magnetic field angle φ. This kind of spontaneous symmetry lowering is called nematicity, and there is still no good theory to explain it as for now. One possible scenario (scenario 1) is based on the superconducting gap symmetry Δ_{4x} or Δ_{4y} with gap nodes [5,6]. Since only superconducting states with odd parity are related to topological superconductivity and only Δ_4 has relationship with nematicity, nodes in the Δ_4 gap is believed to influence the electronic structure and finally causes the nematicity. Another scenario (scenario 2) is the elongation of one of three a-axes, which should be equivalent in the ideal crystal structure. A former study confirmed the elongation of one of *a*-axes by high-resolution X-ray diffraction measurements [7]. By combining scenario 1 and scenario 2, six different directions can be the principal axis of nematicity. Since the Δ_4 gap can have gap nodes along three equivalent directions for both Δ_{4x} (*a*-axis) and Δ_{4y} (a^*-axis), one of these six axes will be selected as the principal axis of nematicity. Although sophisticated theory and state-of-the-art experiments could clarify the essence of nematic superconductivity, presence of uncontrolled inhomogeneities of the sample cannot be ignored. Direct visualization of nematicity or nematic domains will be highly important since we cannot distinguish whether the result is caused by the meaningful physical source or just simply by the nematic domains with different nematic orientations. Crystals with high degrees of homogeneity are highly required for the reliable measurement.

In the present study, we focused on Sr-doped Bi_2Se_3, and explored the presence of abnormal current paths observed in the in-plane magnetic-field-angle dependence of resistivity [8]. Possible origins of abnormal current paths will be discussed.

2. Experimental Methods

2.1. Crystal growth

Single crystals $Sr_xBi_2Se_3$ (x = 0.17, 0.08, 0.04, 0.02, and 0) were grown by the two-step method as described in Ref. [8]. In short, Sr pieces and Bi_2Se_3 powder (prepared in advance) were thoroughly mixed and sealed into a quartz tube and heated in an electrical furnace. After reacting at 1000°C for 24 h, the sample was cooled down to 650°C at a rate of 3 °C/h. Then the sample was slowly cooled down to room temperature after shutting down the furnace (furnace cooling).

2.2. Measurements

Magnetization of the sample was measured by a superconducting quantum interference device (SQUID) magnetometer (MPMS-XL5, Quantum Design). An X-ray diffractometer (Smartlab, RIGAKU) was used to determine the direction of the *a*-axis of the sample, and it was cut into rectangular shape with longer edge along the *a*-axis. Temperature dependence and in-plane magnetic-field-angle dependence of resistivity were measured by using physical property measurement system (PPMS, Quantum Design) with a horizontal sample rotator. AC four-probe method was applied with AC current amplitude of 1 mA. In these measurements, current was passed along the *a*-axis of the sample with dimensions of ~1.5 mm × 0.5 mm × 0.1 mm.

3. Results and Discussion

3.1. In-plane magnetic-field-angle dependence

In this study, $Sr_xBi_2Se_3$ with a wide range of x were measured (x = 0.17: superconducting, x = 0.08, 0.04, 0.02, and 0: non-superconducting). figure 1 (a) shows temperature dependence of magnetization of the sample with x = 0.17, showing superconductivity with an onset of diamagnetism at T_c = 2.9 K.

For transport measurements, we prepared four contacts for current and voltage by sputtering Au onto the sample as shown in figure 1(c). Angular dependence of the change in magnetoresistivity ($\Delta\rho$) at 2.0 K and 8.5 T is shown by red squares in figure 1 (b). Here, $\Delta\rho$ is $\rho - \rho_0$ with ρ_0 being the resistivity at zero field evaluated from the fitting described below. By using Fourier series expansion, we try to analyse $\rho(\varphi)$: the resistivity is decomposed into a constant term, $\cos(\varphi - \alpha_1)$ term , and $\cos(2(\varphi - \alpha_2))$ term as shown by equation (1):

$$\rho(\varphi) = \rho_0 + \rho_1\cos(\varphi - \alpha_1) + \rho_2\cos(2(\varphi - \alpha_2)). \quad (1)$$

The $\cos(2(\varphi - \alpha_2))$ term is interpreted as originated from magnetoresistance, nematicity, or misalignment of the field from the in-plane direction. On the other hand, in the usual transport measurements, the $\cos(\varphi - \alpha_1)$ term should not exist. One possible origin of the $\cos(\varphi - \alpha_1)$ term is the Hall effect induced by nonideal current path. In order to have Hall voltage, a transport current must flow along the *c*-axis between the voltage contacts as shown by the orange solid line with arrows in figure 1(c) [8]. Such an unintentional current along the *c*-axis can appear when the current is bent by macroscopic defects in the sample.

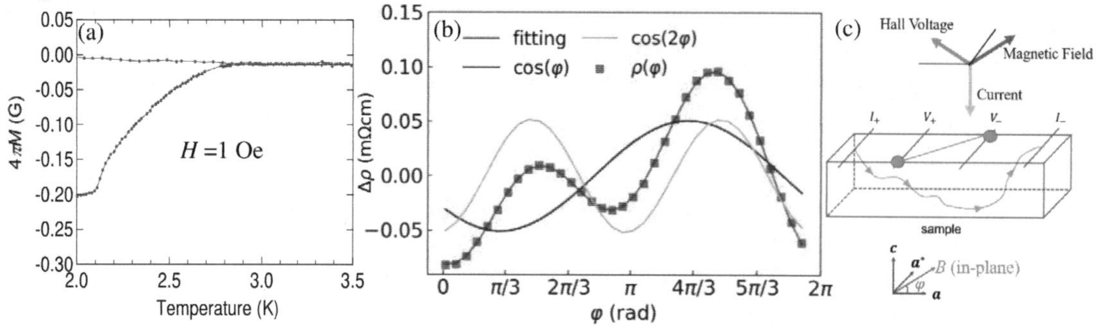

Figure 1. (a) Temperature dependence of magnetization for $Sr_{0.17}Bi_2Se_3$. (b) In-plane magnetic-field-angle (φ) dependence of change in resistivity $\Delta\rho = \rho - \rho_0$ in $Sr_{0.17}Bi_2Se_3$ at 2 K and 8.5 T, with $\rho_0 = 1.168$ mΩcm. (c) Possible current flow in the $Sr_{0.17}Bi_2Se_3$ single crystal. The current path along the *c*-axis exists, which may be bent by obstacles. Green circles are possible effective contact points of voltage contacts.

Similar in-plane magnetic-field-angle dependence measurements were also done in samples with smaller x that do not show superconductivity. Figure 2 (a), (b), (c), and (d) show resistivity as function of φ in samples with $x = 0.08$, 0.04, 0.02, and 0, respectively. Their angular dependence cannot be fitted reasonably well with terms only up to cos2φ. Instead, we need to use the following equation including cos3φ and cos4φ terms:

$$\rho(\varphi) = \rho_0 + \rho_1\cos(\varphi - \alpha_1) + \rho_2\cos(2(\varphi - \alpha_2)) + \rho_3\cos(3(\varphi - \alpha_3)) + \rho_4\cos(4(\varphi - \alpha_4)). \quad (2)$$

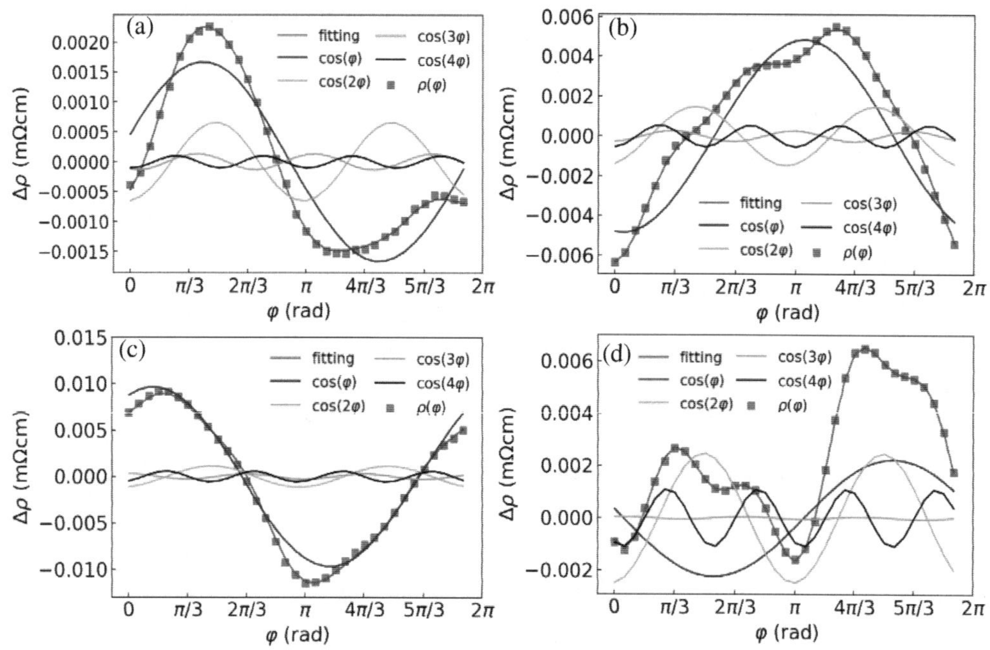

Figure 2. Plots of in-plane magnetic-field-angle dependence of magnetoresistivity $\Delta\rho = \rho - \rho_0$ at 2 K and 8.5 T in $Sr_xBi_2Se_3$ with (a) $x = 0.08, \rho_0 = 0.141$ mΩcm, (b) $x = 0.04, \rho_0 = 0.374$ mΩcm, (c) $x = 0.02, \rho_0 = 0.471$ mΩcm and (d) $x = 0, \rho_0 = 0.588$ mΩcm. The fitting curve and all cosφ, cos2φ, cos3φ and cos4φ terms are also plotted.

In figure 2(a)-(d), angular dependences of all terms are also plotted. Although the appearance of nematicity can induce two-fold symmetry in many physical quantities in $Sr_xBi_2Se_3$, the three-fold symmetry of the crystal structure can still influence the physical quantities and eventually generate a new $\cos(3(\varphi - \alpha_3))$ term in the in-plane magnetic-field-angle dependence measurement. On the other hand, physical origins of $\cos 4\varphi$ remains elusive.

3.2. In-plane magnetic-field-angle dependence

In order to examine all the scenarios, magnetic field dependence of resistivity in samples with $x = 0.08$, 0.04, 0.02, and 0 are plotted in figure 3 (a)-(d). Based on these data, we analyzed the magnetic field dependence of amplitudes of all terms in figure 4 (a)-(d). It is clear that both $\cos(\varphi - \alpha_1)$ and $\cos(2(\varphi - \alpha_2))$ terms show stable relationships with magnetic field, while amplitudes of $\cos(3(\varphi - \alpha_3))$ and $\cos(4(\varphi - \alpha_4))$ terms are much smaller than those of $\cos(\varphi - \alpha_1)$ and $\cos(2(\varphi - \alpha_2))$ terms and fluctuate.

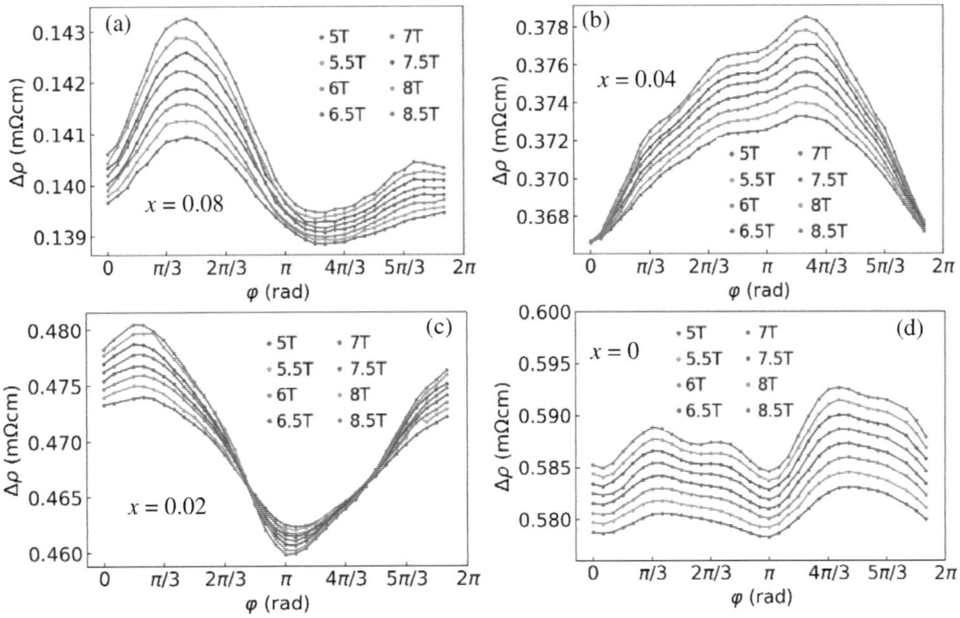

Figure 3. In-plane magnetic-field-angle dependence of magnetoresistivity at 2 K and between 5 T and 8.5 T in $Sr_xBi_2Se_3$ with x = (a) 0.08, (b) 0.04, (c) 0.02, and (d) 0.

In order to obtain detailed magnetic field dependence of amplitudes of $\cos(\varphi - \alpha_1)$ and $\cos(2(\varphi - \alpha_2))$ terms, $\rho_1(H)$ and $\rho_2(H)$ are plotted in double-logarithmic scale in figure 5 (a) and (b), respectively. Linear fittings of data in figure 5(a) and (b) give power-law exponents k_1 and k_2 of $\rho_1(H)$ ($\propto H^{k_1}$) and $\rho_2(H)$ ($\propto H^{k_1}$), which are summarized in Table 1. From these values, we can see that ρ_1 is always proportional to H, while the magnetic field dependence of ρ_2 changes from linear for the sample with small x into quadratic for the sample with large x. The latter fact is consistent with the scenario that $\cos(2(\varphi - \alpha_2))$ term is originated from magnetoresistance. Actually, Sultana *et al.* found that in pure Bi_2Se_3 single crystal, the magnetoresistance is linear to magnetic field at high fields (large than 5 T) for in-plane fields [9]. This interpretation is further supported by the fact that α_2 stays close to $\pi/2$ for all the samples. The difference between the transverse and longitudinal magnetoresistance will make magnetoresistance be minimum at $\varphi = 0$ and π, which also means α_2 should be $\pi/2$. It should be noted that the $\cos(2(\varphi - \alpha_2))$ term can also be caused by the misalignment of magnetic field from the plane. However, in such a case, α_2 is expected to take random values. figure 5 (c) shows x dependence of the phase of the $\cos(\varphi)$ and $\cos(2\varphi)$ terms. If our interpretation of Hall effect origin of $\cos(\varphi)$ terms is correct, the Hall voltage should be zero at $\varphi = 0$ and π, which means α_1 should stay $\pi/2$. However, α_1 fluctuates a lot with x. Such a fluctuation can be partly explained by the shift of effective locations of the two voltage terminals, V_+ and V_- as shown by green circle in figure 1(c). Namely, if locations of the effective voltage terminals are not

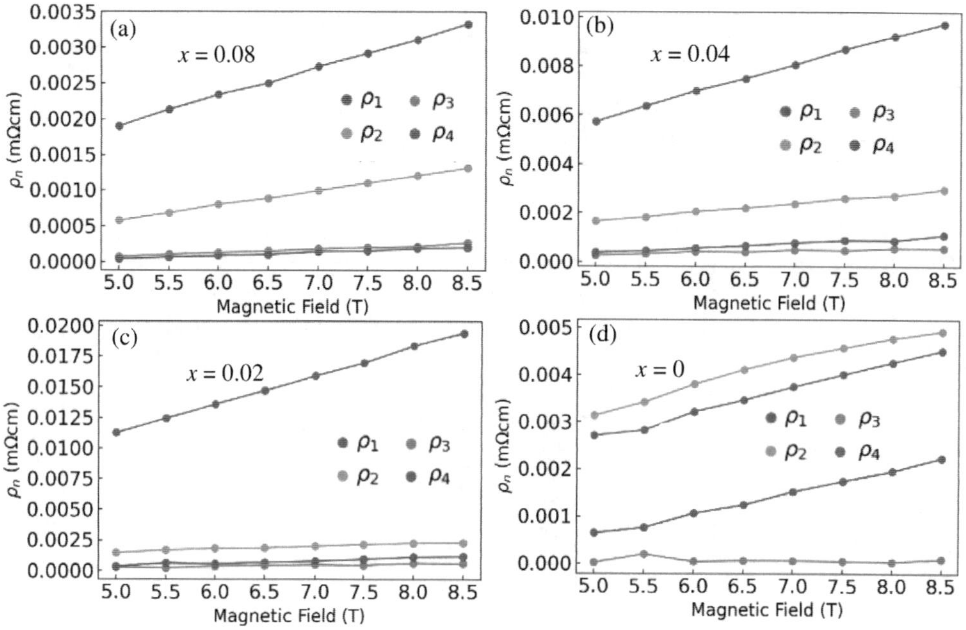

Figure 4. In-plane magnetic-field dependence of the amplitudes of all angular components in equation (2) at 2 K in $Sr_xBi_2Se_3$ with x = (a) 0.08, (b) 0.04, (c) 0.02, and (d) 0.

parallel to the *a*-axis, the Hall voltage can be zero at non-zero φ. The normalized amplitude of all terms at 2 K and 8.5 T for samples with different x are summarized in Table 2. Normalized amplitudes of $\cos(\varphi)$ and $\cos(2\varphi)$ terms roughly decrease with decreasing x. The absence of $\cos(3\varphi)$ and $\cos(4\varphi)$ terms in the sample with $x = 0.17$ is due to very large normalized amplitudes of $\cos(\varphi)$ and $\cos(2\varphi)$ terms.

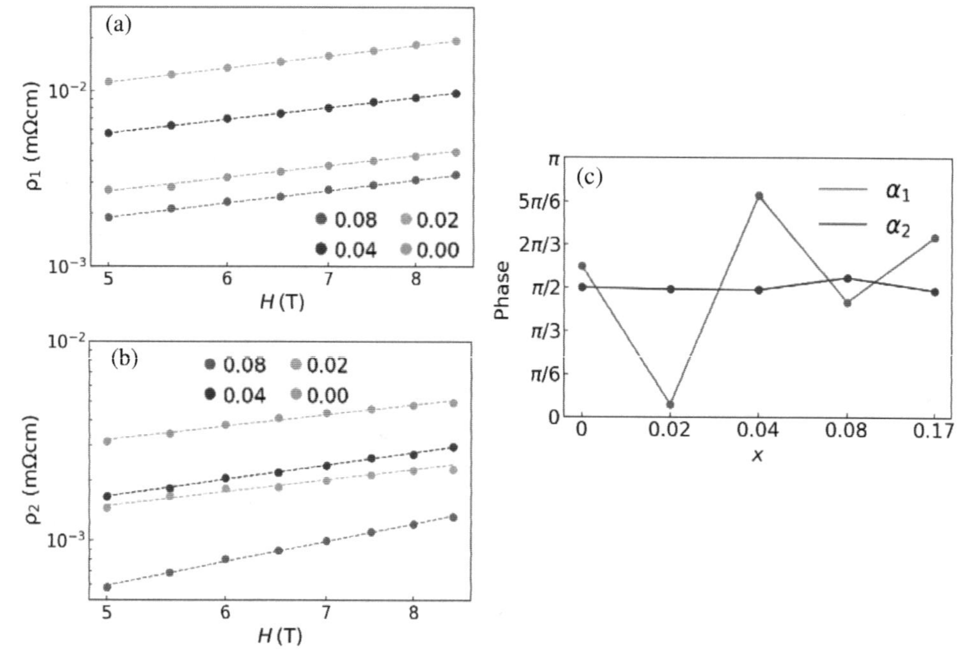

Figure 5. Double-logarithmic plots of (a) ρ_1 and (b) ρ_2 in samples with different x values. (c) Evolution of phase factors of the $\cos(\varphi)$ and $\cos(2\varphi)$ terms with x.

Table 1. The fitting parameter k_1 and k_2.

x	0.08	0.04	0.02	0.00
k_1	1.038	0.995	1.025	0.996
k_2	1.520	1.067	0.891	0.863

Table 2. The normalized amplitude of all terms in equation (2) in $Sr_xBi_2Se_3$ at 2 K and 8.5 T.

x	0.17	0.08	0.04	0.02	0.00
ρ_1/ρ_0	7.89%	2.37%	2.61%	4.11%	0.76%
ρ_2/ρ_0	7.88%	0.93%	0.79%	0.48%	0.84%
ρ_3/ρ_0	0.00%	0.19%	0.15%	0.12%	0.01%
ρ_4/ρ_0	0.00%	0.14%	0.29%	0.24%	0.37%

4. Summary

$Sr_xBi_2Se_3$ single crystals with $0 \leqq x \leqq 0.17$ were grown by self-flux method, and in-plane anisotropy of transport properties were investigated. In-plane magnetic-field-angle (φ) dependence of resistivity in $Sr_xBi_2Se_3$ with $x = 0.17$ is dominated by cos(φ) and cos(2φ) terms. However, φ dependence of resistivity becomes complicated in non-superconducting samples with smaller x, which can be fitted by including terms up to cos(4φ). The cos(2φ) term can be interpreted to be originated from magnetoresistivity or nematicity. The cos(φ) term can only be explained by the presence of the *c*-axis current, possibly induced by sample inhomogeneities. The cos(3φ) term may reflect the intrinsic three-fold symmetry of the system, while the origin of the cos(4φ) term remains elusive. Finally, to better understand the unusual in-plane field-angle dependence of resistivity, direct visualization of current paths using magneto-optical imaging would be helpful as has been attempted in Bi_2Se_3 [4].

References

[1] Kane C L and Mele E J 2005 *Phys. Rev. Lett.* **95** 226801
[2] Moore J E and Balents L 2007 *Phys. Rev. B* **75** 121306
[3] Kane C L and Mele E J 2005 *Phys. Rev. Lett.* **95** 146802
[4] Jash A, Ghosh S, Bharathi A and Banerjee SS 2022 *Bull Mater Sci* **45** 17
[5] Fu L and Berg E 2010 *Phys. Rev. Lett.* **105** 097001
[6] Hashimoto T, Yada K, Yamakage A, Sato M and Tanaka Y 2013 *J. Phys. Soc. Jpn.* **82** 044704
[7] Kuntsevich A Y, Bryzgalov M A, Prudkoglyad V A, Martovitskii V P, Selivanov Y G and Chizhevskii E G 2018 *New J. Phys.* **20** 103022
[8] Matsumae T, Pyon S and Tamegai T 2022 J. *Phys.: Conf. Ser.* **2323** 012005
[9] Sultana R, Awana G, Pal B, Maheshwari P K, Mishra M, Gupta G, Anurag G, Thirupathaiah S and Awana V P S 2017 *J. Supercond. Nov. Magn.* **30** 2031

The in- and out-of-plane magnetisation of highly underdoped $YBa_2Cu_3O_{6+x}$ single crystals

I Kokanović[1,2,*] and J R Cooper[1]

[1]Cavendish Laboratory, University of Cambridge, Cambridge CB3 OHE, U.K.
[2]Department of Physics, Faculty of Science, University of Zagreb, Zagreb, Croatia
*Email: kivan@phy.hr

Abstract. Previously we have shown how measurements of static magnetic susceptibility $\chi_c(T)$ of $YBa_2Cu_3O_{6+x}$ single crystals for magnetic fields applied along the *c*-axis, and $\chi_{ab}(T)$ for fields in the *ab*-plane, can give useful information about their thermodynamic properties which are still being hotly debated. SQUID magnetometry above the superconducting (s/c) transition temperature T_c is used for larger crystals, while piezolever torque magnetometry gives $\chi_c(T)$-$\chi_{ab}(T)$ for tiny crystals. Here we present new data for more heavily under-doped crystals with hole concentrations per CuO_2 plane, p =0.058 to 0.073. We again find that the T-dependent anisotropy well above T_c arises from the pseudogap and the g-factor anisotropy, while at lower T there are Gaussian s/c fluctuations with a strong cut-off. This is possibly a different region where neutron scattering studies give evidence for competition between incommensurate magnetic short-range order and superconductivity. We have studied crystals with three values of x, measuring $\chi_c(T)$ and $\chi_{ab}(T)$ immediately after fixing x by quenching on to a copper block and again after allowing sufficient time at room temperature for the Cu-O chains to order. Thus, we report data for three pairs of underdoped (UD) crystals with T_c ranging from 13 to 36 K, namely UD13 and UD20, UD15 and UD30, UD23 and UD36. As found previously for polycrystalline samples, ordering the Cu-O chains increases T_c and substantially reduces the Curie term. These are isotropic, in contrast to the Pauli-like susceptibility from the CuO_2 planes where there is g factor anisotropy, and this allows us to make a powerful novel analysis. At higher T, $\chi_c(T)$- $\chi_{ab}(T)$ varies as a + bT where the bT term arises from the pseudogap. At lower T, deviations from this behaviour are analysed in terms of 2D or 3D Gaussian s/c fluctuations to obtain in-plane coherence lengths and upper critical fields as T→0.

1. Introduction

Understanding the normal state of cuprate superconductors is still an important and topical problem in condensed matter physics. The anomalous normal state properties of underdoped single crystals, in particular the pseudogap, remain the subject of an intense debate that has been made even more interesting by the relatively recent discovery of charge density waves in a number of moderately underdoped cuprates e.g. [1]. One fundamental thermodynamic property is the temperature (T) dependent, static magnetic susceptibility $\chi(T)$. In two papers [2, 3] we have shown how measurements of static magnetic susceptibility $\chi_c(T)$ of $YBa_2Cu_3O_{6+x}$ single crystals for magnetic fields applied along the *c*-axis, and $\chi_{ab}(T)$ for fields in the *ab*-plane, can give new and useful information about their thermodynamic properties. SQUID magnetometry above the superconducting (s/c) transition temperature T_c has been used for larger crystals [2], while piezolever torque magnetometry gives

$\chi_c(T)$- $\chi_{ab}(T)$ for tiny crystals [3]. The main results in these two papers are that the T-dependent anisotropy well above T_c arises from the pseudogap, and the g-factor anisotropy, while at lower T there are Gaussian s/c fluctuations with a strong cut-off above approximately1.1 T_c in agreement with Rullier-Albenque et al [4]. Here we present some new data for highly under-doped crystals with hole concentrations per CuO_2 plane, p =0.058 to 0.073 with T_c ranging from 13 to 36 K. This is an interesting region where neutron scattering studies show spin-glass magnetic short-range order and superconductivity [5, 6], followed by charge density wave order setting in above p = 0.08, see Fig. 4 of Ref. [1]. However, at present we do not see clear differences in our data in these two regions. We have studied crystals with three values of x, measuring $\chi_c(T)$ and $\chi_{ab}(T)$ immediately after fixing x by quenching on to a copper block and again after allowing sufficient time at room temperature for the Cu-O chains to order. In this doping range, as found by others [7] for polycrystalline samples, ordering the Cu-O chains increases T_c and substantially reduces the Curie term. The $\chi_D(T)= \chi_c(T) - \chi_{ab}(T)$ data show a weakly T-dependent linear region at higher T where $\chi_D(0)$ is 1.3-1.6 x10^{-4}emu/mol. We discuss deviations from this linear behavior in terms of 2D or 3D Gaussian superconducting fluctuations.

2. Results and Discussion

The crystals of $YBa_2Cu_3O_{6+x}$ were grown by K. Iida using the top-seeded melt processing technique [8] and high purity (5N) starting materials.

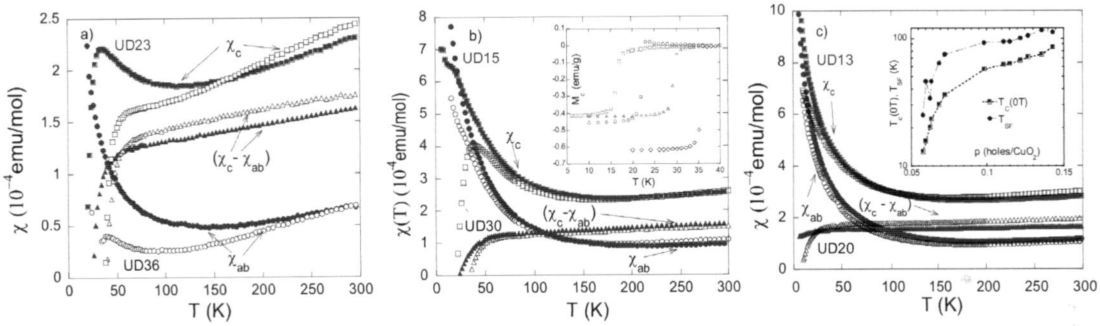

Figure 1. (a-c) Temperature dependence of the static magnetic susceptibility $\chi_c(T)$ (□) and $\chi_{ab}(T)$ (o) for H parallel the c axis and the ab plane respectively, together with the anisotropy $\chi_c(T) - \chi_{ab}(T)$ (Δ) for six $YBa_2Cu_3O_{6+x}$ crystals measured in an applied magnetic field of 5 T. The blue symbols show data taken within 30 minutes after quenching and the black symbols after several months at room temperatute. The inset to Fig. 1 b) shows the superconducting transition of the same crystals measured with H//c on warming in 10 Gauss using a SQUID magnetometer after cooling in zero field. The insert to Fig. 1 c) shows T_c in zero applied field and the onset temperature of superconducting fluctuations T_{SF} vs. hole concentration p for the present samples and those from Ref. [2] on a logarithmic vertical scale.

Sharp superconducting transition temperatures measured on warming in a field of 10 G applied parallel to the c axis after zero field cooling to well below T_c are listed in Table 1. In the following we identify each sample by its T_c value rounded down to the nearest integer, with prefixes UD meaning underdoped. In order to change x, a crystal was held in a small gold boat, annealed in various flowing gases [9] removed quickly from the furnace and quenched on a copper block. It was first mounted and cooled in the SQUID magnetometer within half of hour after quenching and then after annealing at room temperature for several months. The samples were fixed inside a long plastic straw that was the sample holder by inserting two pieces of another straw. These were made by slitting two halves of another straw along their length to fit snugly inside the outer straw. The absence of the inner pieces, or their overlap, in the sample region, over at most +/- 1mm, would give a T-independent offset in the susceptibility data. For the sample weights 50-100 mg used here +/- 1mm corresponds to a systematic shift in χ of +/- $3x10^{-6}$ emu/mole. Checks for the absence of possible ferromagnetic contamination were made by performing M-H scans at 300 and 100 K as well as 30 K above T_c. Any intercepts at H= 0 were less than $2.6x10^{-5}$ emu corresponding to $\Delta\chi$ =0.3/50000 emu/mole at 5 T for the crystals measured, and have been subtracted from the data reported here. Weight changes before and after annealing, and separate thermo-gravimetric studies were used to determine the oxygen content relative to the value for the as-grown crystal. The oxygen content of the as-grown crystal was fixed by comparison with literature data [10]. Namely, the crystal with T_c = 13.5 K was taken to have x=0.35 because immediately after quenching this crystal had a very weak superconducting response at 4 K. Weight changes with time were also used to verify that the equilibrium oxygen content associated with any particular gas mixture and temperature had been achieved. Within the experimental error of ±0.03, in x, $T_c(x)$ of our annealed crystals agree very well with those from UBC group Ref. [10] and are summarized in Table 1. The temperature dependence of the static magnetic susceptibility $\chi_c(T)$ // to the c axis and $\chi_{ab}(T)$ // to the ab plane for H = 5 T and the anisotropy data $\chi_D(T) = \chi_c(T) - \chi_{ab}(T)$ are shown in Fig. 1 for the six $YBa_2Cu_3O_{6+x}$ crystals. The inset to Fig. 1 (b). shows the superconducting transition of the $YBa_2Cu_3O_{6+x}$ crystals for H // c measured on warming in 10 Gauss after zero field cooling.

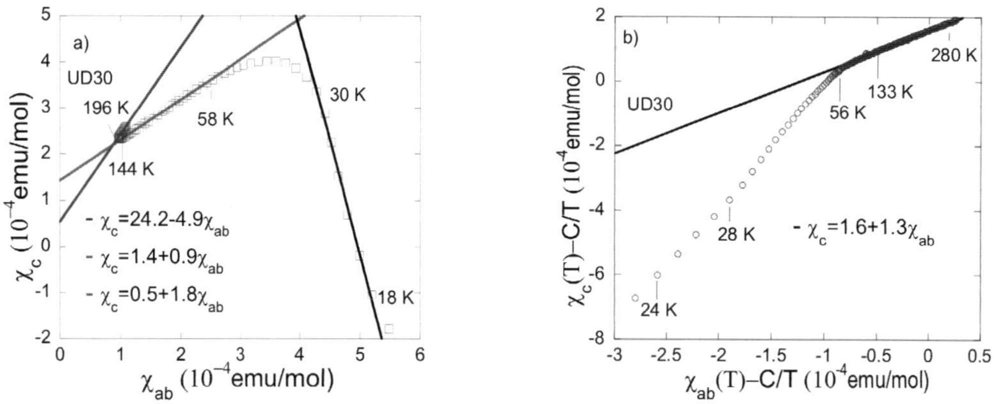

Figure 2. (a) The static magnetic susceptibility $\chi_c(T)$ versus $\chi_{ab}(T)$ for H = 5 T, H//c axis and H//ab plane respectively, of $YBa_2Cu_3O_{6+x}$ single crystal (UD30). The solid lines show linear fits. Arrows show some typical temperatures. The error in the slopes is ± 0.1. **(b)** The static magnetic susceptibility $\chi_c(T) - C/T$ versus $\chi_{ab}(T) - C/T$ for H = 5 T. The solid lines show linear fits with the parameters listed. Arrows show some typical temperatures.

It can be seen from Fig. 1 that $\chi(T)$ is T-dependent. All the crystals measured show clear downturns in $\chi_D(T)= \chi_c(T) - \chi_{ab}(T)$ associated with superconducting fluctuations closer to T_c. The onsets of these downturns, or equivalently the derivative plots in Fig.4, define the onset of superconducting fluctuations at T_{SF} values that are plotted vs. p in the inset to Fig. 1 c). $\chi_D(T)$ has a weakly T-dependent linear region at higher T where $\chi_D(0)$ is 1.3-1.6 $x10^{-4}$emu/mol. With increasing temperature $\chi_c(T)$ and $\chi_{ab}(T)$ reach minimum values in the temperature range above T_c due to the Curie term, C/T. It can be seen from Fig. 1 that the Curie contribution to $\chi(T)$ decreases with annealing the crystals at room temperature. Importantly there is no Curie term in $\chi_D(T)$ for all crystals. Thus, the Curie term is isotropic. Because the anisotropy is T-dependent we believe that it cannot arise from Van Vleck paramagnetism, which involves intermediate states well away from the Fermi energy and would therefore be T-independent even for T as high as 1000 K, but that it must arise from g-factor anisotropy as discussed below [2, 3, 11, 12]. In Figs. 2 and 3 we have plotted $\chi_c(T)$ versus $\chi_{ab}(T)$ and

χ_c(T)-C/T versus χ_{ab}(T)-C/T, with T as an implicit parameter, for the UD30 and UD36 crystals, respectively. All of the χ_c(T) versus χ_{ab}(T) and χ_c(T)-C/T versus χ_{ab}(T)-C/T plots in Figs. 2 and 3 shows a linear relation in the high temperature range. In Figs. 2 b) and 3 b) we have chosen C so that the plots are linear over as wide a range of T as possible in contrast to the complicated behaviour in Figs. 2 a) and 3 a). The linearity confirms that C is isotropic and when C/T is subtracted, the slope is in the range corresponding to $(g_c/g_{ab})^2$=1.3-1.5, is the same as those for other crystals with smaller values of C. Thus, these plots provide a novel and reliable way of determining C in situations where there can be other sources of T-dependence, for example from s/c fluctuations or the pseudogap.

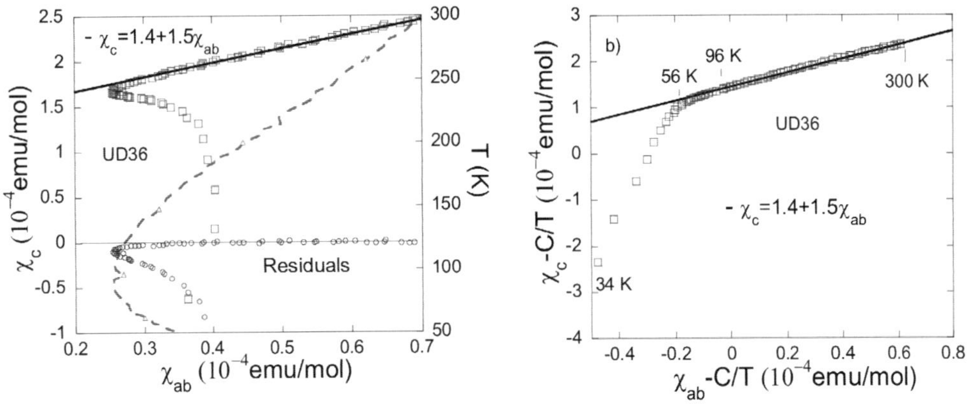

Figure 3. (a) The static magnetic susceptibility and residuals χ_c(T) versus χ_{ab}(T) for H = 5 T parallel the c axis and the ab plane respectively, of the $YBa_2Cu_3O_{6+x}$ single crystal (UD36). **(b)** The static magnetic susceptibility χ_c(T) − C/T versus χ_{ab}(T) − C/T for H = 5 T . The solid lines show linear fits with the parameters listed. The error in the slopes is ± 0.1. Arrows show some typical temperatures. The red dashed line shows T vs χ_{ab} - right hand scale.

As in our earlier work [2, 3] for a quantitative analysis of the magnetic susceptibility data shown in Fig. 1 we consider several contributions to the spin susceptibility for both χ_c(T), χ_{ab}(T) summarised by:

$$\chi(T) = \chi_{PG}(T) + \chi_{FL}(T) + C/T + \chi_{VV}(T) + \chi_{core}(T) \qquad (1)$$

Here χ_{PG}(T) shows the effect of the pseudogap on χ(T), in the model introduced by Loram [13] in which there is a triangular notch of width $2E_G$ in the Fermion density of states (DOS) that is centered on the Fermi energy. We emphasize that this unusual DOS is an empirical way of describing the strong electron correlations in hole-doped cuprates. For a long time we have argued that E_G is a characteristic energy or crossover temperature rather than a well-defined phase transition temperature. $E_G = 0$ for $p \geq 0.19$ and varies as $E_G = E_0(1 - p/0.19)$ for $p \leq 0.19$, where p is the number of added holes per planar Cu. $E_0 \approx 1200$K, approxmately equal to the exchange energy between Cu spins in the antiferromagnetic state at p = 0. As discussed previously [2, 3] this empirical model gives an excellent description of the magnetic susceptibility and electronic entropy of several families of hole-doped cuprates over a wide range of p and T [14, 15]. For the DOS given above, the static (Pauli-like) spin susceptibility χ_{PG} is given by:

$$\chi_{PG} = N_0\mu_B^2\left\{1 - \frac{2k_BT}{E_g}\ln\left[cosh\left(\frac{E_g}{2k_BT}\right)\right]\right\} \qquad (2)$$

For T<0.3 E_g/k_B, the limiting form is:

$$\chi_{PG}(T) = A\frac{2k_BT}{E_g}ln2 \qquad (3)$$

where $A=N_0\mu_B^2$. In the temperature range closer to T_c when H is applied parallel to the c-axis the Gaussian superconducting fluctuation (GF) contribution to the magnetisation along the c axis is [16]:

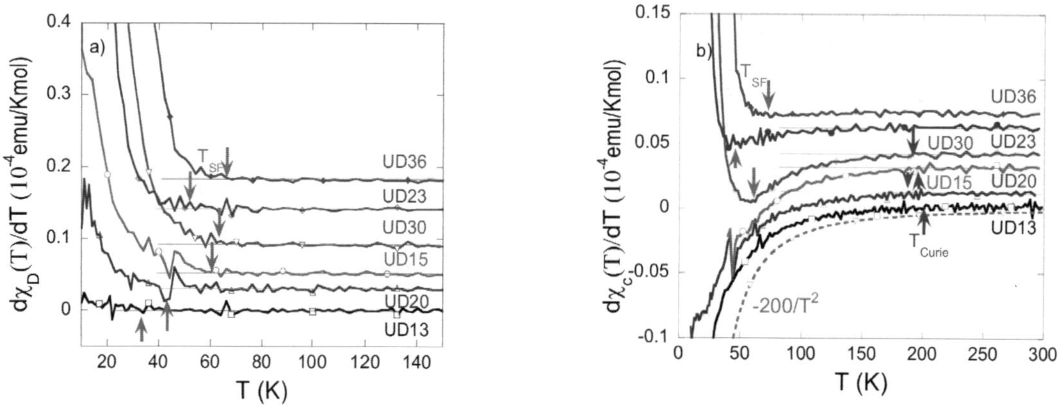

Figure 4. (a) The temperature derivative of static magnetic susceptibility $\chi_D(T)= \chi_c(T) - \chi_{ab}(T)$ at H = 5 T for $YBa_2Cu_3O_{6+x}$ single crystals: UD36, UD23, UD30, UD15, UD20 and UD13. **(b)** The temperature derivative of static magnetic susceptibility $\chi_c(T)$ for the same crystals. The solid lines are a guide for the eye. The arrows show onset temperatures T_{SF} and T_{Curie} where the superconducting fluctuations and Curie contribution start to be visible. The derivatives were calculated by subtracting neighboring points with a spacing of ~2 K, without any smoothing. For clarity the data are shifted vertically.

$$M_c^{FL} = -\frac{\pi k_B TH}{3\Phi_0^2}\frac{\xi_{ab}^2}{s\sqrt{1+\left(\frac{2\xi_{ab}}{\gamma s}\right)^2}} \qquad (4)$$

Here $\gamma = \xi_{ab}/\xi_c$ is the anisotropy, $\xi_{ab} = \xi_{ab}(0)/\varepsilon^{1/2}$ and $\xi_c = \xi_c(0)/\varepsilon^{1/2}$, are the T-dependent GL coherence lengths // and ⊥ to the layers, $\varepsilon = \ln(T/T_c)$, s = 1.17 nm is the distance between CuO_2 bi-layers, Φ_0 is the flux quantum for pairs and k_B is Boltzmann's constant. Eq. 4 is valid in the low field limit when $H \leq \Phi_0/(2\pi\xi_{ab}^2)$ i.e. the magnetic length $l = (\Phi_0/2\pi H)^{1/2}$ is much greater that $\xi_{ab}(T)$ the in-plane coherence length, and then the susceptibility $\chi_c^{FL} \equiv M_c^{FL}/H$ does not vary with H. For H⊥c, $\chi_{ab}^{FL}=0$ in the two-dimensional (2D) limit $s \geq \xi_c(T)$ and in the opposite 3D limit $\chi_{ab}^{FL} = \chi_c^{FL}/\gamma$.

In order to see the onset temperature of superconducting fluctuations (T_{SF}) and the Curie contribution (T_{Curie}) to the magnetic susceptibility, in Figs. 4 (a) and (b), we have plotted the temperature derivative of the static magnetic susceptibilities $\chi_D(T)$ and $\chi_c(T)$, measured at H = 5 T for UD36, UD23, UD30, UD15, UD20 and UD13 crystals. The solid lines are guides to for the eye. The arrows show onset temperatures where the superconducting fluctuations and Curie contribution start to be visible. One important result of the present work is that these T_{SF}, shown in Figs. 4 (a) and (b) are 30-35 K higher than T_c. T_{SF} increases with T_c, as shown in Fig.1 c), and is considered to be the onset temperature of Aslamasov-Larkin type superconducting fluctuations [3, 16].

To analyse the magnetic susceptibility data quantitatively, we have been able to fit the $\chi_D(T)$ data with a nearly constant factor A, changing ξ_{ab}, adding a temperature independent term ($\chi_0=\chi_{VV}+\chi_{core}$) and using $T_{PG}=E_g/k_B = 1200(1-p/0.19)$, and T_c at 5 T which is up to 5 K less than that at H=0. This latter result is consistent with the simple formula for the effect of a magnetic field on $M_c^{FL}(T)$ given in Ref. [3]. The values of the pre-factor A as well as the values of the $\xi_{ab}(0)$ obtained from these fits are listed in Table 1. Adequate fits to the magnetic susceptibility anisotropy data of the measured samples plotted in Fig. 1 can only be achieved by using all terms in Eq. 1 and adding a significant constant term χ_0, ranging from 1.3 10^{-4} emu/mol to 1.6 10^{-4} emu/mol as summarized in Table 1. The extracted values of the $\xi_{ab}(0)$ increase with underdoping. Our interpretation ascribes the temperature dependence of the $\chi_D(T)$ in the temperature range above T_c up to T_c+35 K to Gaussian fluctuations [3, 16–19] and in contrast to other authors [20, 21] the T-dependence of $\chi_D(T)$ at higher T is attributed to the pseudogap combined with the g factor anisotropy.

Table 1. Summary of results. The x values in the first column are determined from weight losses of the same crystals on annealing and during thermogravimetric analysis in flowing argon. The critical temperature T_c was taken as the midpoint of the superconducting transition, as determined by measuring the field-warming magnetisation at 10 Oe after zero-field cooling. C, A, χ_0 and ξ_{ab} are the parameters of Eq. 1 needed fit to the T-dependence of magnetic susceptibility data χ_c - χ_{ab} shown in Fig. 1. T_{PG} is the pseudogap temperature which is constrained to follow the empirical law T_{PG}=1200(1-p/0.19) for p<0.19 [13]. The values of p are obtained from Ref. 10. $H_{c2}(0)=\Phi_0/(2\pi\, \xi^2_{ab}(0))$, T_1-T_2 is the temperature range over which the values of C are obtained.

$YBCO_x$ x_{meas}	T_c (K)	10^4C (emuK/mol)	T_1-T_2 (K)	10^4A (emu/mol)	$10^4\chi_0$ (emu/mol)	T_{PG} (K)	$\xi_{ab}(0)$ (nm)	$H_{c2}(0)$ (T)	p /CuO_2
0.42	36.2	28±1	60-130	1.49	1.52	739	3.5	26.9	0.073
0.40	30.1	160±3	60-200	1.50	1.43	770	4.2	18.7	0.069
0.42	23.3	49±2	60-150	1.46	1.34	796	4.2	18.7	0.064
0.37	20.2	150±3	50-190	1.42	1.59	802	5.3	11.7	0.063
0.40	15.7	200±3	50-200	1.45	1.45	821	6.1	8.9	0.060
0.37	13.0	165±3	40-200	1.51	1.32	833	7.5	5.9	0.058

3. Conclusions

We have presented some new data for highly under-doped crystals of $YBa_2Cu_3O_{6+x}$ with hole concentrations per CuO_2 plane, p =0.058 to 0.073. We have studied crystals with three values of x, measuring $\chi_c(T)$ and $\chi_{ab}(T)$ immediately after fixing x by quenching on to a copper block and again after allowing sufficient time at room temperature for the Cu-O chains to order and analyzed the data for three pairs of underdoped crystals with T_c ranging from 13 to 36 K, namely UD13 and UD20, UD15 and UD30, UD23 and UD36. As found previously for polycrystalline samples [7] ordering the Cu-O chains increases T_c and substantially reduces the Curie term. In contrast to the Pauli-like susceptibility from the CuO_2 planes, the Curie terms are isotropic. Plots of $d\chi_c(T)/dT$ vs. T give an unexpected result, the Curie terms disappear above a p-dependent onset temperature T_{Curie} of order 200 K as shown in Fig. 4 (b). For all six samples $\chi_D(T)=\chi_c(T)-\chi_{ab}(T)$ has a weakly T-dependent linear region at higher T where $\chi_D(0)$ is 1.3 - 1.6 x10^{-4}emu/mol. We have discussed deviations from this linear behaviour in terms of 2D or 3D Gaussian superconducting fluctuations.

Acknowledgments
This work was supported by the Croatian Science Foundation under the project IP 2018 01 8912.

References

[1] Wu T, Mayaffre H, Kramer S, Horvatić M, Berthier C, Hardy W N, Liang R, Bonn D A and Julien M H 2011 *Nature* **477** 191
[2] Kokanović I and Cooper J R 2016 *Phys. Rev. B* **94** 075155
[3] Kokanović I, Hills D J, Sutherland M L, Liang R and Cooper J R 2013 *Phys. Rev. B* **88** 060505(R)
[4] Rullier-Albenque F, Alloul H and Rikken G 2011 *Phys. Rev. B* **84** 014522
[5] Haug D, Hinkov V, Suchaneck A, Inosov D S, Christensen N B, Niedermayer Ch, Bourges P, Sidis Y, Park J T, Ivanov A, Lin C T, Mesot J and Keimer B 2009 *Phys. Rev. Lett.* **103** 017001
[6] Haug D, Hinkov V, Sidis Y, Bourges P, Christensen N B, Ivanov A, Keller T, Lin C T and Keimer B 2010 *New J. Phys.* **12** 105006
[7] Biscaras J, Leridon B, Colson D, Forget A and Monod P 2012 *Phys. Rev. B* **85** 134517
[8] Shiohara Y and Endo A 1997 *Mater. Sci. Eng. R* **19** 1
[9] 7 days in 1% O2/N2 at 650 C for UD13 followed by a 2 months anneal at room temperature (RT) to give UD20, 8 days in 1% O2/N2 at 630 C for UD23 followed by a seven month anneal at RT to give UD30, 7 days in 1% O2/N2 at 620 C for UD23 followed by a seven month anneal at RT to give UD36. The masses of the crystals used were 0.0834g (UD13), 0.0654g (UD15), 0.0541g (UD23).
[10] Liang R, Bonn D A and Hardy W N 2006 *Phys. Rev. B* **73** 180505
[11] Walstedt R E, Bell R F, Schneemeyer L F, Waszczak J V and Espinosa G P 1992 *Phys Rev B* **45** 8074
[12] Abragam A and Bleaney B 1986 *Electron Paramagnetic Resonance of Transition Ions* (Dover, New York) sect. 7.16.
[13] Loram J W, Luo J, Cooper J R, Liang W Y and Tallon J L 2001 *J. Phys. Chem. Sol.* **62** 59
[14] Loram J W, Mirza K A, Cooper J R and Tallon J L 1998 *J. Phys. Chem. Sol.* **59** 2091
[15] Loram J W, Luo J, Cooper J R, Liang W Y and Tallon J L 2001 *J. Phys. Chem. Sol.* **62** 59
[16] Larkin A and Varlamov A 2005 "*Theory of Fluctuations in Superconductors*",Clarendon Press, Oxford (U.K.)
[17] Ussishkin I, S. L. Sondhi S L and Huse D A 2002 *Phys. Rev. Lett.* **89** 287001
[18] Pourret A, Aubin H, Lesueur J, Marrache-Kikuchi C A, Bergé L, Dumoulin L and Behnia K 2006 *Nature* Physics **2** 683
[19] Kokanović I, Cooper J R and M. Matusiak 2010 *Phys. Rev. Lett.* **102** 187002
[20] Xu Z A, Ong N P, Wang Y, Kakeshita T and Uchida S 2000 *Nature* (London) **406** 486
[21] Rullier-Albenque F, Tourbot R, Alloul H, Lejay P, Colson D and Forget A 2006 *Phys. Rev. Lett.* **96** 067002

Peak Effect Induced by Heavy-ion Irradiation in *2H*-$NbSe_2$ Single Crystals

Wenjie Li[1], Sunseng Pyon[1], Satoru Okayasu[2] and Tsuyoshi Tamegai[1]

[1]Department of Applied Physics, The University of Tokyo, 7-3-1 Hongo, Bunkyo-ku, Tokyo 113-8656, Japan

[2]Advanced Science Research Center, Japan Atomic Energy Agency, Tokai, Ibaraki 319-1195, Japan

E-mail: liwenjie@g.ecc.u-tokyo.ac.jp

Abstract. Peak effect has been studied in *2H*-$NbSe_2$ single crystals with splayed columnar defects introduced by 320 MeV Au irradiation. Through the magnetic relaxation experiments, correspondence between the peak effect and magnetic relaxation has been observed. However, by calculating the critical current density (J_c) much before the first measurement using a SQUID magnetometer based on the relaxation rate *S*, it is found that the peak effect is present even at the very early stage of relaxation, indicating that it is caused by the magnetic field dependence of pinning force.

1. Introduction

Columnar defects introduced by heavy-ion irradiations are widely used as effective pinning centers to improve the J_c of superconductors due to their similar structures to vortices [1-3]. In addition to strong enhancement of the J_c, other new phenomena have also been discovered in superconductors with columnar defects. One of the representative phenomena is that the J_c shows a non-monotonic change with the increase of magnetic field. In other words, a peak effect is formed at a specific magnetic field. The peak effect is observed in $YBa_2Cu_3O_{7-\delta}$ crystals when the field is applied parallel to columnar defects ($\theta_H = \theta_{CD}$, θ_H is the angle between sample *c*-axis and magnetic field, θ_{CD} is the angle between sample *c*-axis and columnar defects) [1]. Recently, a similar peak effect has also been observed in the study on conventional superconductor $NbSe_2$ [4]. In Ref. [4], by applying magnetic field in various directions, it was found that the peak effect is quickly suppressed when the magnetic field is applied away from the direction of columnar defects. For $|\theta_{CD} - \theta_H| > 9°$, the peak disappears completely. In our previous study on $NbSe_2$ with tilted columnar defects ($\theta_{CD} = +30°$) and different B_Φ (B_Φ is a dose equivalent matching field, 1 T = $5\times10^{10}/cm^2$), we found that the peak effect is only formed when B_Φ is larger than 0.5 T [5]. In addition to the peak effect induced by tilted columnar defects, a similar peak effect has also been observed in superconductors with splayed columnar defects. For example, in iron-based superconductor $Ba_{1-x}K_xFe_2As_2$, when the magnetic field was applied parallel to the average direction of the two kinds of columnar defects, a peak effect located at ~$1/3B_\Phi$ was observed [3,6]. In our previous study on $NbSe_2$ with splayed columnar defects, we also observed a peak effect [7]. The peak field increases with increasing B_Φ. This feature indicates that this peak effect is induced by a different mechanism from the peak effect which is observed in pristine $NbSe_2$ around H_{c2} [8]. It is widely believed that the latter peak effect is originated from the transition from an ordered phase to a disordered phase around H_{c2} [9]. This interpretation is supported by the fact introduction of point defects (disorder) shifts the peak field to lower magnetic field. In the study of Co doping effects on $NbSe_2$ (Co can be treated as point defects), the peak field was decreased monotonically with increasing Co content [10]. Such a peak shift is obviously different from what we observed in the case of $NbSe_2$ introduced with splayed columnar defects. Therefore, it is necessary to conduct additional experiments to understand the mechanisms of the peak effect in $NbSe_2$ with splayed columnar defects.

Magnetic relaxation in superconductors is a phenomenon related to the dynamics of vortices influenced by pinning centers, which is under the thermal activation and quantum fluctuations a part of the pinned vortex creeps to the nearby pinning centers [11]. This process consumes energy and therefore suppresses J_c. Through the study of magnetic relaxation, we can confirm whether magnetic relaxation can induce the peak effect or not. For that purpose, we conducted magnetic relaxation measurements on $NbSe_2$ single crystals introduced with splayed columnar defects at different temperatures and magnetic fields.

2. Experimental details

$2H$-$NbSe_2$ single crystals used in the present study were synthesized by using the iodine vapor transport method [5]. The total charge of starting material is 2g, which was placed in a two-zone furnace with high and low temperatures set at 800 ºC and 750 ºC, respectively. Single crystals were obtained at high temperature part. The single crystal used in this study has a dimension of 784×589×10 μm^3. Columnar defects were introduced by irradiating 320 MeV Au ions using the tandem accelerator at JAEA in Tokai, Japan. By controlling the angle between irradiation beam and the crystal c-axis, splayed columnar defects (B_Φ = 2 T + 2 T, θ_{CD} = ±10º) were introduced. Magnetization measurements were conducted by using a commercial SQUID magnetometer (MPMS-XL5, Quantum Design). J_c [A/cm^2] is estimated by the Bean model [12], which is

$$J_c = \frac{20\Delta M}{a\left(1-\frac{a}{3b}\right)} \qquad (a<b), \tag{1}$$

where ΔM [emu/cm^3] is the difference of magnetization when the external field is swept down and up. a [cm] and b [cm] are sample width and length, respectively. As for the magnetic relaxation measurements, the data were collected by decreasing the set field value step by step so that magnetic relaxation of the superconducting magnet would not affect the evaluation of relaxation. For each target magnetic field, the relaxation is measured for 1 hour. According to the definition of the normalized magnetic relaxation rate

$$S = \left|\frac{d\ln M}{d\ln t}\right|, \tag{2}$$

S value for $NbSe_2$ with splayed columnar defects was estimated at different temperatures and magnetic fields.

3. Results and discussion

Figure 1 shows the magnetic field dependence of magnetization for $NbSe_2$ single crystals with splayed columnar defects introduced at θ_{CD} = ±10º, each at B_Φ = 2 T at several temperatures. The clear peak effect can be observed at temperatures between 2 K and 5 K. Even at 6 K, a weak peak effect can be observed as shown in the inset of Fig. 1. It can be seen that the peak field (H_p) gradually shifts to lower magnetic field with increasing temperature. This shift of H_p induced by temperature is consistent with our previous study on $NbSe_2$ single crystals with splayed columnar defects (320 MeV Au, θ_{CD} = ±5º, B_Φ = 2 T + 2 T) [7]. Actually, for superconductors with columnar defects, another peak effect at low fields has often been reported, which is the self-field peak effect. The formation of the self-field peak

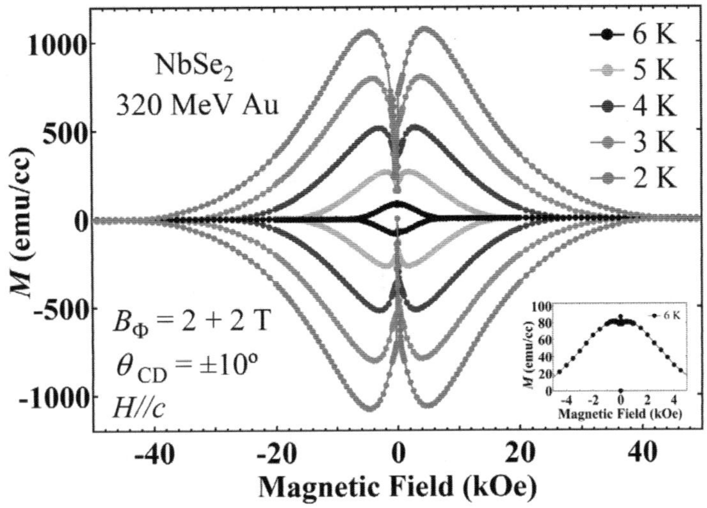

Figure 1. Magnetic field dependence of magnetization for $NbSe_2$ introduced with splayed columnar defects (B_Φ = 2T + 2 T, θ_{CD} = ±10°) by 320 MeV Au irradiation. Inset shows a blow-up of the data at T = 6 K.

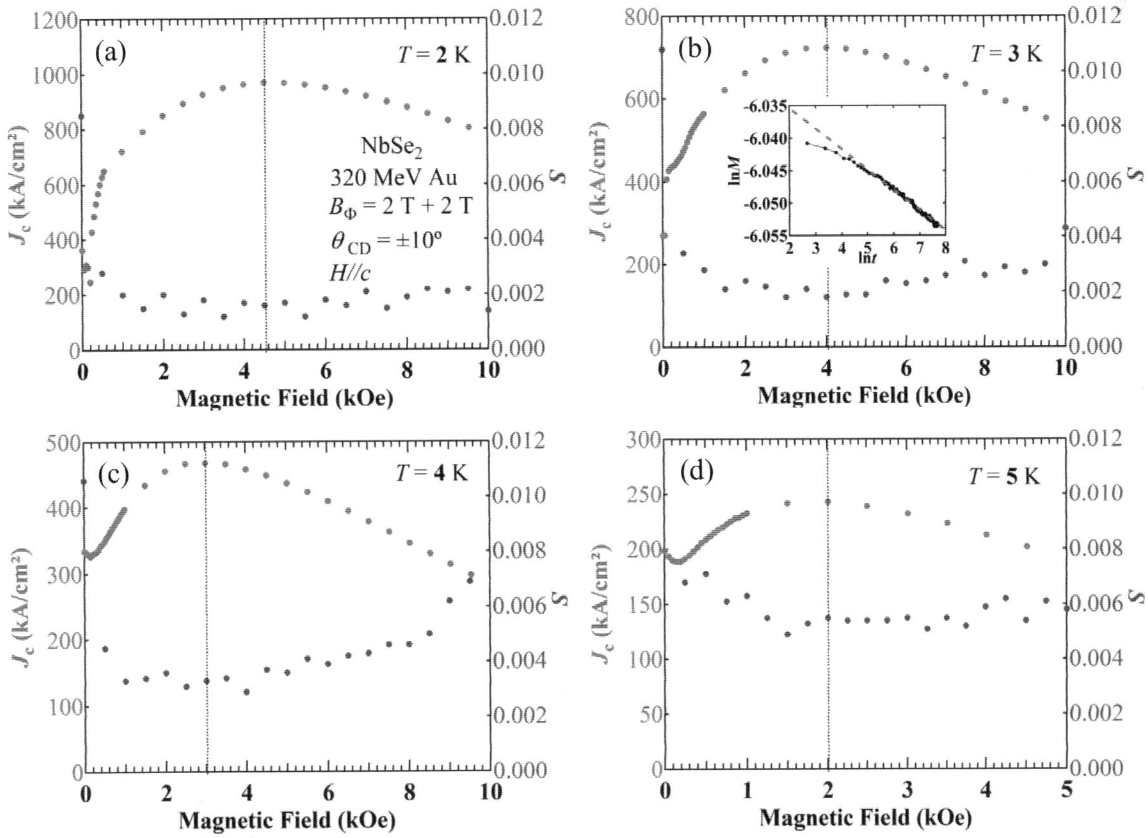

Figure 2. Magnetic field dependence of J_c and relaxation rate for $NbSe_2$ introduced with splayed columnar defects (B_Φ = 2 T + 2 T, θ_{CD} = ±10°) by 320 MeV Au irradiation at T = 2 K (a), T = 3 K (b), T = 4 K (c), and T = 5 K (d). The dashed line indicates the peak field. An example of magnetic relaxation at T = 3 K and H = 500 Oe is shown as an inset of (b). The red broken line is used to evaluate the normalized relaxation rate.

effect is related to the curvature of the vortices at low magnetic fields [2]. At low magnetic fields, vortices at the edge of the superconductor can be curved. The curved vortices cannot be well pinned by straight columnar defects. As the magnetic field increasing, those curved vortices will be straightened again, and they can be pinned better by columnar defects. This is the reason why in the low magnetic field range, J_c increases with increasing the magnetic field. The peak field of the self-field peak effect (H_{sf}) can be estimated based on J_c and sample thickness d; $H_{sf} \sim J_c d$. The calculated H_{sf} for the sample here is ~ 1000 Oe ($T = 2$ K). It is clear that this value is much smaller than the peak field of $H_p \sim 4600$ Oe at $T = 2$ K in Fig. 2(a). This result suggests that the peak effect observed in $NbSe_2$ with splayed columnar defects has different mechanisms from the self-field peak effect.

The magnetization shown in Fig. 1 is measured at some time after the preparation of the critical state at $t = 0$ s. So, J_c values shown in Fig. 2 are not "critical current density" in a strict sense. It means that any structures in $J_c - H$ curve can be induced either by the magnetic field dependence of the true critical current density at $t = 0$ s or by the magnetic field dependence of the relaxation process. As for the magnetic relaxation, there are two representative models to explain it [11]. The first is the Anderson-Kim model [13-14], in which the pinning barrier $U(J)$ is considered to follow a linear dependence on current, which is

$$U = U_0\left[1 - J / J_{c0}\right]. \tag{3}$$

Assuming Arrhenius type dependence on the relaxation time

$$t = t_0 \exp\left(U / kT\right), \tag{4}$$

where t_0 is an effective hopping attempt time for vortices, temporal evolution of J_c can be derived as

$$J_c = J_{c0}\left[1 - \frac{kT}{U_0}\ln\left(\frac{t}{t_0}\right)\right]. \tag{5}$$

By combining formula (5) with the definition of the normalized relaxation rate at Eq. (2), it can be obtained that the normalized relaxation rate has a linear relationship with the temperature;

$$S = \frac{kT}{U_0}. \tag{6}$$

However, Eq. (6) is only be applied to conventional superconductors with negligible relaxation rate. For superconductors with strong thermal fluctuations, it is important to take into account elastic deformations of vortices leading to diverging pinning potential $U(J)$ in the small J limit called collective creep. In such case, the pinning barrier $U(J)$ follow the relation of

$$U = U_0\left[\left(J_{c0} / J\right)^{\mu} - 1\right]. \tag{7}$$

It leads to relaxation of J_c not proportional to ln$t$;

$$J = \frac{J_{c0}}{\left[1 + \left(\mu k_B T / U_0\right)\ln\left(t / t_0\right)\right]^{1/\mu}}. \tag{8}$$

Based on Eq. (8), the normalized relaxation rate S can be calculated as

$$S = \frac{k_B T}{U_0 + \mu k_B T \ln\left(t / t_0\right)}. \tag{9}$$

At low enough temperatures, collective creep model converges into Anderson-Kim model and at high temperature it predicts a temperature independent normalized relaxation with

$$S = \frac{1}{\mu \ln\left(t / t_0\right)}. \tag{10}$$

In this formula, μ is a collective creep exponent. For example, μ = 1/7, 5/2 (3/2), and 7/9 correspond to creep for single vortices, small bundles, and bundles much larger than the London penetration length, respectively. Since a typical value for ln (t/t_0) is ~20-30, the S value is in the range of 0.02~0.04 [11]. In high temperature superconductors, the measured S is consistent with this predicted value [15]. In the case of $NbSe_2$, the measured S in this study is much smaller than that in high temperature superconductors. The maximum value of S in the present study is ~0.006. Similar values of the normalized relaxation rate have also been observed in $NbSe_2$ introduced with tilted columnar defects [4].

Figure 2 shows the magnetic field dependence of J_c and the normalized relaxation rate in $NbSe_2$ irradiated by 320 MeV Au with splayed columnar defects at different temperatures. An example of the relaxation curve is shown in the inset of Fig. 2(b). From the obtained results, it can be seen that at temperatures such as T = 3 K, 4 K, and 5 K the correspondence between the peak effect and the normalized relaxation rate can be observed. Namely, in the vicinity of the peak effect, the normalized relaxation rate gets a minimum value. This correspondence may suggest that the peak effect is caused by the field dependence of magnetic relaxation. Actually, such a dynamical origin of the peak effect has been demonstrated in $Bi_2Sr_2CaCu_2O_{8+y}$ by the direct relaxation measurements in the early stage of relaxation using a Hall probe [16]. We calculate the J_c before the relaxation occurs based on temporal evolution of J_c expressed by Eq. (5) with the determined value of the normalized relaxation rate. About the calculation, choosing a time much closer to $t = 0$ s is better. In our case, we assume that the J_c is measured at 15 s after the creation of the critical state at $t = 0$ s, and calculate J_c at $t = 1$ μs. The J_c at $t = 1$ μs is calculated by $J_c\ (t = 1\ \mu\text{s}) = (1 + \ln(15/10^{-6}) \times S) \times J_c\ (t = 15\ \text{s})$. As shown by the dashed line in Fig. 3, the J_c value does increase. However, the structure of peak effect is almost unchanged even at the early stage of relaxation, which means that the global structure of the peak effect cannot be strongly affected by only considering the magnetic relaxation. Despite the presence of corresponding changes in J_c and S, we can conclude that the magnetic relaxation is not the origin of the peak effect in $NbSe_2$ with splayed columnar defects.

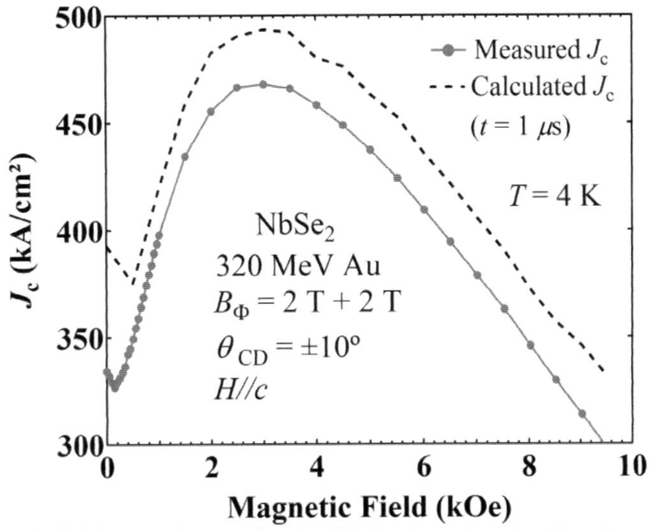

Figure 3. Magnetic field dependence of J_c for $NbSe_2$ introduced with splayed columnar defects (B_Φ = 2 T + 2 T, $\theta_{CD} = \pm 10°$) by 320 MeV Au irradiation at T = 4 K. The dashed line is the calculated J_c at t = 1 μs after the creation of the critical state based on the normalized relaxation rate data.

4. Summary

The magnetic relaxation experiments have been conducted on $NbSe_2$ single crystals with splayed columnar defects. Although a correspondence between the peak effect and the field dependence of normalized relaxation rate has been observed, the peak effect is observed even in the magnetic field dependence of J_c at very early stage of relaxation, indicating that it is caused by the magnetic field dependence of pinning force.

References

[1] Civale L, Marwick A D, Worthington T K, Kirk M A, Thompson J R, Krusin-Elbaum L, Sun Y, Clem J R and Holtzberg F 1991 *Phys. Rev. Lett.* **67** 648
[2] Tamegai T, Taen T, Yagyuda H, Tsuchiya Y, Mohan S, Taniguchi T, Nakajima Y, Okayasu S, Sasase M, Kitamura H, Murakami T, Kambara T and Kanai Y 2012 *Supercond. Sci. Technol.* **25** 084008
[3] Park A, Pyon S, Ohara K, Ito N, Tamegai T, Kambara T, Yoshida A and Ichinose A 2018 *Phys. Rev. B* **97** 064516
[4] Eley S, Khilstrom K, Fotovat R, Xiao Z L, Chen A, Chen D, Leroux M, Welp U, Kwok W K and Civale L 2018 *Sci. Rep.* **8** 13162
[5] Li W, Pyon S, Okayasu S and Tamegai T 2021 *J. Phys.: Conf. Ser.* **1975** 012003
[6] Takahashi A, Pyon S, Kambara T, Yoshida A and Tamegai T 2020 *J. Phys. Soc. Jpn.* **89** 094705
[7] Li W, Pyon S, Okayasu S and Tamegai T 2022 *J. Phys.: Conf. Ser.* **2323** 012017
[8] Merithew R D, Rabin M W, Weissman M B, Higgins M J and Bhattacharya S 1996 *Phys. Rev. Lett.* **77** 3197
[9] Paltiel Y, Zeldov E, Myasoedov Y, Rappaport M L, Jung G, Bhattacharya S, Higgins M J, Xiao Z L, Andrei E Y, Gammel P L and Bishop D J 2000 *Phys. Rev. Lett.* **85** 3712
[10] Iavarone M, Capua R Di, Karapetrov G, Koshelev A E, Rosenmann D, Claus H, Malliakas C D, Kanatzidis M G, Nishizaki T and Kobayashi N 2008 *Phys. Rev. B* **78** 174518
[11] Yeshurun Y, Malozemoff A P and Shaulov A 1996 *Rev. Mod. Phys.* **68** 911
[12] Bean C P 1964 *Rev. Mod. Phys.* **36** 31
[13] Anderson P W 1962 *Phys. Rev. Lett.* **9** 309
[14] Kim Y B, Hempstead C F and Strnad A R 1963 *Phys. Rev.* **131** 2486
[15] Malozemoff A P and Fisher M P A 1990 *Phys. Rev. B* **42** 6784
[16] Tamegai T, Iye Y, Oguro I and Kishio K 1993 *Physica C* **213** 33

Journal of Physics: Conference Series **2545** (2023) 012006 doi:10.1088/1742-6596/2545/1/012006

Turbulent Structure and Characterization of '11'-type Iron-based Superconductors by Magneto-optical Imaging

T Ren[1], Y Sun[2], S Pyon[1] and T Tamegai[1]

[1]Department of Applied Physics, The University of Tokyo, 7-3-1 Hongo, Bunkyo-ku, Tokyo 113-8656, Japan

[2]Department of Physics, Southeast University, Nanjing 211189, China

E-mail: rentong@g.ecc.u-tokyo.ac.jp

Abstract. Magneto-optical imaging is employed to investigate the vortex penetration modes in '11'-type iron-based superconductors (IBSs) $Fe(Te_{0.6}Se_{0.4})$ and FeSe single crystals. In $Fe(Te_{0.6}Se_{0.4})$, vortices penetrate mostly homogeneously from all edges, forming an unstable turbulent interface with pre-existing anti-vortices, where 'Meissner hole' emerges as those have been reported in '122'-type IBSs. In FeSe, however, the penetration is restricted to numerous mesoscopic stripe-like regions. Further experiments revealed an inhomogeneous vortex propagation pattern, where vortices are most likely to be channeled along the twin boundaries.

1. Introduction

Vortex motions in type-II superconductors like vortex flow and pinning never cease to intrigue physicists. The major problem is the dissipation that results from moving vortices, which caps the maximum critical current density in use. Therefore, studying vortex dynamics and pinning mechanisms have been our main interests. This research is focused on two vortex phenomena. One is the so-called macroturbulence [1, 2, 3, 4], which is described by the following scenario. Fluid-like vortices and anti-vortices with sheer velocity parallel to their interface should induce Kelvin-Helmholtz instability, perturbing the border like turbulence on liquid surfaces. Such perturbed, meandering vortex-antivortex interface is formerly reported in $YBa_2Cu_3O_{7-\delta}$ [5], MgB_2 [6], and IBSs [7, 8]. From a microscopic perspective, vortices at the interface are bent to form loops with diameters larger than the minimum diameter determined by the equilibrium between pinning force and line tension. Such vortex loops, forming torus-like structure in remagnetized superconductors, is called Meissner hole (MH) [5]. At the skin layer of MH flows a local excess current which is twice as large as critical current in the same area. The other phenomenon is the mesoscopic domains in FeSe and $BaFe_2(As_{1-x}P_x)_2$ induced by spin and orbital degrees of freedom [9]. Such domains have similar dimensions to another in-plane structural domain, namely twin boundaries (TBs), which have been reported in IBSs such as $Ba(Fe_{1-x}Co_x)_2As_2$ [10] and FeSe [11] below the structure transition. The latter was proven to act as intrinsic pinning potential for vortices [12]. Nevertheless, scanning tunneling microscopy (STM) study revealed that TBs could noticeably suppress the superconducting gap along its trajectory. In both scenarios, vortices are accepted to accumulate along TBs, leading to macroscopic inhomogeneous penetration.

In this paper, we start from discussion on the vortex turbulence found in Fe($Te_{0.6}Se_{0.4}$), followed by an attempt to create a similar situation in FeSe, only to find an inhomogeneous fractal penetration pattern instead.

2. Experimental Methods

Single crystals of Fe($Te_{0.6}Se_{0.4}$) were grown from high purity Fe (99.99%) powder, Te (99.999%), and Se (99.999%) grains with ratio 2 : 1 : 1 of 5 g in total. The mixture were heated up to 1050 °C in an evacuated quartz tube for two days, followed by slow cooling to 650 °C within a week. Crystals with typical dimensions of 1 mm × 1 mm × 0.1 mm were picked out for use. To unveil the superconductivity of the as-grown crystals, necessary oxygen annealing was performed to remove excess Fe. In detail, single crystals were sealed in O_2&Ar atmosphere with 2.5% oxygen (molar ratio) with respect to Fe, followed by heating up to 400 °C and held for 1 hour. Crystals prepared in this manner have $T_c = 14$ K. More systematic annealing effect on Fe($Te_{0.6}Se_{0.4}$) can be found in ref. [13, 14, 15]. Chemical vapor transport method (Ref. [16]) was used to grow FeSe single crystal ($T_c \sim 8.5$ K), where $AlCl_3$ and KCl with a ratio of 2 : 1 were used as flux. The assembly was loaded into a dual-zone furnace for one month, with temperature gradient set from 400 °C to 250 °C.

In the characterization of vortex penetration, we employed the differential magneto-optical (MO) imaging, where integrated backgrounds were subtracted from the integrated target image [17, 18]. By doing so, it is possible to obtain high-resolution MO images with suppressed fluctuations caused by the environments. All samples were cooled to the target temperature under zero field, followed by the application of vertical magnetic fields induced by a copper coil. Depending on the situation, additional horizontal field was also applied to remove the domain formation of the garnet.

3. Result and Discussion

3.1. Meissner hole in Fe($Te_{0.6}Se_{0.4}$) single crystal

Figure 1(a) shows a typical MH created in a ($Ba_{0.67}Rb_{0.33}$)Fe_2As_2 single crystal in our former experiments [8]. To reproduce one in Fe($Te_{0.6}Se_{0.4}$), a piece of crystal of dimensions ~1,000 × 600 × 20 μm^3 was zero-field cooled to 3.1 K, followed by a -1,000 Oe virgin penetration. Thus, antivortices partly occupied the crystal from all edges. After that the magnet was switched to 350 Oe, introducing positive vortices to form a vortex-antivortex interface (see Fig. 1(b)). As was expected, the border became curly and show meandering texture. Profiles (Figs. 1(c), (d)) taken along red dashed lines in Figs. 1(a) and (b) display characteristic induction steps (black arrows) induced by local excess current, or Meissner current [7, 8]. In contrast to the MH in ($Ba_{0.67}Rb_{0.33}$)Fe_2As_2, MH in Fe($Te_{0.6}Se_{0.4}$) appears to be more inhomogeneous, and partly exhibit branch-like penetration (rectangle area in Fig. 1(b)). Such branch-like penetration pattern could be attributed to the rugged edge from which vortex initially enter the crystal. Inverse Biot-Savart conversions [7] yield Figs. 1(e) and (f) mapping the in-plane current density for Figs. 1(a) and (b), respectively. Highlighted Meissner current in ($Ba_{0.67}Rb_{0.33}$)Fe_2As_2 (Fig. 1(e)) is mostly continuous and even in brightness, which is reasonable because MH should always be located in the mid-plane of the crystal. Meissner current in Fe($Te_{0.6}Se_{0.4}$) (Fig. 1(f)), on the contrary, is shattered and uneven. Based on the fluctuation of intensity of Meissner current, an alternative hypothesis would be that the MH is no more fixed centrally, but instead wobbling in a three-dimensional path.

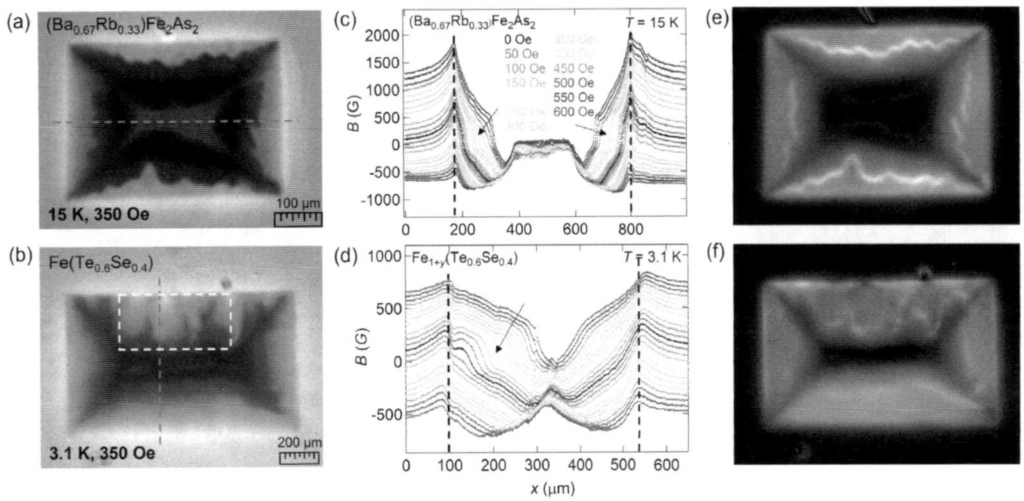

Figure 1. MO images of remagnetized (a) $(Ba_{0.67}Rb_{0.33})Fe_2As_2$ and (b) $Fe(Te_{0.6}Se_{0.4})$ single crystals at 15 K and 3.1 K, respectively. Novel branch-like MH is framed in white rectangle in (b). Induction profile taken along red dashed lines in (a) and (b) are plotted in (c) and (d), respectively. Black arrows indicate the induction step caused by local excess current, and vertical dashed lines mark sample edges. In-plane current density mapping (e) and (f) are converted from (a) and (b) using inverse Biot-Savart law, respectively.

3.2. Inhomogeneous vortex penetration in FeSe single crystal

Despite initial intention of observing vortex turbulence in FeSe single crystal, we found it impossible to generate a clear Meissner hole due to a unique vortex penetration behavior in this material. As shown in Fig. 2 (a), after zero-field cooling and application of 100 Oe field,

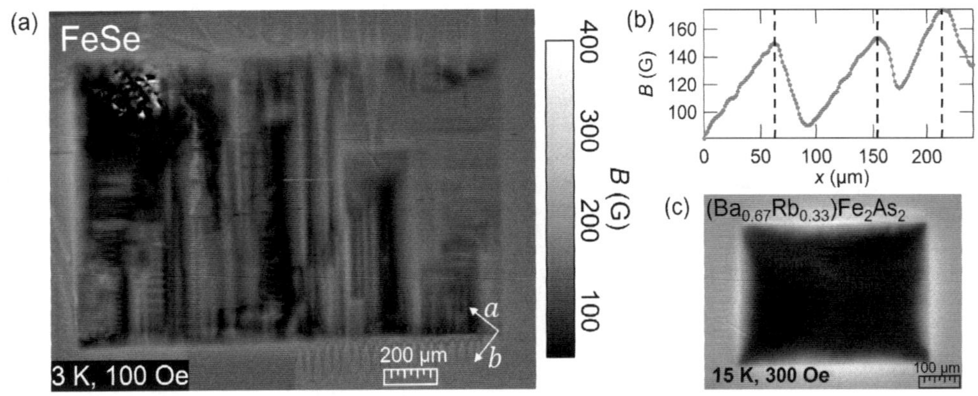

Figure 2. MO images of (a) FeSe single crystal at 3 K and (c) $(Ba_{0.67}Rb_{0.33})Fe_2As_2$ crystal at 15 K being penetrated by 100 Oe and 300 Oe external field, respectively. White coordinates indicate (100) and (010) directions in the orthorhombic phase. Induction profile (b) taken along the red line in (a).

stripe-like domains run along (110) or $(1\bar{1}0)$ directions of orthorhombic unit cell of FeSe at 3 K

[19]. These directions are determined by in-plane ϕ scan using X-ray diffraction with reference to the peak (104). The criss-cross domains have dimensions varying from 10 μm to 100 μm in width, with a maze-like texture sharing geometric similarity with the optical observations of TBs in IBSs [10, 20]. Such linear vortex penetration has also been reported in susceptometry measurements with a scanned superconducting quantum interference device (SQUID), where enhanced superfluid density is confirmed along TBs in underdoped $Ba(Fe_{1-x}Co_x)_2As_2$ [21], and individual vortices are visualized to be stopped by TBs [12]. What is more, Fig. 2(b) depicts an asymmetric vortex distribution aside the domain centers (vertical lines). Vortex density gradient on the right side is noticeably larger than that on the left side, implying that vortices accumulate more densely at the right side of the domain center. This is compatible with the analytical calculation of current density distribution in a superconducting stripe under perpendicular field, concluding that in critical states vortices could penetrate deeper into a broader area [22, 23].

To get a dynamic landscape of the penetration process in FeSe. We swept field up to 240 Oe on another sample (see Figs. 3(a)-(e)). Visible inhomogeneity starts from 30 Oe (Fig. 3(b)),

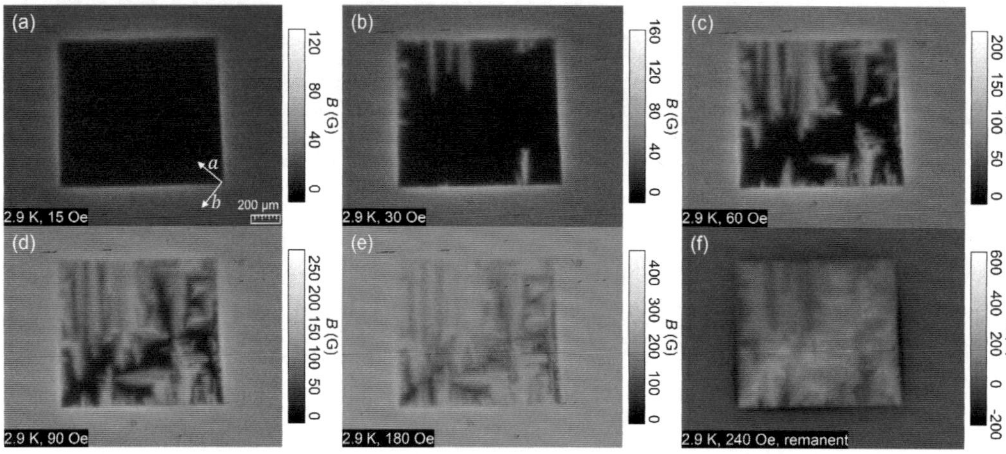

Figure 3. MO images of a zero-field cooled crystal under an ascending field (a) 15 Oe, (b) 30 Oe, (c) 60 Oe, (d) 90 Oe, and (e) 180 Oe. (f) MO images of the crystal at its remanent state after application of 240 Oe.

where vortices from edges intrude the interior of sample along (110) or (1$\bar{1}$0) directions. This distinctly differs from the typical penetration pattern in '122' IBSs like $(Ba_{0.67}Rb_{0.33})Fe_2As_2$ (Fig. 2(c)), where the penetration boundary is smooth and homogeneous in all directions. As vortices propagate deeper into the sample, the root channels tend to branch, but still are restricted to (110) (or (1$\bar{1}$0)) direction (Figs. 3(c), (d)). Finally, vortices fill the crystal with a maze-like fractal pattern (Fig. 3(e)). During the whole process, vortices in FeSe behave like they are being channeled by TBs, and they are stopped by other vortex channels. This provides strong evidence to the pinning mechanisms of TBs in IBSs. In Fig. 3(f) we took a remanent state MO image of FeSe after application of 240 Oe field, only to find a blurred maze pattern, which might be caused by the formation of local vortex turbulence.

It is also worth mentioning that scanning tunneling spectroscopy (STS) revealed suppressed superconductivity along TBs [11], which offers an alternative explanation that vortices tend to penetrate along TBs. Nevertheless, this scenario is contradictory to a more recent study claiming the absence of superconducting gap modification across TBs despite possible twist of the order parameter [24].

4. Summary
We synthesized high-quality single crystals of Fe($Te_{0.6}Se_{0.4}$) and FeSe, and performed magneto-optical observation of vortex behavior in these materials. Following our former research on ($Ba_{1-x}Rb_x$)Fe_2As_2, we find Meissner hole in Fe($Te_{0.6}Se_{0.4}$) with novel boundary configuration. On the other hand, vortex penetration in FeSe is demonstrated to be distinctly inhomogeneous, channeling along twin boundaries. We also provide a proof of asymmetric accumulation of vortices along twin boundaries.

Acknowledgments
This work is supported by a Grant in Aid for Scientific Research (A) (17H01141) from the Japan Society of Science (JSPS), and Support for Pioneering Research Initiated by Next Generation (SPRING) of Japan Science and Technology Agency (JST).

Reference
[1] Fisher L M, Goa P E, Baziljevich M, Johansen T H, Rakhmanov A L and Yampol'skii V A 2001 *Phys. Rev. Lett.* **87** 247005
[2] Fisher L M, Bobyl A, Johansen T H, Rakhmanov A L, Yampol'skii V A, Bondarenko A V and Obolenskii M A 2004 *Phys. Rev. Lett.* **92** 037002
[3] Baggio C, Howard M and Saarloos W v 2004 *Phys. Rev. E* **70** 026209
[4] Dvash E E, Shapiro I, Rosenstein B, Logoboy N and Shapiro B Y 2012 *J. Low Temp. Phys.* **167** 39
[5] Vlasko-Vlasov V, Welp U, Crabtree G, Gunter D, Kabanov V and Nikitenko V 1997 *Phys. Rev. B* **56** 5622
[6] Johansen T H, Baziljevich M, Shantsev D V, Goa P E, pe rin Y M G, Kang W N, Kim H J, Choi E M, Kim M S and Lee S I 2002 *EPL* **59** 599
[7] Mohan S, Tsuchiya Y, Nakajima Y and Tamegai T 2011 *Phys. Rev. B* **84** 180504
[8] Ren T, Pyon S and Tamegai T 2022 *J. Phys. Conf. Ser.* **2323** 012016
[9] Shimojima T, Motoyui Y, Taniuchi T, Bareille C, Onari S, Kontani H, Nakajima M, Kasahara S, Shibauchi T, Matsuda Y and Shin S 2021 *Science* **373** 1122
[10] Prozorov R, Tanatar M A, Ni N, Kreyssig A, Nandi S, Bud'Ko S L, Goldman A I and Canfield P C 2009 *Phys. Rev. B* **80** 174517
[11] Song C L, Wang Y L, Jiang Y P, Wang L, He K, Chen X, Hoffman J E, Ma X C and Xue Q K 2012 *Phys. Rev. Lett.* **109** 137004
[12] Kalisky B, Kirtley J R, Analytis J G, Chu J H, Fisher I R and Moler K A 2011 *Phys. Rev. B* **83** 064511
[13] Sun Y, Taen T, Tsuchiya Y, Shi Z X and Tamegai T 2012 *Supercond. Sci. Technol.* **26** 015015
[14] Sun Y, Tsuchiya Y, Taen T, Yamada T, Pyon S, Sugimoto A, Ekino T, Shi Z and Tamegai T 2014 *Sci. Rep.* **4** 1
[15] Sun Y, Shi Z and Tamegai T 2019 *Supercond. Sci. Technol.* **32** 103001
[16] Böhmer A E, Hardy F, Eilers F, Ernst D, Adelmann P, Schweiss P, Wolf T and Meingast C 2013 *Phys. Rev. B* **87** 180505
[17] Soibel A, Zeldov E, Rappaport M, Myasoedov Y, Tamegai T, Ooi S, Konczykowski M and Geshkenbein V B 2000 *Nature* **406** 282
[18] Yasugaki M, Itaka K, Tokunaga M, Kameda N and Tamegai T 2002 *Phys. Rev. B* **65** 212502
[19] Sun J P, Matsuura K, Ye G Z, Mizukami Y, Shimozawa M, Matsubayashi K, Yamashita M, Watashige T, Kasahara S, Matsuda Y, Yan J Q, Sales B C, Uwatoko Y, Cheng J G and Shibauchi T 2016 *Nat. Commun.* **7** 12146
[20] Tanatar M A, Kreyssig A, Nandi S, Ni N, Bud'ko S L, Canfield P C, Goldman A I and Prozorov R 2009 *Phys. Rev. B* **79** 180508
[21] Kalisky B, Kirtley J R, Analytis J G, Chu J H, Vailionis A, Fisher I R and Moler K A 2010 *Phys. Rev. B* **81** 184513
[22] Brandt E H, Indenbom M V and Forkl A 1993 *EPL* **22** 735
[23] Brandt E H 1995 *Phys. Rev. Lett.* **74**(15) 3025
[24] Watashige T, Tsutsumi Y, Hanaguri T, Kohsaka Y, Kasahara S, Furusaki A, Sigrist M, Meingast C, Wolf T, Löhneysen H v, Shibauchi T and Matsuda Y 2015 *Phys. Rev. X* **5** 031022

Journal of Physics: Conference Series **2545** (2023) 012007 doi:10.1088/1742-6596/2545/1/012007

Transition Temperature of a Superconductor under a Helical Magnetic field

Masaru Kato

Department of Physics and Electronics, Osaka Metropolitan University, 1-1 Gakuencho, Nakaku, Sakai, Osaka 599-8531, Japan

Mkato76@omu.ac.jp

Abstract. We consider a bilayer system with a conventional s-wave superconductor and a chiral helimagnet. If we assume that structure of an external magnetic field from the helimagnet is same as the magnetic moment structure in the helimagnet, a helical magnetic field is applied to the superconductor. Our previous simulation showed that vortex structures depend on strongly on structure of the helical magnetic field as well as the amplitude of the field. We study how pitch of the helical magnetic field affects the transition temperature. For this purpose, we formulate the linearized Ginzburg-Landau equation for the helical magnetic field.

1. Introduction

A superconductor under an external field shows several interesting phenomena. A vortex state is one of such phenomena [1,2]. Magnetic flux of the vortex is quantized to $_0 = hc/2e$, due to single-valuedness of the macroscopic wave function of superconductivity. Here, h is the Planck constant, c is the speed of light and -e is the electric charge of an electron. For applications of the superconductor, controlling, especially pinning of vortices is important. Dynamics of vortices is also interesting problem [3].

Usually, the external field is assumed to be uniform and the vortex is parallel to the field. Previously, we considered a double layer of a superconductor and a chiral helimagnet. In the chiral helimagnet, order of magnetic moment is expressed as.

$$\boldsymbol{S}(x) = S\left(0, \sin\theta(x), \cos\theta(x)\right) \quad (1)$$

Here the angle (x) is given as,

$$(x) = 2\sin^{1}\left[\operatorname{sn}\left(\frac{\sqrt{H}}{k}x\right)\right] + \quad (2)$$

where H is a normalized applied magnetic field and k is the modulus of Jacobi's elliptic function $\operatorname{sn}(x|k)$. k is determined by following relation,

$$\frac{}{4\sqrt{H}} = \frac{E(k)}{k} \quad (3)$$

Where $E(k)$ is the complete elliptic integral of the second kind. is determined from interactions of magnetic moments. If we can assume the structure of the magnetic field in the superconductor from

Journal of Physics: Conference Series **2545** (2023) 012007 doi:10.1088/1742-6596/2545/1/012007

the helimagnet is that of magnetic moments, then we have a superconductor under a helical magnetic field. Vortex structures in such superconductor are not trivial. We studied these vortex states, solving the Ginzburg-Landau (GL) with the finite element method (FEM) [5-8]. In two-dimensional (2D) computation where only perpendicular magnetic field is applied to the superconductor, we found that vortex and anti-vortex coexist and when applied magnetic field H is increased, vortices that are parallel to the applied magnetic field increase and anti-vortices decrease. In such 2D model, only vortices and anti-vortices appear and then next we investigated a three-dimensional (3D) model but without the applied field i.e. $H = 0$. Then the external field from the helimagnet is given as,

$$\boldsymbol{H} = H\left(0, \sin kx, \cos kx\right) \tag{4}$$

where $\frac{2}{k}$ is a pitch of the helix (figure 1). In the 3D model, we can expect 3D structures of vortices. Also, we expected that vortices may be parallel to the external field (equation (4)). However, our numerical results in which the helical magnetic field rotated by 2 from an end of the superconductor to another end, showed the vortices are slanted to $\pm x$ direction. The field is within the yz plane, therefore directions of vortices and the field are not parallel with each other [8]. This result was different from our expectation, therefore we investigated the origins of this result. We considered there are two origins of this result. First one is effect of boundaries. This is because, in previous model, positions where vortices appear are close to the positions where the field goes diagonally from one edge to another edge. In order to investigate the effect of the boundary, we studied a superconductor under a slanted uniform magnetic field [9]. In this case we found two parallel vortices, which are also parallel to the field. So, we can rule out the effect of the boundary. Another origin is the vortex-vortex interaction between non-parallel vortices. Interaction between parallel (anti-parallel) vortices is repulsive (attractive), respectively. But interaction between non-parallel vortices is not trivial. In order to investigate how interaction depends on the angle between vortices, we studied a superconductor under slightly rotated field [10]. We found that even in the slightly rotated field, directions of vortices are not parallel to the external field. Further, four vortices form a complicated structure. We think pitch of the helical magnetic field may affect the superconducting state, especially vortex states.

In this study, as a first step of study of the effects of the pitch of the helical magnetic field, we investigate the superconducting transition temperature under helical magnetic field. In section 2 we show our models and in section 3 we show numerical results. Section 4 is devoted to summary.

2. Models

In order to obtain the superconducting transition temperature in the mean-field approximation, we solve the linearized GL equation (LGL). The LGL equation is given as,

$$\frac{1}{2m^*}\left(\frac{\hbar}{i}\nabla + \frac{2e}{c}\boldsymbol{A}\right)^2 \psi = -\alpha\psi \tag{5}$$

where is a superconducting order parameter, and $2e$ and m are charge and mass of the Cooper pair, respectively. The magnetic vector potential $\boldsymbol{A}$ is related to a helical magnetic field as follows,

$$\boldsymbol{H} = H\left(0, \cos kx, \sin kx\right) = \nabla \times \boldsymbol{A} \tag{6}$$

because the supercurrent tends to zero at the transition temperature and then the flux density $\boldsymbol{B} = \nabla \times \boldsymbol{A}$ equals to the external magnetic field. $= {}_0\left(1 \quad \frac{T}{T_{c0}}\right)$ is the coefficient in the GL equation and T_{c0} is the transition temperature at zero field. After we introduce the coherence length $\xi(T)^2 = -\frac{\hbar^2}{2m^*\alpha}$, the LGL equation becomes,

$$\left(\frac{1}{i}\nabla+\frac{2\pi}{\Phi_0}\boldsymbol{A}\right)^2\psi=\frac{1}{\xi(T)^2}\psi=\frac{1-\frac{T}{T_c}}{\xi_0^{\ 2}}\psi \tag{7}$$

Here ξ_0 is the coherence length at zero temperature.

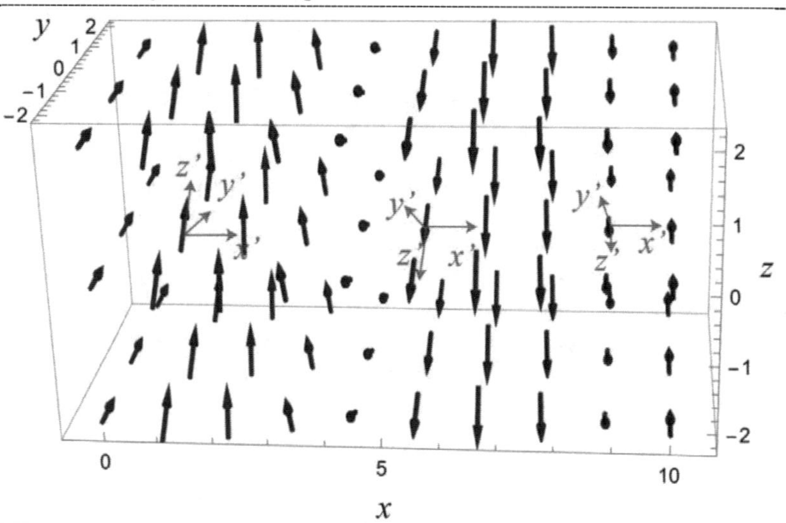

Figure 1. Schematic diagram of a helical magnetic field. Cartesian coordinate (x,y,z) and helical coordinate (x',y',z'). The helical coordinate depends on the x-coordinate.

In order to solve equation (7), we consider two method. First one is transformation of coordinate system from the Cartesian coordinate to a helical coordinate (figure 1) defined as,

$$\begin{cases} x'=x \\ y'=y\cos kx+z\sin kx \\ z'=-y\sin kx+z\cos kx \end{cases} \tag{8}$$

In this coordinate, the magnetic field is express as

$$\boldsymbol{H}=H\boldsymbol{a}_2 \tag{9}$$

which is uniform. Here the unitary vectors in this coordinate system defined as [11],

$$\boldsymbol{a}_1=\left(1,-k\left(y'\sin kx+z'\cos kx\right),k\left(y'\cos kx-z'\sin kx\right)\right) \tag{10}$$

$$\boldsymbol{a}_2=\left(0,\cos kx,\sin kx\right) \tag{11}$$

$$\boldsymbol{a}_3=\left(0,-\sin kx,\cos kx\right) \tag{12}$$

Because this coordinate system is a general curvilinear coordinate, these unitary vectors are not orthogonal with each other and we must consider reciprocal unitary vectors. Then, the LGL equation is not simple, although the magnetic field is uniform. So, in this paper we take second approach.

2.1. Vector potentials in the Cartesian coordinate

For the vector potential in usual Cartesian coordinate, first we assume following one,

$$\boldsymbol{A}=\left(0,-\frac{H}{k}\cos kx,-\frac{H}{k}\sin kx\right) \tag{13}$$

which gives $\nabla\times\boldsymbol{A}=H\left(0,\cos kx,\sin kx\right)$. Then the LGL equation becomes as,

$$\left\{\left(\frac{1}{i}\frac{}{x}\right)^2+\left[\frac{1}{i}\frac{}{y}\quad\frac{2}{_0}\frac{H}{k}\cos kx\right]^2+\left[\frac{1}{i}\frac{}{z}\quad\frac{2}{_0}\frac{H}{k}\sin kx\right]^2\right\} \quad = \frac{1\quad\frac{T}{T_c}}{^2_0} \tag{14}$$

Let $(x,y,z)=\exp\left(ik_y y+ik_z z\right)\ (x)$, and we get,

$$\frac{\mathrm{d}^2}{\mathrm{d}x^2}\quad 2\frac{2}{_0}\frac{H}{k}\left(k_y\cos kx+k_z\sin kx\right)\quad = \left[\frac{1\quad\frac{T}{T_c}}{^2_0}\quad k_y^2\quad k_z^2\quad\left(\frac{2}{_0}\frac{H}{k}\right)^2\right] \tag{15}$$

Another type of the vector potential can be taken as,

$$\boldsymbol{A}=\left(Hz\cos kx\quad Hy\sin kx,0,0\right) \tag{16}$$

which gives $\nabla\times\boldsymbol{A}=H(0,\cos kx\,,\sin kx)$. Then the LGL equation becomes as,

$$\left\{\left[\frac{1}{i}\frac{}{x}+\frac{2}{_0}\frac{H}{k}(z\cos kx\quad y\sin kx)\right]^2+\left(\frac{1}{i}\frac{}{y}\right)^2+\left(\frac{1}{i}\frac{}{z}\right)^2\right\}\quad = \frac{1\quad\frac{T}{T_c}}{^2_0} \tag{17}$$

Let $(x,y,z)=\exp\left[\quad i\frac{2}{_0}\frac{H}{k}(z\sin kx+y\cos kx)\right]\exp\left(ik_y y+ik_z z\right)\ (x)$, and we get equation (15). Therefore, for two vector potentials, we get same equation. So we solve equation (15).

We use following scaling

$$x = {_0}x$$

$$k=\frac{k}{_0}$$

$$\boldsymbol{A}=\boldsymbol{A}\ \frac{_0}{2\ _0} \tag{18}$$

$$H=H\ \frac{_0}{2\ _0^2}$$

Also we take transformation of coordinate as,

$$k\,x\qquad =2x \tag{19}$$

where $\tan\quad=\frac{k_z}{k_y}$. Equation (17) becomes as,

$$\frac{\mathrm{d}^2}{\mathrm{d}x^{\ 2}}+\left\{\frac{4}{k^{\ 2}}\left[1\quad\frac{T}{T_c}\quad k_y^{\ 2}\quad k_z^{\ 2}\quad\left(\frac{H}{k}\right)^2\right]\quad 2\frac{4H'}{k^{\ 3}}\sqrt{k_y^{\ 2}+k_z^{\ 2}}\cos(2x\quad)\right\}\quad =0 \tag{20}$$

This is the Mathieu equation [12,13], which is defined as,

$$\frac{\mathrm{d}^2\ (x)}{\mathrm{d}x^2}+(a\quad 2q\cos 2x)\ (x)=0 \tag{21}$$

In equation (20), the characteristic number a of the Mathieu equation is given as,

$$a=\frac{4}{k^{\ 2}}\left[1\quad\frac{T}{T_c}\quad k_y^{\ 2}\quad k_z^{\ 2}\quad\left(\frac{H'}{k'}\right)^2\right] \tag{22}$$

Also, amplitude $2q$ of the periodic potential is given as,

$$q=\frac{4H}{k^{\ 3}}\sqrt{k_y^{\ 2}+k_z^{\ 2}} \tag{23}$$

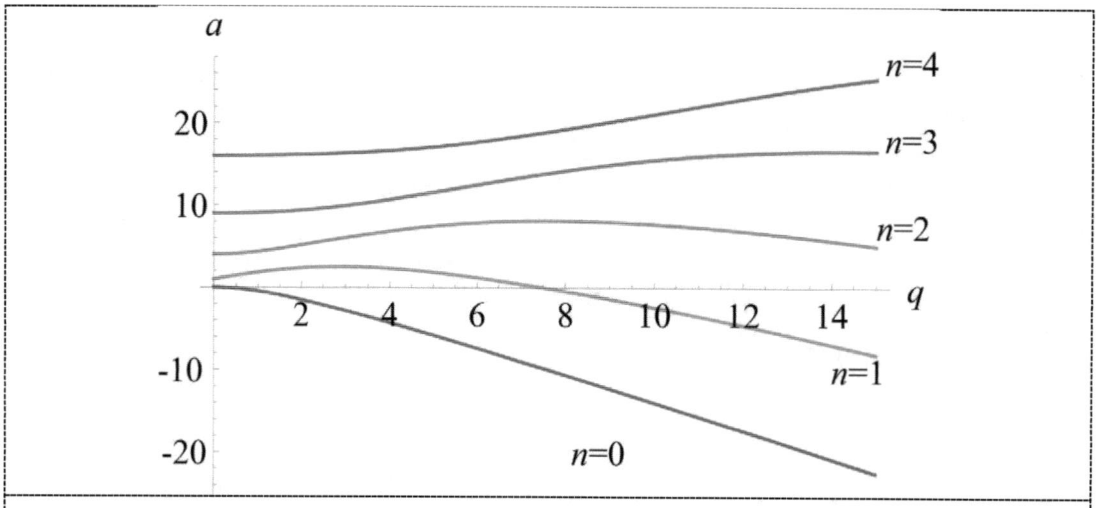

Figure 2. Characteristic number $a(q,n)$ as a function of q for n=0, 1, 2,3 and 4.

3. Numerical Results

Solutions of the Mathieu equation (equation (21)) are called the Mathieu function and periodic functions are denoted as, $\mathrm{ce}_n(r,q)$ and $\mathrm{se}_n(r,q)$ (n is an integer), which are related cosine and sine functions, respectively. As shown in equation (22), the characteristic number a is related to the transition temperature. We show q and n dependence of $a(q,n)$ in figure 2. The lowest branch of $a(q,n)$ ($n=0$) is related to the transition temperature. From this q dependence of $a(q,n)$, we obtain the transition temperature. In figure 3, we show the transition temperature as a function of the amplitude of the field for $k\xi_0$ = 0.2, 0.4, 0.6, and 0.8, where we set $k_x = k_y = 0$. This result shows the transition temperature depends not only on the magnitude of the field but also the pitch of the helix.

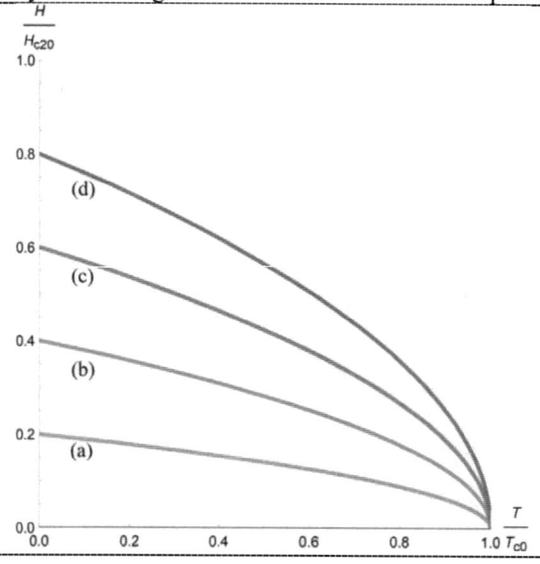

Figure 3. Transition temperature as a function of the magnitude of the field for $k\xi_0$=0.2 (a), 0.4 (b), 0.6 (c), 0.8 (d).

4. Summary

In this paper we have deduced equations for the transition temperature of a superconductor under a helical magnetic field from the LGL equation. Two types of the vector potential give same equation. This equation for the transition temperature becomes the Mathieu equation. The integer order periodic Mathieu functions are related to our problem. From the characteristic number of the Mathieu equation, we have obtained the transition temperature as a function of the magnitude of the field. The transition temperature depends on the pitch of the helical magnetic field.

In the real helimagnets, the pitch under zero external field is fixed for each material. Typical period of the helical magnetic structure is ~50nm [14]. This value is comparable to the penetration depth of the superconductor. In order to investigate how the helical magnetic field affects the superconducting state, we consider pitch dependence of the transition temperature. So larger or smaller value of the pitch may be unphysical. But we believe that this study will clarify effects of the chiral magnetic field on the superconductivity.

We expect that the order parameter structure in the superconducting state under helical magnetic field depends on this pitch of the field, especially vortex structures depend on the pitch as well as the magnitude of the field. To study these dependences is a future problem.

References

[1] Abrikosov A A 1957 *Zh. Teor. Fiz.* **32** 1442 (1957 *Sov. Phys. JETP* **5** 1174)
[2] Huebener R P 2001 *Magnetic Flux Structures in Superconductors* (Berlin: Springer-Verlag)
[3] Blatter G, Feigel'Man M V, Geshkenbein V B, Larkin A I and Vinokur V M 1994 *Rev. Mod. Phys.* **66** 1125
[4] Kishine J. Inoue K and Yoshida Y 2005 *Prog. Theor. Phys. Suppl.* **159** 82
[5] Fukui S, Kato M, Togawa Y 2016 *Supercond. Sci. Technol.* **29** 125008
[6] Fukui S, Kato M, Togawa Y and Sato O 2018 *J. Phys. Soc. Jpn* **87** 084701
[7] Fukui S, Kato M, Togawa Y and Sato O 2018 *J. Phys.: Conf. Ser.* **1054** 012027.
[8] Fukui S, Kato M, Togawa Y and Sato O 2021 *Physica C* **589** 1353918
[9] Yokoji H and Kato M 2021 *J. Phys.: Conf. Ser.* **1975** 012001
[10] Yokoji H and Kato M, in preparation.
[11] Stratton J D 1941 *Electromagnetic Theory* (NewYork and London: McGraw-Hill)
[12] Whittaker E T and Watson G N 1927 *A Course of Modern Analysis fourth edition* (Cambride: Cambridge University press)
[13] Wiltse J. C. 2012 *Graphs and Tables of the Mathieu Functions and Their First Derivatives* (Bloomington: AuthorHouse)
[14] Togawa Y et al 2015 *Phys. Rev. B* **92** 220412 (R)

Stable operation characteristics and perspectives of the large-current HTS STARS conductor

N Yanagi[1,2], Y Narushima[1,2], Y Onodera[1,2], D Garfias-Dávalos[2], S Ito[3], Y Terazaki[1], S Hamaguchi[1,2], H Chikaraishi[1], S Takada[1,2], N Hirano[1,2], K Takahata[1,2]

[1]National Institute for Fusion Science, 322-6 Oroshi-cho, Toki, Gifu 509-5292, Japan
[2]The Graduate University for Advanced Studies, SOKENDAI, 322-6 Oroshi-cho, Toki, Gifu 509-5292, Japan
[3]Tohoku University, 6-6-01-2 Aramaki-Aza-Aoba, Aoba-ku, Sendai, Miyagi 982-8579, Japan

yanagi@nifs.ac.jp

Abstract. The High-Temperature Superconducting (HTS) magnet option has been explored for fusion reactors as well as for next-generation fusion experimental devices. The Stacked Tapes Assembled in Rigid Structure (STARS) conductor uses HTS tapes with simple stacking without twisting and transposition. A practically applicable STARS conductor is presently being developed with an operating current of 18 kA at 20 K temperature and ~15 T magnetic field. The conductor is required to have a high current density of 80 A/mm^2. For the second stage of the conductor development, internal electrical insulation is applied between the copper stabilizer casing and the outer stainless-steel jacket, and a 6-m conductor sample was fabricated in a solenoid coil shape with a 600-mm diameter and three turns. The coiled sample was tested in 8 T, 20 K using a facility equipped with a maximum 13-T, 700-mm bore solenoid coil. Excitation up to the rated current of 18 kA was successfully attained with stable operation. The characteristics of the conductor observed during the excitation test are discussed.

1. Introduction

The High-Temperature Superconducting (HTS) magnet is a feasible option for fusion reactor magnets. Development of large-current HTS conductors has progressed at National Institute for Fusion Science (NIFS) to apply to the helical fusion reactor FFHR [1-5], in parallel to many other projects for developing HTS magnets in the world [6]. Using HTS conductors, it would be possible to make magnets with a magnetic field of >16 T at an elevated temperature of >20 K, which are clear advantages compared to Low-Temperature Superconducting (LTS) conductors. The higher cryogenic stability is also expected as an essential advantage, which may choose a higher current density for a winding conductor. The Stacked Tapes Assembled in Rigid Structure (STARS) conductor concept has been developed [3-5]. The eminent feature of the STARS conductor is that Rare-Earth Barium Copper Oxide (REBCO) tapes are simply stacked without twisting and transposition. The stack of tapes is embedded in a copper stabilizer casing. The conductor exterior is a stainless-steel jacket for mechanical reinforcement. Previously in 2014, a transport current of 100 kA was successfully observed by a prototype conductor sample, having a 3-m total length, and the 300-mm central region placed in a magnetic field of 5 T, with a 20 K temperature control using an electrical heater [3, 7, 8]. This prototype test demonstrated the feasibility of the concept of the STARS conductor.

In the next phase of development, a 20-kA-class nominal current is a target to be used for the next-generation fusion experimental devices including the helical type and others. For this, three types of HTS conductors with different internal configurations have been developed at NIFS. One of them is the STARS conductor [9], and the other two options are FAIR [10, 11] and WISE [12, 13] conductors, which also use REBCO tapes. For all these conductors, the crucial target is to achieve a high current density of 80 A/mm^2 for the coil winding. For the first stage of the conductor development, short (1-3 m) samples of each conductor have been fabricated and tested in liquid nitrogen (temperature: 77 K) with no external magnetic field. The critical currents were measured and compared with expectations. Then, the samples (reassembled in 2-m lengths) were situated in the large superconductor testing facility with 9-T split coils, and a temperature-controlled helium gas supply of 4.2-50 K. The maximum current was limited to 20 kA in this test. The STARS conductor was successfully confirmed to have a stable operation at 8 T with 20 K. The critical current was measured at 11.2 kA for 40 K, 8 T and at 13.0 kA for 40 K, 6 T. Since the STARS conductor had first passed this test, it was then decided to test it with a longer length using a larger testing facility equipped with a maximum 13-T solenoid coil and a 700-mm bore (with a sample chamber of 687 mm in diameter) facility. Details of this experimental result are described in this paper.

2. STARS conductor with internal electrical insulation

As described in [3-5], one of the innovative concepts included in the STARS conductor design is internal electrical insulation between the copper stabilizer casing and the stainless-steel jacket. The stainless-steel jacket of neighboring conductors would be welded together after the winding process. Thus, the stainless-steel jacket would serve as the mechanical reinforcement structure for the entire windings in a coil, a similar role to the "radial plate" used in the toroidal field coils for ITER. This way of including the structure would be useful, especially for the three-dimensional helical coil windings, in which the inclusion of a separate structure like the radial plate is not easy. The STARS conductor with internal electrical insulation has been designed firstly as shown in figure 1, which gives almost a similar layout as that without insulation [9], having an outer dimension of 22.5 mm by 10 mm. Note that the cross-sectional area of this choice would be the same as that of the aluminum-stabilized NbTi/Cu conductor used in the LHD helical coils with 18 mm by 12.5 mm [14]. Two sample conductors having 1-m and 3-m lengths were fabricated with internal electrical insulation. In these samples, there are three rows of each 15 stacks of 4-mm-wide REBCO tapes (FESC-SCH04 of Fujikura). The stainless-steel jacket was formed into two pieces and sealed by Laser Beam Welding (LBW). With the former sample fabricated for a 3-m length, without internal electrical insulation as the first trial, an initial test of LBW showed that the maximum temperature during the welding was ~44 centigrade [9], which was sufficiently below the recommended limiting value of 200 centigrade for REBCO tapes. This time, a similar procedure was employed so that the resultant temperature increase due to LBW inside the copper stabilizer casing was supposed to be in the same range.

A 1-m conductor sample with internal insulation (as shown in figure 1(a)) was then fabricated and tested in liquid nitrogen without a magnetic field, followed by a 3-m sample also tested in liquid nitrogen. The critical current was evaluated to be ~3,930 A for both samples. It should be noted that the presently observed critical current agrees well with the value observed formerly for a conductor sample with no internal electrical insulation [9]. This fact indicates that the fabrication process of the STARS conductor has been established and robust. The experiment has been carried out with a couple of cooling cycles (warmups and cooldowns). There has been no degradation of the measured critical currents (within the measurement error of ~1%).

After this experiment, the internal structure of the conductor has been slightly changed to have a single row of 15 stacks of 12-mm-wide REBCO tapes as shown in figure 1(b). The reason for selecting the previous three-row option was to reduce the edge-wise strain during the helical winding by allowing a slight movement (sliding) of each stack. By adopting a sophisticated winding method to minimize the edge-wise strain as discussed in [15], it has been determined that the sliding of stacks is not necessary. It is also noted that the screening currents induced in the REBCO films increases by employing wider tapes. We consider this would not cause a serious problem because (1) the AC losses (hysteresis losses) could be permitted due to the slow charging rate in the helical reactor, which would be operated mostly in a DC mode (since there is no need to induce plasma current in the helical reactor) and (2) the error magnetic field generated by the screening current does not affect the confining field seriously provided that the periodicity around the torus (ten times along the helical coils) is maintained [16]. This internal configuration would be employed for the STARS conductor design hereafter.

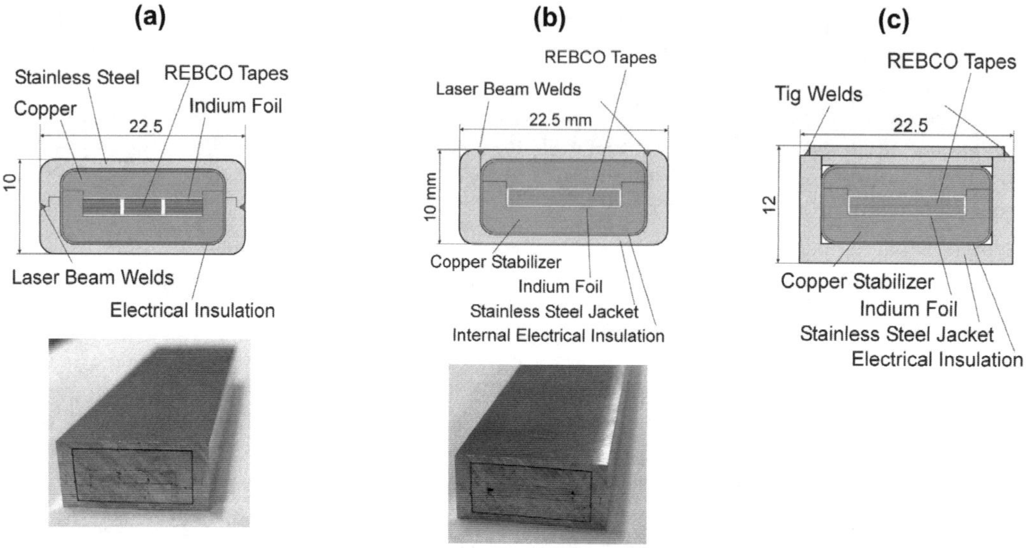

Figure 1. (a) Schematic cross-sectional layout of the STARS conductor and a photograph of a mockup sample with three rows of 15 stacks of 4-mm-wide REBCO tapes, employed in the 1-m and 3-m samples with internal electrical insulation. (b) Schematic cross-sectional image of the STARS conductor sample and a photograph of a mockup with a single row of 15 stacks of 12-mm-wide REBCO tapes with internal electrical insulation. (c) Schematic cross-sectional layout of the STARS conductor employed for the 6-m conductor sample.

3. Design and fabrication of a 6-m STARS conductor sample with a solenoid structure

In the next phase of the STARS conductor development, a 6-m sample was designed and fabricated to be tested in a larger testing facility at NIFS equipped with a solenoid coil magnet producing the maximum magnetic field of 13 T on a sample coil. It should be noted that a 2-m conductor test using the 9-T split coil system was omitted for the present sample, as the inclusion of the internal electrical insulation may not give an essential difference to the performance of the conductor. In the 13-T facility, the magnet has a bore of 700 mm in diameter. Inside the bore, a vacuum chamber with an inner diameter of 687 mm is situated for installing a sample conductor formed into a coil shape. In the vacuum chamber, a cooling pipe supplies helium gas with temperature control in the range of 4.2-50 K. With a 6-m conductor length, the sample was formed into a solenoid coil with a 600-mm diameter so that it would fit into the sample chamber, as is shown in figure 2. The actual conductor used for this coil sample was finally designed as shown in figure 1(c) with a cross-section of 22.5 mm by 12 mm, which was tentatively selected for this sample. The three-turn solenoid winding has been formed by first fabricating the copper casing and stainless-steel jackets by machining with each 120-degree piece. The copper casing was then assembled using tentative support jigs made of resin, with precise sizes, fabricated by additive manufacturing. Next, REBCO tapes (FESC-SCH12 of Fujikura, width: 12 mm) were wound into the groove of the assembled copper casing. After the winding was completed with 15 REBCO tapes, the lid part of the copper stabilizer casing was attached, and the copper casing was wrapped by two turns of Kapton tape as the internal electrical insulation. Then, pieces of stainless-steel jackets were attached around the Kapton layer and they were welded together by TIG welding instead of LBW. As the direction of the magnetic field produced by the sample coil is the same as that by the background magnet, a hoop force should appear radially outward. Thus, the sample coil was installed in a 30-mm stainless-steel ring for mechanical support.

For this 6-m STARS conductor sample wound in a 600-mm diameter solenoid coil, we chose a fabrication method by assembling the copper stabilizer casing and the stainless-steel jacket prepared in pieces. We note that with a bending radius of 300 mm and 2 mm thickness of the stack of REBCO tapes, the bending strain reaches ~0.3% maximum. Though this strain could be still accepted for a REBCO tape, the problem is more in the coil feeder region where the minimum bending radius of ~100 mm gives a bending strain of ~1%, which is not tolerable. In this respect, the present coil sample is different from the actual condition of the STARS conductor for its use with a long length and in a large coil winding, such as of the helical coils. However, it is considered that the present sample should play an important role to examine the feasibility of the STARS conductor with a sufficient length. We should note that in the previous test of the prototype STARS conductor with a 100 kA current capacity and 3-m length, only a 300-mm portion was kept at 20 K under a magnetic field of <8 T [3]. In the former test of the 20-kA class STARS conductor without internal electrical insulation, the whole sample of 2-m long was kept at 20 K, but the magnetic field was applied again only in the same 300-mm region [9]. Thus, the present experiment with a 6-m STARS conductor has become a big step by cooling the whole conductor at 20 K under a high magnetic field. The temperature control was provided for the whole sample with indirect cooling by supplying helium gas into a copper tube mechanically attached to the sample as shown in figure 2.

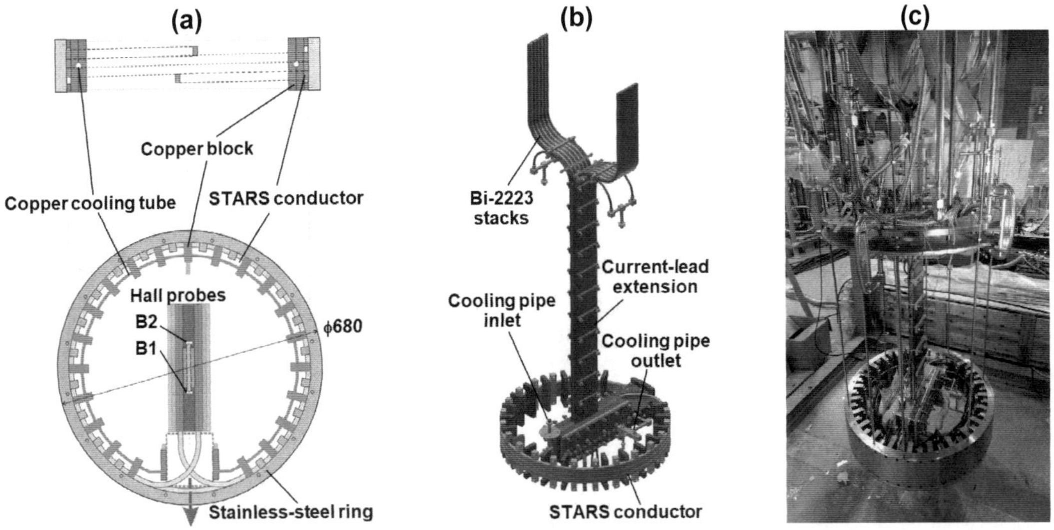

Figure 2. (a) Schematic drawing of the 6-m STARS conductor sample with side and plan views. Positions of the two Hall probes (B1 and B2) are indicated. The arrow pointing down indicates the direction of movement of the whole sample due to the imbalance of electromagnetic forces. (b) Three-dimensional view of the sample without showing the stainless-steel supporting ring. (c) Photograph of the sample during the attachment work to the current leads of the 13-T conductor testing facility.

The coil feeder sections are the most difficult part for the present sample formed in a coil shape. It was chosen to connect the sample coil to the current leads by having a pair of "current-lead extensions" which is shown in figure 2(b). The difficulty is that this joint was located in a high magnetic field region. Here the low resistive mechanical lap joint technique developed at Tohoku University [17-20] was applied, which employs a staircase-like structure for the stack of REBCO tapes and indium foils on the tapes. The "current-lead extensions" were formed by copper plate bent into a "Y" shape and Bi-2223 HTS tapes were embedded in grooves. We note that a low-melting temperature metal, U-Alloy 60 (melting temperature: 60 centigrade) was used for "soldering".

4. Experiment of the 6-m STARS conductor sample

The 6-m conductor sample was assembled and installed into the sample chamber of the 13-T magnetic field, 700-mm-bore, 4.2-50 K temperature controlled, 50-kA current conductor testing facility. The detailed specification of this facility is described in [21]. Figure 3 shows the first excitation test result of the STARS conductor sample. The current was increased stepwise above 10 kA to 18 kA. During this operation, no large rise in the voltage signal was observed, which indicates that the excitation was carried out stably without any disturbing event such as a normal-transition and a subsequent quench. The pair of voltage taps V1 was attached to the copper blocks which were connected to the conductor terminals. The joint resistance at a joint was evaluated to be ~50 nΩ. It is understood that these low-resistive mechanical lap joints played an important role in the success of the present experiment by ensuring no temperature increase in the sample. The magnetic field was measured by the two Hall probes B1 and B2. A numerical calculation obtains close values with the measured values with both B1 and B2, by considering the locations of the external magnet as well as the winding path of the STARS conductor sample with the solenoid coil shape including the coil lead sections. Presently there is a ~0.7% discrepancy, especially for the magnetic field component produced by the external coils. Some factors should be considered to have a better agreement, such as the exact position of the sample coil considering the thermal contraction by cooling regarding the relative position between the external magnet and the sample coil. It is considered that screening currents are induced in the REBCO films of the STARS conductor sample by the application of the external magnetic field. This may also affect the measured magnetic field and the detailed analysis is our future study. During the excitation of the sample coil, there is a possibility that the sample coil moves by the electromagnetic force in the direction indicated in figure 2(a). If we assume that the sample coil moves by a maximum of 14 mm, the magnetic field at the Hall probe positions should change by ~±0.02 T. A precise analysis is future work.

In this experiment, the critical current was measured at a condition with an external magnetic field of 6 T and the inlet temperature of 40 K. In this case, the voltage rose sharply at the sample current of 11.8 kA as shown in figure 4. This value was considerably lower than the expected critical current of 13.0 kA, which was formerly observed for a 2-m conductor sample with no internal insulation, as was reported in [9]. In the present experiment, though the inlet temperature was well controlled at ~40 K, the outlet temperature was as high as ~45 K. The reason for this non-uniform temperature distribution may be explained by a low vacuum condition with a pressure of ~0.1 Pa due to a leak in the sample chamber. The critical current measurement will be carried out again in the second test planned in the near future by fixing the leakage. In the present critical current measurement, it was observed that the voltage rose abruptly so that the gradual increase of the voltage, accompanied by a flux-flow resistance before reaching the critical current, was difficult to observe due to the relatively large noise ratio. For the next experiment, it is planned to install a measuring cable that is co-wound with the sample solenoid coil to reduce the inductive noise signal.

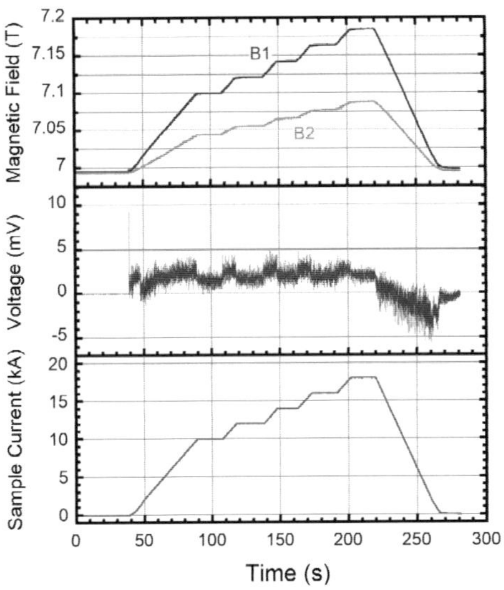

Figure 3. Temporal waveforms of the magnetic field intensities measured by Hall probes (top), the voltage measured by voltage taps across the whole coil sample (middle), and the current of the sample (bottom). The temperature was kept at ~20 K.

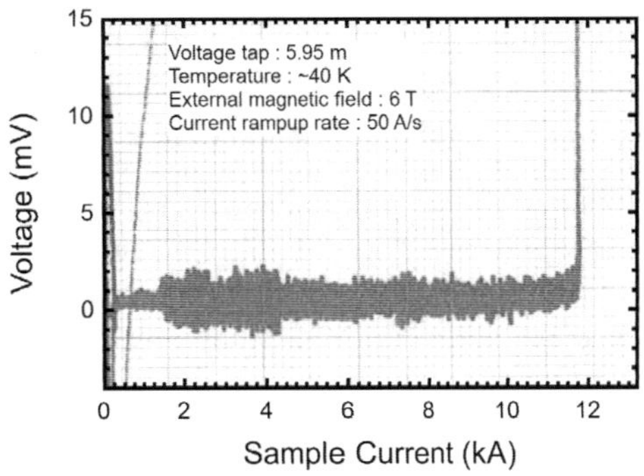

Figure 4. Voltage observed for a whole sample conductor as a function of the sample current during a critical current measurement conducted at 6 T with an inlet temperature of ~40 K. There was non-uniform temperature distribution.

5. Discussion

In the present experiment, a series of repetitive excitations were also carried out with a fast (1 kA/s) ramp-up and down rate of the sample current, and up to 100 times excitations. No apparent change was observed in the excitation characteristics during and after the repetitive excitations. Figure 5 shows waveforms of the magnetic field strengths observed in the 100-time repetition excitation test, depicted for the last two excitations. It should be noted that there was a gradual decrease in the magnetic field, for ~20 s, after the sample current decreased to zero. Figure 6 plots this residual magnetic field strength, just at the time that the sample current reached zero, as a function of the ramp rate. It shows a clear linear dependence between the residual magnetic field and ramp rate. We consider that the residual magnetic field is produced by the circulation current flowing across the stacked REBCO tapes, especially through the joint sections at the coil leads where a staircase-like connection was made with a copper block. The existence of a circulation current should be equivalent to the formation of a non-uniform current distribution among REBCO tapes, which may be induced by the imbalance of self-inductances among REBCO tapes. Here we consider a simple model consisting of two parallel coils with self-inductances L_1 and L_2, both connected to an external power supply through terminals. A circuit equation gives a difference of the currents flowing in these two coils, ΔI. When the whole current I_0 supplied by a power supply becomes zero from a finite value with a ramp-down rate of $-dI_0/dt$, ΔI still exists but decays due to the loss generated in the joint resistances connecting each inductance to the terminals. In this phase, ΔI becomes a circulation current flowing through the two coils, which can be expressed as a function of time t as,

$$\Delta I = \frac{L_2 - L_1}{4R}\frac{dI_0}{dt}\exp\left(-\frac{2R}{L_2 + L_1 - 2M}t\right), \tag{1}$$

where R is the joint resistance taking the same value for simplicity for the two coils, and M is the mutual inductance between the coils. It should be noted that equation (1) indicates that the circulation current is proportional to the difference between the two self-inductances and the ramp rate. This feature may explain the linear dependence of the residual magnetic field on the ramp rate observed in the present experiment. The absolute value of the residual magnetic field is being examined as the next step, which will be reported elsewhere. It is also considered that screening currents induced in the REBCO films by the application of the external magnetic field as well as by the sample current may also contribute to the residual magnetic field. However, we consider that its magnitude is limited compared to that generated by the circulation current. It should be noted that the screening current must be irrelevant to the ramp rate, and the measured decay time constant of ~20 s is too short for the screening current, which must be attenuated by the flux-flow resistance.

In this experiment, despite having a non-uniform current distribution, a stable operation has been confirmed. However, the validity of simple stacking of HTS REBCO tapes, without twisting or transposition, for making a large conductor is still under intense discussion. It is being discussed that the AC losses may not become serious, especially for fusion magnets [22]. The non-uniform current distribution is a serious problem, especially for LTS conductors in terms of cryogenic stability. Due to the orders of magnitude larger cryogenic stability margin for HTS, this may not become a serious problem. We confirmed this possibility in a small experiment having stacked five REBCO tapes and current feeding from only one of the tapes [23]. The experimental observation of the voltage in each tape can be well described by a numerical calculation [24]. Extension of calculation to the present sample coil will be reported elsewhere.

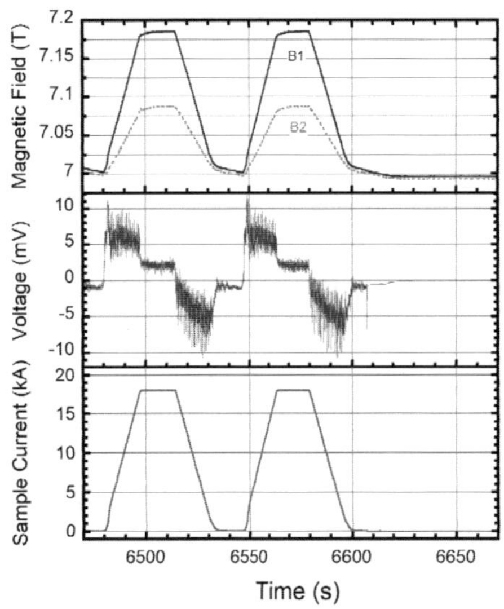

Figure 5. Waveforms of the magnetic field measured by Hall probes (top), voltage across the whole coiled conductor sample including the joint (middle), and current (bottom) observed in the last two excitations of the 100-times repetition excitation test.

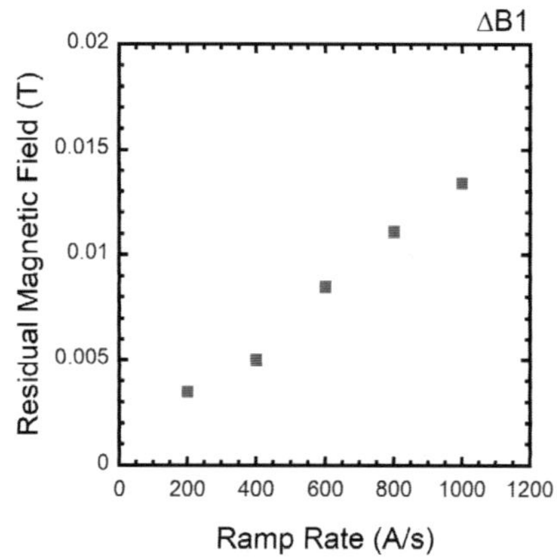

Figure 6. The residual magnetic field, when the transport current reached zero, as a function of the ramp rate for excitations.

6. Summary

The HTS STARS conductor is being developed for application to the next-generation fusion experimental devices The rated current is 18 kA at 20 K and ~15 T, with a target current density of 80 A/mm^2. A 1-m and 3-m conductor samples were fabricated with internal electrical insulation situated between the copper stabilizer casing and stainless-steel jacket. The samples were tested in liquid nitrogen first with no external magnetic field, which confirmed exactly the same result that was observed for a former non-insulation sample. Then, a 6-m conductor sample with internal electrical insulation was fabricated with a 600-mm diameter and a coiled structure of 3 turns. It was tested in 8 T, 20 K using a facility equipped with a maximum 13-T, 700-mm bore solenoid coil. An external magnetic field was applied uniformly to the whole conductor of 6-m length, and temperature control was also provided for the whole sample with indirect cooling by supplying helium gas into a copper tube mechanically attached to the sample. A stable operation up to 18 kA current was confirmed. A series of 100-time repetitive excitations with a 1 kA/s ramp rate was also carried out. For the success of this experiment, the low-resistive mechanical lap joints between the current feeders and the conductor terminals played an important role by ensuring no temperature increase. A residual magnetic field was observed after the transport current became zero, which might be caused by a circulation current among the simply-stacked REBCO tapes due to the imbalance of self-inductance among them, which may correspond to the formation of non-uniform current distribution. The experiment indicates that the STARS conductor with a simple stacking configuration is considered promising to be used in large-scale fusion magnets. A second experiment is planned to make more excitation cycles of >1000 times, and the circulation current will be measured more precisely.

Acknowledgments
The authors thank Yasutaka Maeda and Hideo Kakui from Metal Technology, Co. Ltd. for fabricating the STARS conductor samples. The continuous support for conductor experiments provided by K. Ohba and T. Baba from the Technical Division of NIFS is specially granted. One of the authors (N. Y.) thanks Xiongyi Huang from ASIPP for a fruitful discussion on the internal electrical insulation of the STARS conductor concept.

References
[1] Sagara A et al 2014 *Fusion Engineering and Design* **89** 2114
[2] Sagara A et al 2017 *Nuclear Fusion* **57** 086046
[3] Yanagi N et al 2015 *Nuclear Fusion* **55** 053021
[4] Yanagi N et al 2016 *Cryogenics* **80** 243
[5] Yanagi Net al 2019 *Journal of Fusion Energy* 38 147
[6] Bruzzone P et al 2018 *Nuclear Fusion* **58** 103001
[7] Terazaki Y et al 2014 *IEEE Transactions on Applied Superconductivity* **24** 4801305
[8] Terazaki Y et al 2015 *IEEE Transactions on Applied Superconductivity* **25** 4602905
[9] Yanagi N et al 2022 *Plasma and Fusion Research* **17** 2405076
[10] Mito T et al 2020 *Journal of Physics Communications* **4** 035009
[11] Onodera Y et al 2020 *Journal of Physics: Conference Series* **1559** 012118
[12] Matsunaga S et al 2020 *IEEE Transactions on Applied Superconductivity* **30** 4601405
[13] Narushima Y et al 2022 *Plasma Fusion Research* **17** 2405006
[14] Yanagi N et al 1999 *IEEE Transactions on Applied Superconductivity* **9** 1113
[15] Narushima Y et al 2020 *Plasma Fusion Research* **15** 1405076
[16] Yanagi N et al 2011 *Fusion Science and Technology* **60** 648
[17] Ito S et al 2014 *Plasma Fusion Research* **9** 4602305
[18] Ito S et al 2018 *Fusion Engineering and Design* **136** 239
[19] Ito S et al 2019 *Fusion Engineering and Design* **146** 590
[20] Ito S et al 2021 *Nuclear Fusion* **61** 115002
[21] Imagawa S et al 2015 *Plasma Fusion Research* **10** 3405012
[22] Uglietti D et al 2020 *Cryogenics* **110** 103118
[23] Meulenbroeks T A J et al 2019 *Journal of Physics: Conference Series* **1293** 012065
[24] Garfias-Dávalos D et al 2022 *Plasma Fusion Research* **17** 2405066

Characteristics and causes of voltage observed at the current feeder of high-temperature superconducting WISE conductor

Y Narushima[1,2], D Garfias-Davalos[2], Y Onodera[1], S Hamaguchi[1,2], H Chikaraishi[1,2], N Yanagi[1,2], T Baba[1], J Miyazawa[1,2]

[1] National Institute for Fusion Science, 322-6 Oroshi-cho, Toki, Gifu, 509-5292, Japan
[2] The Graduate University for Advanced Studies, SOKENDAI, 322-6 Oroshi-cho, Toki, Gifu, 509-5292, Japan

narushima.yoshiro@nifs.ac.jp

Abstract. The HTS (high-temperature superconducting) conductor is a feasible candidate for constructing magnets for next-generation fusion devices because of its higher critical current in a high magnetic field. A new concept of the HTS-WISE (Wound and Impregnated Stacked Elastic tapes) conductor has been studied aiming to apply the fusion reactor magnet. Here, the WISE-U conductor is composed of stacked thirty REBCO tapes (10 mm width, 65 μm thickness, I_c = 370 A @77 K, s.f.) wrapped by a stainless-steel coil tube which is inserted into the metal pipe. The 4 m-long REBCO tapes are folded with a radius of curvature of 35 mm in a hairpin-like structure. A low-melting-point metal U-Alloy 60 whose melting point is 60°C is poured into the pipe for impregnation to make the non-insulation conductor. The REBCO tapes and the current feeder made of oxygen-free copper were also impregnated with the U-Alloy 60 to connect. This fabrication method has the advantage of being easier to fabricate than the technique of connecting each tape using indium foil. The energization test results showed that a maximum current value of 16.9 kA was recorded at B = 5 T and T = 30 K, however, a burnout occurred in the current feeder before the critical current was determined. Then, the improved WISE conductor has been designed and tested which showed a maximum of 19kA was reached in the self-field and 20K. However, burnout still occurred in the current feeder section. In those experiments, the superconducting section has not been damaged at all. If this burnout had been avoided, a higher current-carrying capacity could have been obtained. Identifying the cause of burnout and improving the current feeder is required.

1. Introduction

In the environment of the nuclear fusion reactor, the high magnetic field for the confinement of the plasmas and the neutron irradiation from the fusion reaction of the plasma particles cannot be avoided. In such conditions, the REBCO (Rare-Earth Barium Copper Oxide) is expected to be applied to fusion magnets, due to the high mechanical rigidity of the REBCO tapes provided by the use of Hastelloy as a substrate. The low amount of silver in REBCO tapes is less susceptible to radioactivation by neutron irradiation, giving it an advantage over silver-rich HTS conductors like BSCCO. On the other hand, relatively large anisotropy concerning the direction of the magnetic field exists in the REBCO tapes. Therefore, when a magnetic field has applied perpendicular to the REBCO tape surface, the characteristics of the critical current are generally degraded. Compared to low-temperature superconductors (LTS), high-temperature superconductors (HTS) have the advantage of a higher critical current value under high magnetic fields as well as a higher critical temperature, which allows for a wider range of operating conditions as reported by Bednorz and Müller in 1986 [1]. From the viewpoint of making an HTS cable, the re-distribution of the current would be difficult in the stacked REBCO tapes due to the disturbance by the high resistance substrate layer. Those problems become even more serious in the condition of the time-varying operation of the current, which causes the polarization of the current profile leading to a risk of current concentration on certain tapes. To solve those problems, many efforts have been done. To make the anisotropy of the current path average, the CORC (Conductor on Round Core) conductor is composed of wrapping multiple tapes. Although the perpendicularly applied magnetic field to the HTS tape determines the critical current of a whole conductor, the CORC concept appeals the flexibility and allows for a small bending radius [2]. Another approach for the equal distribution of the current is that a meander-shaped cut of HTS tapes is stacked called the Roebel conductor, which is capable of minimizing AC losses [3].

To realize the HTS magnet applying to the helical fusion reactors, the current-carrying characteristics under the condition of low temperature and high magnetic fields should be clarified. Helical fusion devices produce only the external magnetic field to confine the plasma and do not need to make the plasma current for the poloidal magnetic field. Because a significant change of the magnetic field like in tokamaks does not occur, a concerned AC loss is not expected and an application of a stacked conductor is available. In National Institute for Fusion Science (NIFS) has studied three types of HTS conductors with different concepts (STARS (Stacked Tapes Assembled in Rigid Structure) [4], FAIR (FSW, Al-alloy, Indirect-cooling, REBCO) [5], and WISE (Wound and Impregnated Stacked Elastic tapes) [6]), which brings the benefits of diversity to the study of HTS conductors. In this article, we report the characteristics of the WISE conductor. We have tested the WISE conductor under the condition of a temperature $T = 30$ - 50 K and a magnetic field $B = 5$ - 8 T. The experimental conditions were established based on the magnetic fields and temperatures that the superconducting coils in a Helical fusion reactor would encounter. Although these conditions are not identical to those expected in FFHR-b3 ($T = 20$ K, $B = 19$ T) [7], we have tried to make them as close as possible using the current testing equipment.

The following section briefly explains the experimental setup, including the WISE conductor and test facility. Section 3 shows the experimental result. We will summarize in Section 5 following the discussion in section 4.

2. Experimental setup

The WISE-U conductor shown in figure 1 (a) as the overall figure contains 4 m-long REBCO tapes (Shanghai Superconducting Technology Co., Ltd. 10 mm width, 65 µm thickness, and critical current $I_c = 370$ A @$T = 77$ K, S. F.) with 30 layers. The stacked tape was inserted into a stainless steel flexible tube (figure 1 (b)), which was then inserted into a U-shaped pipe with a curvature radius of $R =$ 35 mm (figure 1 (c)). At this time, the stacked tape is flexible and can be formed into such a structure. This is one of the characteristics of WISE conductors. Finally, it was impregnated with low-melting-point metal after the stacked tapes are inserted in the straight 2 m length aluminum pipes. Due to the shape of the WISE conductor seeming to be a 'U' figure, this is named WISE-U. The stacked REBCO tapes are impregnated with the

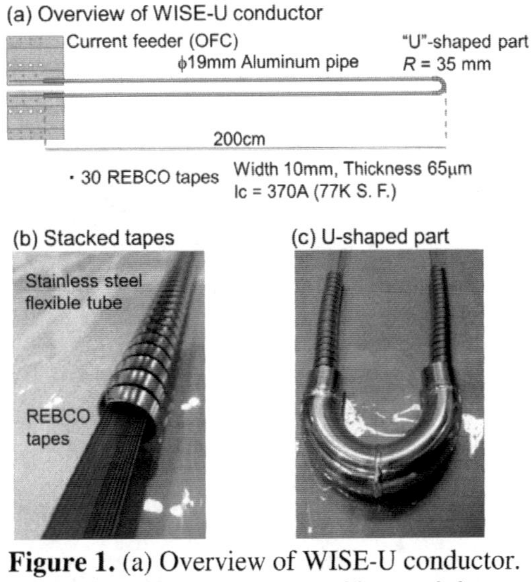

Figure 1. (a) Overview of WISE-U conductor. (b) Stacked HTS tapes inserted into stainless steel flexible tube. (c) Detail of the U-shaped part.

low-melting-point metal, U-Alloy 60, to be fixed. This test facility requires two parallel straight conductors, and they usually need to be connected at the tip, which is the area with the highest magnetic field. This is because it is difficult for a general conductor to be bent into a small curvature. However, connecting at the tip can cause joule heating due to connection resistance, which can lead to issues with temperature control and accurate measurement. On the other hand, WISE-U conductors can be bent into a small curvature due to their flexibility before impregnation. Therefore, it is not necessary to connect them at the tip, and this can help to avoid joule heating issues associated with tip connections.

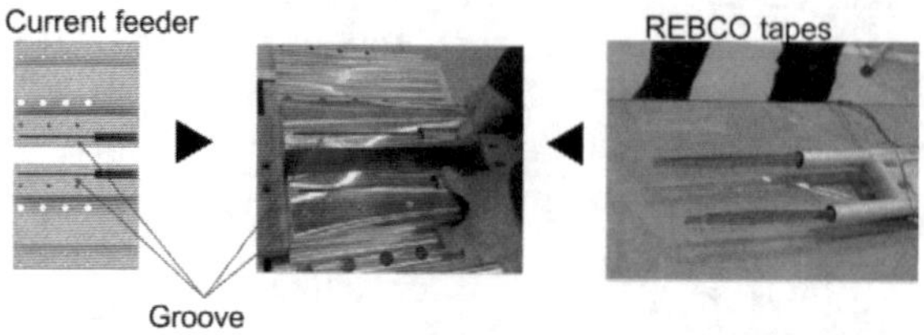

Figure 2. Assembly of current feeder and REBCO tapes. The liquid U-Alloy 60 is poured into the groove to create an electrically connected joint.

The connection between current leads and REBCO tapes of the WISE-U conductor adopts the same method as low-melting-point metal U-Alloy 60 impregnation of the superconducting section. After the REBCO tape is inserted into the groove of the current feeder (figure 2), the entire current feeder is heated to 100°C. Following the liquid U-Alloy 60 is poured into the groove. At this time, the REBCO tape and groove surfaces are flux-cleaned so that the U-alloy adheres firmly as if soldered. This technique is expected to be simple to fabricate, requiring no crimp torque control and no indium foil to ensure sufficient current flow between the HTS tape and the current leads.

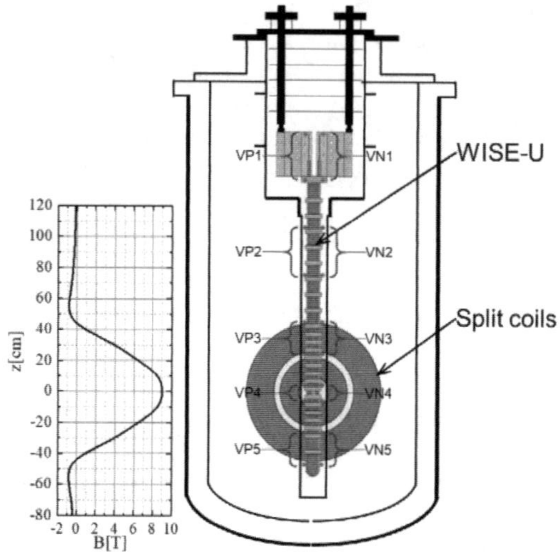

Figure 3. WISE-U conductor inserted in test facility. Red points on WISE-U indicate the voltage taps. Profile of magnetic field is shown on the left.

For energization tests at low temperatures (T = 30 - 50 K) and high magnetic fields (B = 5 - 8 T), the WISE-U conductor was installed in the test facility which can produce temperatures up to T = 4 K using liquid/gas helium (figure 3 right). And a split coil, consisting of two sets of circular coils, can generate a magnetic field up to B = 9 T. The magnetic field profile is shown on the left side of figure 3, where the magnetic field strength is more than 80% of the maximum field within z = ±170 mm from the center, and it drops to zero at z = ±440 mm. The voltage taps VP4 and VN4 are located at the position where the magnetic field produced by the split coils is at its maximum. This magnetic field is applied in a direction parallel to that of the REBCO tape in WISE-U.

3. Experimental results

3.1. First trial of energization test of WISE-U conductor

3.1.1 Behavior of voltage on superconducting section

Figure 4 shows the relationship between the voltage and supplied current (I_{ps}) into the WISE-U as an I - V curve. The voltage V on the vertical axis is the normalized voltage measured at the position with the

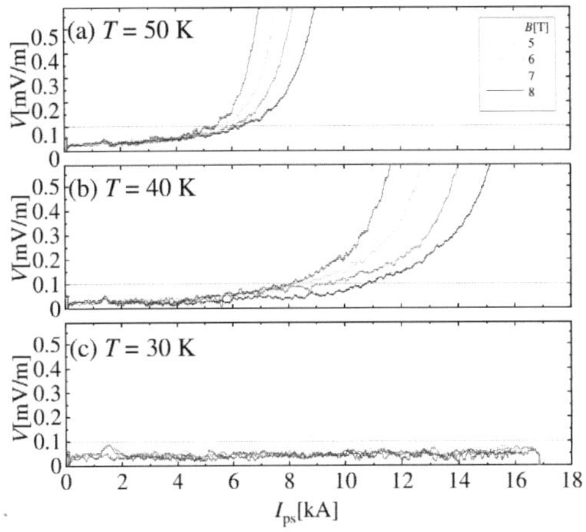

Figure 4. $I - V$ curves of WISE-U conductor in (a) T = 50 K, (b) T = 40 K and (c) T = 30 K, respectively. Voltage is measured at VP4. Magnetic fields B = 5 T, 6 T, 7 T, 8 T are indicated by blue, green, red and black, respectively. Dotted horizontal lines indicate the 0.1 mV/m critical current index.

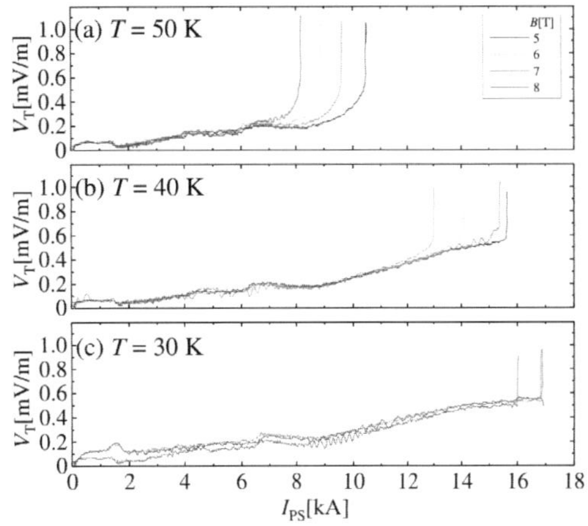

Figure 5. $I - V$ curves at the current feeder of WISE-U conductor at (a) T = 50 K, (b) T = 40 K and (c) T = 30 K, respectively. Magnetic fields B = 5 T, 6 T, 7 T, 8 T are indicated by blue, green, red and black, respectively.

strongest external magnetic field produced by the split coil of the test facility. The voltage displayed in figure 4 corresponds to the voltage measured at the location z = 0, which is labeled as "VP4" and corresponds to the position where the magnetic field is the strongest, as shown in figure 3. As the magnetic field is strongest at this location, it is more likely to generate a voltage here than at other locations. Therefore, analyzing the voltage at the z = 0 location is a useful approach for evaluating the performance of the WISE-U conductor. The critical current I_c obtained as the current value when the voltage reaches 0.1 mV/m, indicated by the dashed line, ranges from I_c = 5.4 kA to 6.5 kA at T = 50 K (figure 4 (a)) and from I_c = 8.1 to 10.7 kA at 40 K (figure 4 (b)). Almost zero voltages are shown in the low current range whereas nonlinear increment appears approaching the critical current. It can be seen that the critical current values increase with decreasing temperature and magnetic field. These experimental results show that the WISE-U conductor has the typical features of the superconductor. On the other hand, the case of T = 30 K is a special case, in which the quench detector works and the energization is terminated before the critical current value is reached. In this case, a maximum current value of 16.4 kA was obtained in a magnetic field of B = 8 T, which corresponds to a current density of j_E = 60 A/mm^2 and a maximum experienced electromagnetic force of 131 kN/m. In addition to the results presented above, Ref. [8] also discusses in detail the energization cycle evolution of n - values, the comparison of experimentally obtained critical current values with theoretical calculations, and the electromagnetic forces experienced.

3.1.2 Behavior of voltage on current feeder part

The behavior of the reference voltage which is the total voltage in the entire length of the WISE-U conductor input to the quench detector to operate is shown in figure 5. The quench detector was set to operate at a total voltage of 0.5 mV/m. In the case of T = 30 K shown in figure 5 (c), the total voltage increases when the I_{PS} is beyond 10kA whereas the voltage at the strongest external magnetic field (figure 4 (c)) maintains V < 0.1 mV/m.

The voltages in the current feeder and superconducting sections at T = 30 K and B = 5 T are shown in figure 6. The WISE-U conductor is also drawn to reference the voltage measurement location. The voltages in the WISE-U superconducting section (Figs. 6 (b)-(e) and (g)-(j)) do not show a sudden

Journal of Physics: Conference Series **2545** (2023) 012009 doi:10.1088/1742-6596/2545/1/012009

increase, indicating that the superconducting section was not damaged from the current and resulting

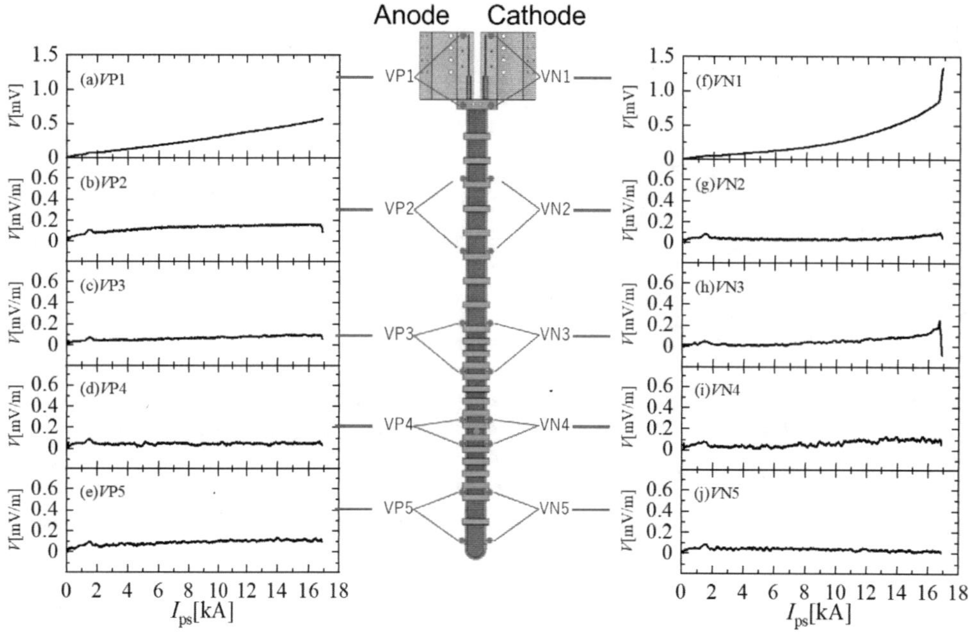

Figure 6. *I-V* curves at (a) and (b) the current feeder, and (b - e) and (g - j) the superconducting sections at $T = 30$ K and $B = 5$ T. The WISE-U conductor is included for reference to indicate the voltage measurement location.

electromagnetic force, remains healthy. On the other hand, the voltage in the current feeder section increases with the current (Figs. 6 (a) and 6 (f)). In particular, the voltage in figure6 (f) shows a spike-like waveform just before the end of the energization. In fact, at the end of this energization, the current feeder was damaged and could not continue energizing.

To clarify what happened in the current feeder, the details of the voltage and resistance in the current feeder section, are shown in figure 7. In the region of I_{ps} < 12kA, the voltage on the cathode side is lower than that on the anode side (figure 7 (a)). The resistance is nearly constant at 30 nΩ on the anode side, while on the cathode side, it increases from 20 nΩ to 30 nΩ (figure 7 (b)). In this range of I_{ps} < 12 kA,

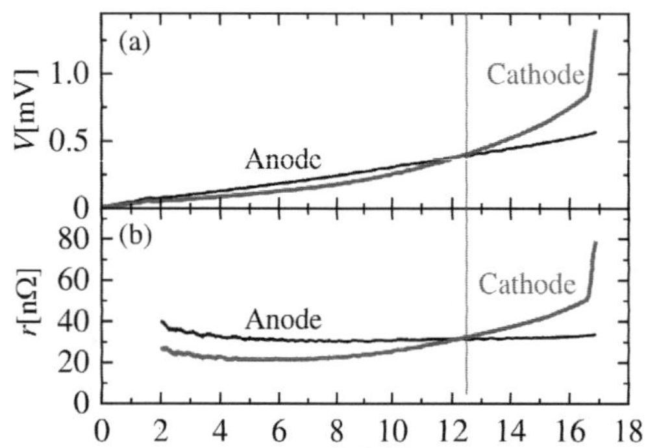

Figure 7. (a) Voltage and (b) resistance in the current feeder. The cathode and anode sides are shown in red and black, respectively. The cross-over point is indicated by a vertical solid line.

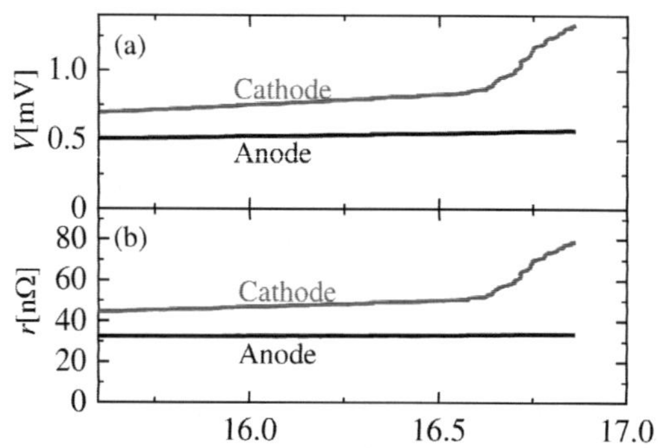

Figure 8. Zoom-up view of (a) voltage and (b) resistance in the current feeder during a sudden increase in voltage state, with I_{ps} increasing from 15.6 kA to 17 kA.

the current feeder was thought to have a margin against heat generation during energization because the electric power could roughly be estimated to be about 4 W. Beyond I_{ps} = 12 kA, the increase in resistance on the anode side is less pronounced, but the resistance and voltage on the cathode side exceed those on the anode side due to nonlinearities. The behavior of voltage and resistance around I_{ps} = 16 kA are shown in figure 8. Both the cathode side and the anode side gradually increase the voltage and resistance up to I_{ps} = 16.6 kA. Consequently, from I_{ps} = 16.6 kA to 16.9 kA, the rate of change of them on the anode side does not change, but the rate of them on the cathode side shows a sharp increase. Finally, the voltage (resistance) increases rapidly, and at I_{ps} = 16.9 kA the voltage reaches 1.3 mV and the resistance 80 nΩ. The rough estimation yields the electric power of 23 W. Finally, the REBCO tape inside the current feeder on the cathode side has been burned out to cut.

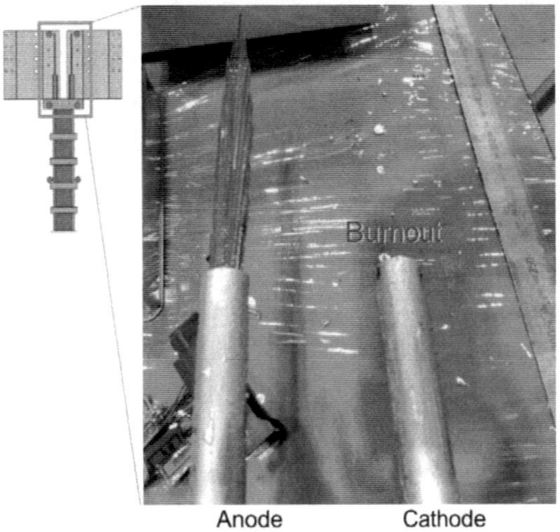

Figure 9. Conductor removed from the current feeder. The left side (anode) REBCO tapes remain, but the right side (cathode) tapes are completely burned out and lost.

3.1.3 Burnout of REBCO tapes in current feeder

After the energization test, the WISE-U conductor was removed from the current feeder to assess the situation and condition of the REBCO tapes on the cathode side. As the result, the stacked REBCO tapes survived on the anode side, but those on the cathode side were burned out as shown in figure 9. The other superconducting section was not damaged. These facts mean that the local quench has occurred inside the current feeder whereas the superconductor keeps its health.

3.2. Second trial of energization test by improved current feeder of WISE-U conductor

3.2.1. Redesign of current feeder of WISE-U conductor

Avoiding or suppressing the Joule heating by eliminating normal conductors as much as possible is one of the essential requirements for improving the overall performance of superconducting conductors. The WISE-U conductor studied here does not use a normal conductor connection at the tip, as shown in figure 1, which can eliminate Joule heating from the superconducting section. This can be realized by the feature of the flexibility of the WISE conductor in the fabrication process. However, in the current lead part of the WISE-U conductor, a normal conductor material must be used, which may generate Joule heating.

The burnout in the current feeder may be due to some degradation in the REBCO tape, which may have caused a local quench due to the positive feedback between the increase in resistance due to current flow and the subsequent Joule heating. The REBCO tape is very vulnerable to bending in the edge direction. The edge-wise strain of REBCO tape is typically limited to less than approximately 0.5% [9, 10]. In WISE-U

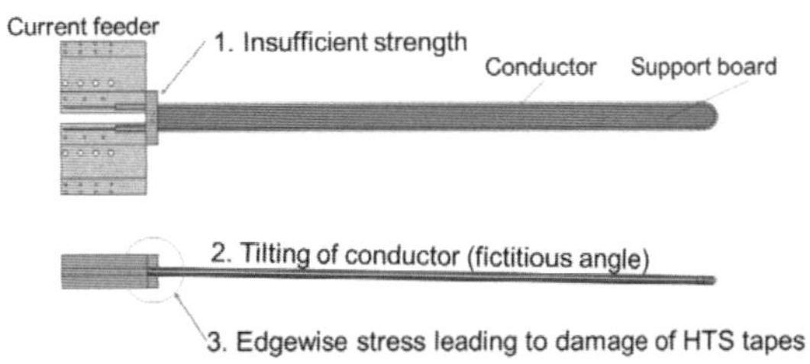

Figure 10. Possible insufficient fixing strength of WISE-U conductor and resulting damage to REBCO tape.

conductors, the edge-wise strain of concern is expected to be concentrated in the current lead area, where the radius of curvature may be extremely small. While it is difficult to determine the actual radius of curvature quantitatively, assuming a relatively large radius of curvature of 1000 mm (which is likely underestimated), the edge strain would still be 0.5%. This value should not be taken lightly. The neck of the WISE-U conductor was not sufficiently fixed, which may have resulted in insufficient fixation accuracy and strength, causing edge-directional torsion to the REBCO tape due to bending as shown in figure 10.

To solve this problem, it is necessary to improve the fabrication accuracy and robustness of the current feeder as shown in figure 11. The point of improvement is to change the structure to fix the current feeder and the support plate accurately and firmly. In addition, the pipe is changed from aluminum which is anodized to insulation to copper to ensure a path for the current to escape in the unlikely event of a quench.

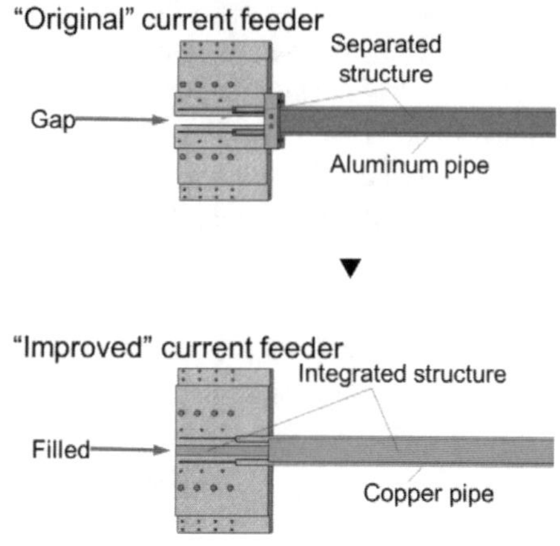

Figure 11. Improved current feeder structure using the integrated support board stucture.

3.2.2. Voltage appearance in the improved current feeder

The energization test has been performed. However, the burnout in the current feeder has re-occurred. Figure 12 shows the waveforms of supplied current, the voltage of the superconducting section of the WISE-U conductor, the voltage at the current feeder, and the temperature at the current feeder under the condition of $T = 20$ K and self-magnetic field. The voltage of the current feeder on the anode side (Red solid line in figure 12 (c)) gradually increases with time. When the supplied current reaches 19kA, a sudden increment of the voltage is observed. Simultaneously, the temperature of the current feeder on the anode side (Red solid line in figure 12 (d)) quickly increases from $T = 20$ K to 30K. Then the supplied current was shut down. After that, the temperature gradually decreases.

4. Discussion

One of the advantages of WISE conductors is the ability to form flexible conductors, which eliminates the need for a normal-conductor connection between superconducting conductors. Therefore, concerns about "finite resistance at the connection with normal conductors" do not exist in principle. Initially, we attempted to use a low-melting-point metal for the connection between the current leads and the REBCO tape by adapting the fabrication method of the WISE conductor. This was expected to simplify the process. However, when we did the energization tests, burnout of the REBCO tape inside the current leads was recognized as an unavoidable problem. The reason for the occurrence of burnout despite the modification of the current feeder is thought here. The structure

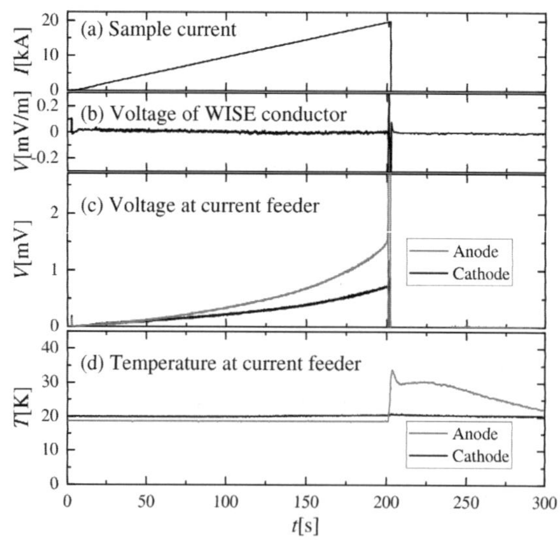

Figure 12. Waveforms of (a) supplied current, (b) voltage of WISE conductor, (c) voltage at current feeder, and (d) temperature at current feeder.

of the current feeder was thought to be worked as an idea to make the current feeder sufficient fabrication to suppress the stress of the REBCO tapes. The stress on REBCO tapes making damages is thought to be avoided. Therefore, it can be thought that the current feeder involves the essential problem including the U-Alloy. The resistivity of the U-Alloy used here is 2x10^{-7} Ωm at 77K, which is enough low for the contact material. One possibility is the finite thickness of the U-Alloy between the REBCO tapes and the current feeder. As shown in figure 2, the stacked REBCO tapes are inserted into the slit of the current feeder, and U-Alloy is poured, which makes the nonuniform thickness of the U-Alloy as shown in figure 13. The current is supplied from the copper current feeder to stacked REBCO tapes via the U-Alloy and the effective resistance should be varied depending on the thickness of the U-Alloy, resulting in the nonuniform supplied current distribution indicated by yellow arrows in figure 13. The supplied current is concentrated at a certain position to lead to the burnout of the REBCO tapes. The reason why the burnout did not occur simultaneously on both may depend on manufacturing differences. For example, the connection resistance can be suppressed if the REBCO tape were put on the surface of the copper leading to the thickness of the U-Alloy could be reduced. In other words, it is possible to totally reduce the connection resistance realizing the uniform current distribution if all REBCO tapes are directly put on the copper.

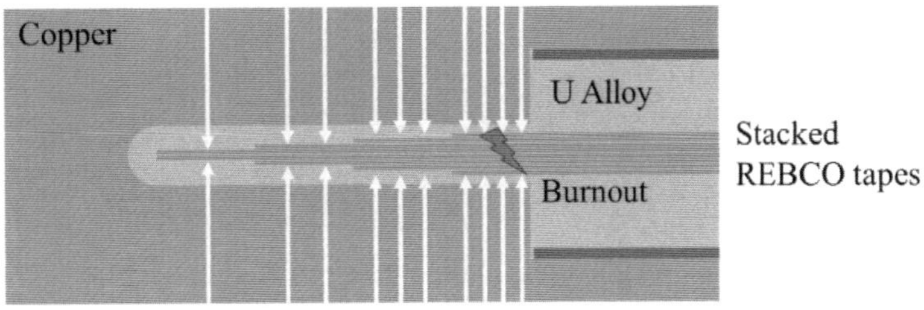

Figure 13. Schematic view of non-uniform current distribution inside the current feeder. The yellow arrows indicate the current path which is concentrated near the burnout point. The red spark in the picture indicates a location where the risk of burning is high.

5. Summary

Energization tests of the WISE conductor which is aimed to be applied in next-generation helical fusion reactors were conducted under conditions of B = 5 T to 8 T and T = 30 K to 50 K. Critical current values of 5.4 A to 10.7 kA were obtained under the condition of T = 40 K and 50 K. The maximum current value of 16.4 kA was obtained in a magnetic field of T = 30 K and B = 8 T, which corresponds to a maximum experienced the electromagnetic force of 131 kN/m. This demonstrates the robustness of the WISE-U conductor and the correctness of the WISE concept. On the other hand, burnout occurred on one of the current feeders. It was thought that the REBCO tape had a weak spot due to the insufficient fabrication of the current feeder structure, which caused the REBCO tape to degrade. The design of the current feeder was improved and the energization test has been done again. However, burnout has occurred again. We thought that there was no problem in the superconducting section of the WISE-U conductor, but a fundamental problem in the current feeder part. The idea to solve the problem is that the structure of the current feeder is improved to make the connection resistance be reduced.

Through this study of the WISE-U conductor, we have identified the importance of the connection resistance between the current feeder and HTS tape. Solving this issue could bring us one step closer to our original goal, which is the application of WISE conductors in the Helical fusion reactor magnets. However, in reality, creating the Helical fusion reactor magnets requires consideration not only of the connection resistance but also of the connection between HTS tapes. This is because it is almost impossible to wind the entire Helical fusion reactor magnets with continuous HTS tapes. It is necessary to connect the HTS tapes while winding them. Various kinds of research have been conducted on the splicing method of HTS tapes [11]. Although the design was focused on improving structural weaknesses, further studies will be carried out to further reduce contact resistance.

Acknowledgments

This study is supported by NIFS budget UFPP001

References

[1] Bednorz J G and Müller K A, 1986 *Z. Phys. B Condensed Matter* **64** 189-93
[2] Weiss J D, Mulder T, Kate H J T and Van der Laan D C 2017 *Supercond. Sci. Technol.* **30** 014002
[3] Hao L, Shen B, Ma J, Yang J, Patel I, Hu J, Tian M, Wei H, Shah A, Wang Q, and Coombs T A 2022 *IEEE Trans. Appl. Supercond.* **32** 5900505
[4] Yanagi N, Ito S, Terazaki Y, Seino Y, Hamaguchi S, Tamura H, Miyazawa J, Mito T, Hashizume H and Sagara A 2015 *Nuclear fusion* **55** 053021
[5] Mito T, Onodera Y, Otsuji M, Hirano N, Takahata K, Yanagi N, Iwamoto A, Hamaguchi S, Takada S, Baba T, Chikumoto N, Kawagoe A and Kawanami R 2021 *IEEE Trans. Appl. Supercond.* **31** 4202505
[6] Matsunaga S, Narushima Y, Onodera Y, Terazaki Y, Miyazawa J, and Yanagi N 2020 *IEEE Trans. Appl. Supercond.* **30** 4601405
[7] Miyazawa J, Goto T, Hamaji Y and Kobayashi M.I. 2021 *Nucl. Fusion* **61** 126062
[8] Narushima Y, Terazaki Y, Onodera Y, Yanagi N, Hirano N, Hamaguchi S, Chikaraishi H, Baba T and Miyazawa J 2022 *Plasma and Fusion Research* **17** 2405006
[9] De Leon M B, Nisay A R, Diaz M A, and Shin H S 2022 *IEEE Trans. Appl. Supercond.* **32** 400605
[10] Narushima Y, Miyazawa J, Matsunaga S, and Yanagi N 2020 *Plasma and Fusion Research* **15** 1405076
[11] Ito S, Hashizume H, Yanagi N, Tamura H 2019 *Fusion Eng. Des.* **146** 590-593

Electromagnetic and Thermal Analysis of HTS AC Cable using *T-A* Formulation

Gaotai Lyu, Yutaka Terao and Hiroyuki Ohsaki

Graduate School of Frontier Sciences, The University of Tokyo, Kashiwa 277-8561, Japan

lyu.gaotai22@ae.k.u-tokyo.ac.jp

Abstract. High temperature superconducting (HTS) cables have been developed into various products and attracted wide attention in industrial applications due to their high critical temperature, current density, zero resistance and overcurrent strength. Recently the triaxial YBCO cables are used in AC applications with the consideration of the flexibility and economy of materials. In this paper, the transient electromagnetic thermal characteristics of a triaxial HTS cable are analysed by the *T-A* formulation and the Thermal-Fluid Coupling Module of the COMSOL software under the condition of alternating current flow. Firstly, the electromagnetic *T-A* formulation and the basic theory of heat transfer are introduced. Transient magnetic fields, critical current distribution, and magnetization losses are discussed under *T-A* modelling. To evaluate the thermal behaviour of the cable under normal current conditions, liquid nitrogen laminar flow is coupled with the heat transfer module, while the heat source data are obtained from the previous *T-A* results.

1. Introduction

Superconductors, with their advantages of zero resistance and complete diamagnetism, make up for the shortcomings of traditional conductors in the industrial field, and superconducting devices have better performance in some application scenarios. In recent years, power cables [1], aircraft motors [2], current limiters [3], and energy storage systems [4] using superconducting technology have been extensively researched and improved in different methods.

High temperature superconductors (HTS) have developed various products since last century because of higher critical temperature, current density, and magnetic field compared with the low temperature superconductors (LTS). With the development of research and application of high temperature superconductors, Rare Earth-Barium-Copper-Oxide (REBCO) conductor as the second-generation HTS has the advantages of higher current density and higher critical current.

In terms of cables, three main wiring methods have been developed, Roebel cable, twisted pair stacked cable (TSTC), and round core conductor cable (CORC) [5]. Round core conductor cables are helically wound tapes around a round core, which offers the advantages of flexibility, mechanical strength and ease of manufacture compared to other wiring methods. At the same time, the analysis and improvement of the winding method of the CORC cable can effectively reduce the magnetization loss.

To analyze the superconducting cables' electrothermal characteristics, researchers from universities mainly utilize the Multiphysics simulation. In this process, the electromagnetic calculation of the power cable uses Maxwell's equations. However, too many vectors will increase the computational difficulty and reduce the computational speed. Therefore, different formulations are proposed, which are used to separate the variables and derive partial differential equations, reducing the computational complexity.

To determine whether cooling parameters such as fluid velocity and cooling channel diameter are adequate under normal operating conditions of laminar flow, the electromagnetic fields during normal operation are analyzed and AC losses are calculated. Using AC losses as input to the thermal and fluid models constructed in the second step, the transient behavior of the cooling process is observed, and the performance of the cooling parameters is analyzed and discussed. The next stage of the established models will be used in transient fault current analysis to evaluate the temperature performance and cooling performance of each layer to provide a reference for HTS cable design.

2. Theoretical model construction

2.1. *T-A formulation*

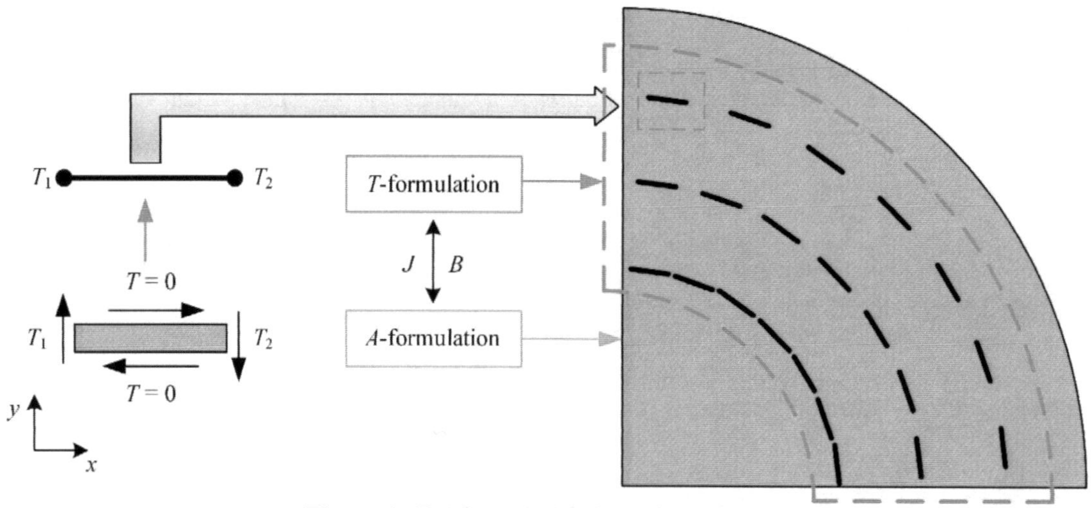

Figure 1. *T-A* formulation for HTS cable

T-A formulation is gradually applied in the field of superconductivity because of its higher efficiency for HTS materials' electromagnetic analysis. As shown in Figure 1, *T* and *A* correspond to the current-related variables on the superconductor and the magnetic field-related variables in the air domain, respectively. These two parts are calculated independently, and then coupled to obtain the electromagnetic field distribution.

The governing equation for *T*-formulation is as follows

$$\boldsymbol{J} = \nabla \times \boldsymbol{T} = \nabla \times (T \cdot \vec{n}) \tag{1}$$

where $\boldsymbol{T}$ is the current vector potential. The relationship between current density $\boldsymbol{J}$ and magnetic flux density $\boldsymbol{B}$ can be derived from Faraday's law

$$\nabla \times \boldsymbol{E} + \frac{\partial \boldsymbol{B}}{\partial t} = 0 \tag{2}$$

where $\boldsymbol{E}$ is the electric field contains current density in terms of the following E-J equation

$$\boldsymbol{E}=\boldsymbol{E}_0 \left(\frac{|\boldsymbol{J}|}{J_c(B)} \right)^n \frac{\boldsymbol{J}}{J_c(B)} \tag{3}$$

where $\boldsymbol{E}_0$ is 1×10^{-4} V/m and the fitting parameter n is 31. Considering the effect of parallel and perpendicular magnetic field to the HTS tapes [6], J_c can be obtained by

$$J_c(B) = \frac{J_{c0}}{\left(1+\sqrt{(kB_{\parallel})^2 + B_{\perp}^2} / B_c\right)^b} \tag{4}$$

where $B_{\parallel}$ and $B_{\perp}$ is the magnetic flux density parallel and the perpendicular to the tape, respectively. J_{c0} is the critical current density without. k is 0.0605, b is 0.7580, and B_c is 103 mT.

The governing equations based on the A-formulation are derived from Ampere's law and are applied to calculate the magnetic field distribution.

$$\boldsymbol{B} = \nabla \times \boldsymbol{A} \tag{5}$$

$$\nabla \times (\frac{1}{\mu_0} \nabla \times \boldsymbol{A}) = \boldsymbol{J} \tag{6}$$

where $\boldsymbol{A}$ represents the magnetic vector potential. From the calculation of T-formulation, the surface current $\boldsymbol{J}_s$ of the conductor can be derived

$$\boldsymbol{n} \times (\boldsymbol{H}_2 - \boldsymbol{H}_1) = \boldsymbol{J}_s \tag{7}$$

where $\boldsymbol{H}_2$ and $\boldsymbol{H}_1$ are the magnetic fields at two sides of the conductor. The magnetization loss [7] can be calculated by

$$Q = 2\int_{T_1/2}^{T_1} \int_S \boldsymbol{E} \cdot \boldsymbol{J} ds dt = \frac{2}{\cos^2 \theta} \int_{T_1/2}^{T_1} \int_S \boldsymbol{E}' \cdot \boldsymbol{J}' ds dt \tag{8}$$

where unit of Q is J/m/cycle, T_1 represents the period, θ is the winding angle, and S is the area of HTS material. The current I can be obtained with Dirichlet boundary conditions

$$I = (T_1 - T_2) \cdot h_{tape} \tag{9}$$

where h_{tape} is the thickness of the HTS tape, T_1 and T_2 are current vector potential in different sides as shown in Figure 1.

2.2. *Thermal and fluid flow modelling*

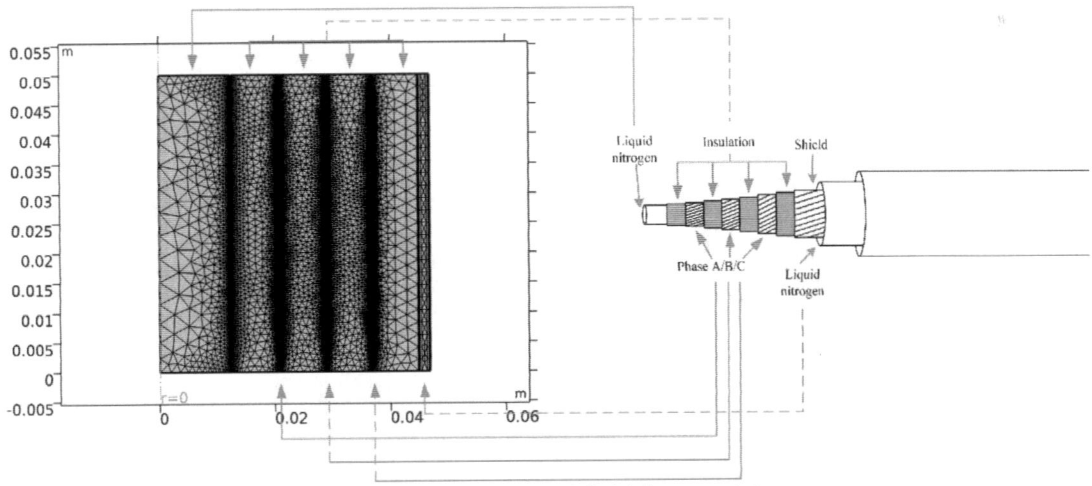

Figure 2. Thermal fluid flow model

In order to select an appropriate fluid analysis process, the flow state of the fluid should be defined at first. Fluids can be classified into laminar, turbulent flow, and mixed states according to their velocity and viscosity. For laminar flow, the fluid flows in an orderly manner along smooth paths in the layers, with no eddies or vortices of the fluid and vertical cross flow, and little mixing occurs. The opposite is the case with turbulent flow, where the flow pattern is more chaotic. Flow regimes are defined using Reynolds number

$$\mathrm{Re} = \frac{\rho u D_H}{\mu} \tag{10}$$

where ρ is the density of the fluid, D_H is the hydraulic diameter of the pipe, u is the average velocity of the fluid, μ is the dynamic viscosity of the fluid. Generally, the fluid Reynolds number below 2300 is laminar flow, above 4000 is turbulent flow, and 2300-4000 is a transitional state of both laminar and turbulent flow. In this paper, the laminar flow module is coupled with the thermal module to discuss and analyze the heating conditions of each layer of the cable and the influence of the fluid flow rate under normal conditions.

To analyze the thermal fluid phenomena of the HTS cable, a 2D model is built as depicted in Figure 2. The model is divided into inner cooling pipe, insulating layer, phase A, phase B, phase C, and outer cooling pipe. In the thermal analysis section, use the COMSOL heat transfer module to solve partial differential equations

$$\begin{gathered}\rho C_p \frac{\partial T_{\mathrm{temp}}}{\partial t} + \rho C_p u \cdot \nabla T_{\mathrm{temp}} + \nabla \cdot \boldsymbol{q} = Q + Q_{vd} \\ \boldsymbol{q} = -k \nabla T_{\mathrm{temp}}\end{gathered} \tag{11}$$

where k represents the heat conductivity of materials, ρ is the thermodynamic density, $\boldsymbol{q}$ is the heat flux by conduction, u is the fluid velocity of liquid nitrogen (LN2), T_{temp} is the temperature, C_p is the heat capacity, Q is the heat source generated by AC loss from three phases and the average heat leakage which is usually 1 W/m in industrial application, while Q_{vd} represents heating under viscous dissipation in the fluid. The boundary condition of heat flux is set by

$$-\boldsymbol{n} \cdot \boldsymbol{q} = h \cdot (T_{\mathrm{ext}} - T_{\mathrm{temp}}) \tag{12}$$

where h is the heat transfer coefficient, T_{ext} is the external temperature far away from the boundary.

3. Simulation results

In the simulation results section, an electromagnetic analysis is carried out first, and the AC losses are calculated for thermal data in thermal fluid evaluation.

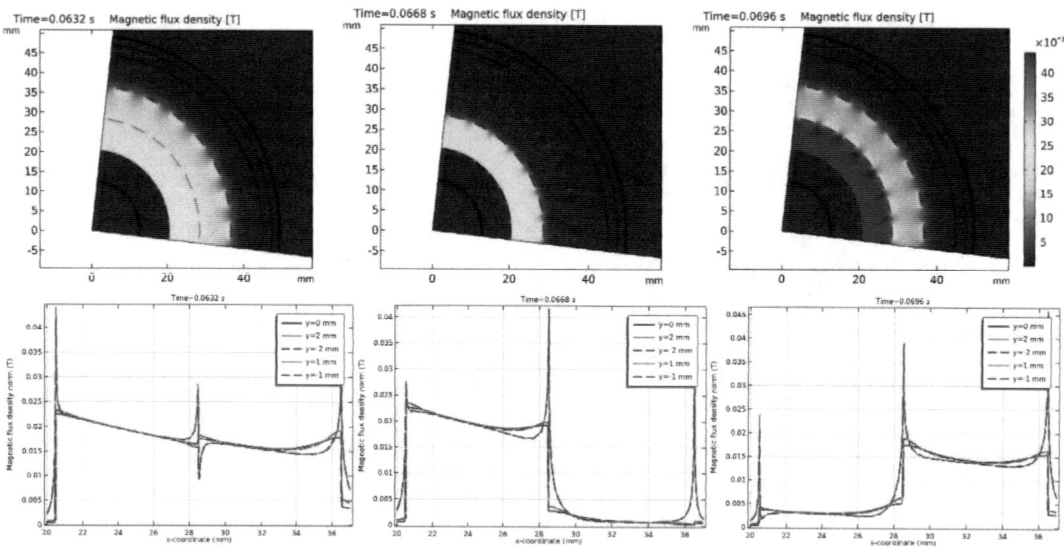

Figure 3. Magnetic flux density from the 1/4 section model

Figure 3 shows transient magnetic flux density and normalized current distribution of the HTS cable from a 1/4 section model. The current is mainly concentrated on both sides of the superconducting tapes. Therefore, the magnetic field has the largest value on both sides of the tapes.

Since the magnetic field generated by three-phase tapes is not only influenced by the phase angle and distance between different phases but also by the different spacing of the tapes at the same phase, the magnetic flux density is relatively lower at peak AC loss of phase C than that of A and B. The magnetic field almost only exists between the A/B/C three-phase, so it will not cause electromagnetic interference to the outside. 0.0632 s, 0.0668 s, and 0.0696 s correspond to the AC loss peaks for phases A, B, and C, respectively. The frequency of the three-phase AC current is 50Hz and the amplitude is about 0.62Ic. The parameters of the cable and superconducting material are listed in Table 1.

As shown in Figure 4, the three-phase AC current is gradually added to the HTS tapes, and the calculation of the AC loss reaches a steady state after about three cycles. Obviously, the magnetic field of each phase affects the losses of other phases, and the losses gradually increase with the increase of the cable radius.

Winding angle of HTS tapes from different layers will influence the magnetic field and magnetization loss. Considering the constraints of material strength and the need to keep AC loss as low as possible, the winding angle is usually designed between 10° and 40°. In this paper, the angle is considered as 20° and the results for equation (8) have been included in Figure 5. Figure 5 depicts the AC loss under different peak value I_p of transportation current at 70 K, while the critical current I_c is about 160 A. In the finite element simulation, it is found that the transportation current value will affect the error of the final calculation result, and the error is relatively small when I_p/I_c is between 0.5 and 0.9.

Table 1. Parameters of HTS cable and tapes

Parameters for tapes	Value	Parameters for cable	Value
Thickness of the YBCO layer	0.1 μm	Thickness of insulation	8 mm
Thickness of Substrate	50 μm	Cooling temperature	70 K
Thickness of copper stabilizer	40 μm	Current amplitude	2800 A
Width of the tape	4 mm	Cable diameter	100 mm
Critical current (70 K)	160 A	Heat leakage	1 W/m

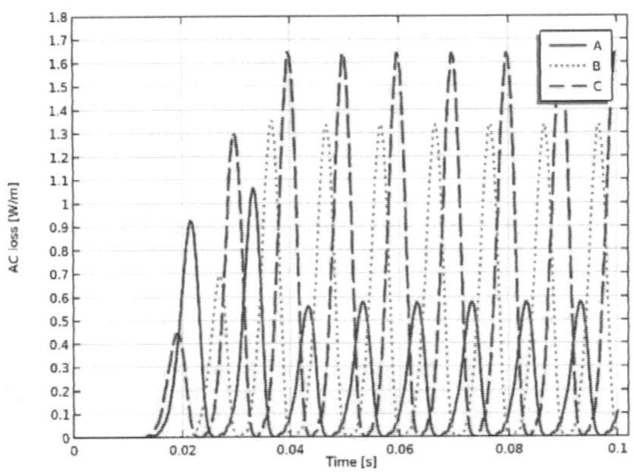

Figure 4. Instantaneous AC loss

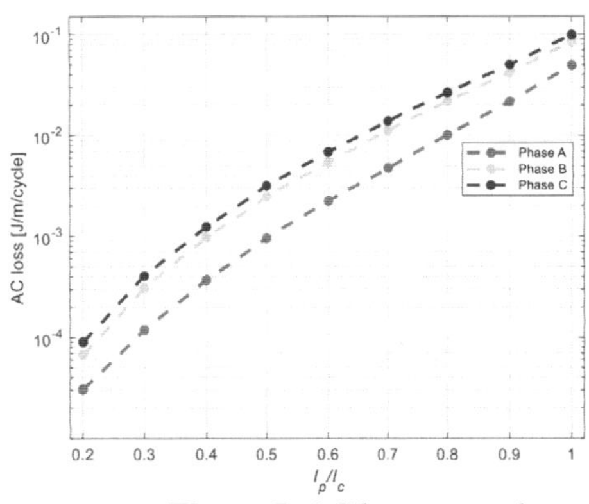

Figure 5. AC loss per cycle

Figure 6 is the stable temperature distribution of the HTS cable at 2 hours, and the maximum temperature in the middle part is about 70.9 K. Thermal insulation is set at the upper and lower boundaries excluding the inner and outer channels. Without consideration of fault condition, current cooling settings are able to ensure the normal operation of HTS cable. As shown in Figure 7, LN_2 flows in opposite directions in the inner and outer channels. Due to the viscosity of the fluid, the central flow velocity is much higher than that near the pipe wall. As the length of the pipe increases, the formation of the flow from the inlet end to the fully developed flow can be seen.

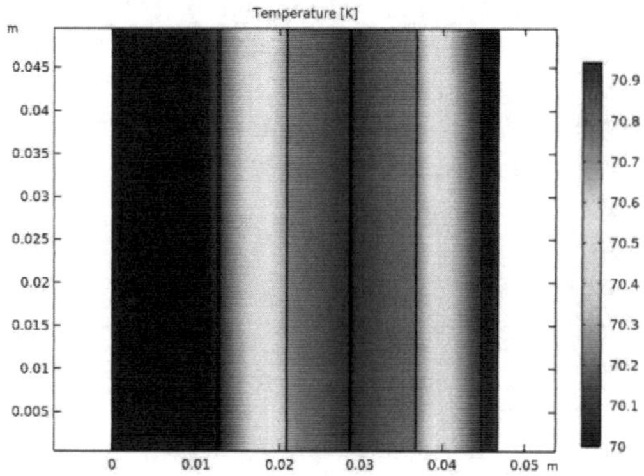

Figure 6. Temperature distribution

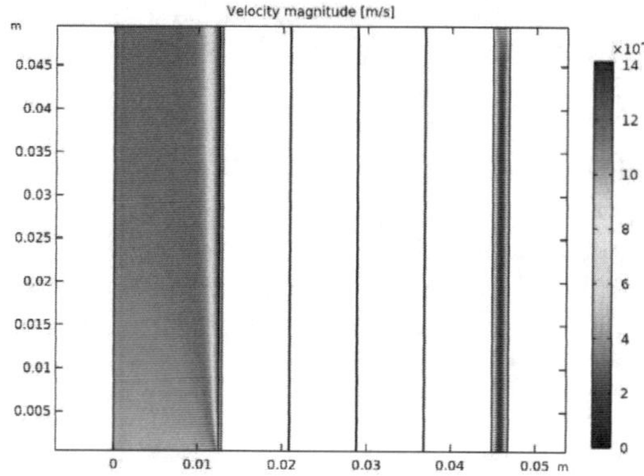

Figure 7. Fluid velocity distribution

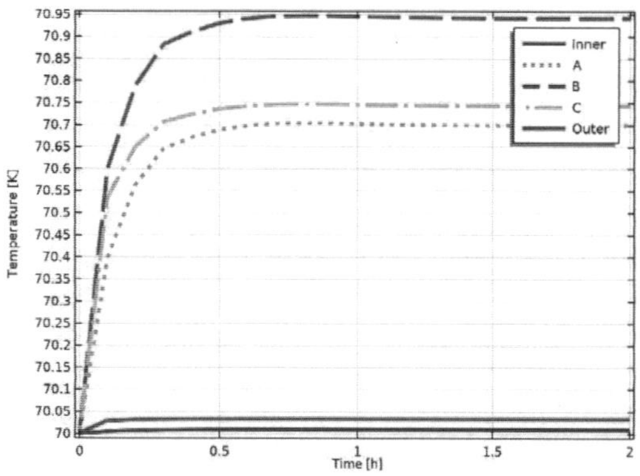

Figure 8. Temperature of cooling channels and each phase with time

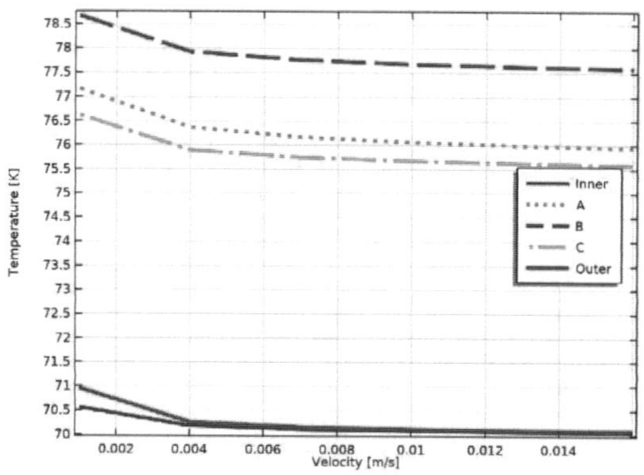

Figure 9. Steady temperature applying different fluid velocity

Figure 8 shows the cooling process for normal operating cable under fluid velocity 0.01m/s. It takes about 1 hour for the temperature to change from the transient state to the steady state, and the AC loss when $I_p = 0.6I_c$ is also considered as the heat source of each phase. As the middle part, the B phase has the highest temperature, about 70.95 K. The temperature of phase C is higher than that of phase A, partly due to heat leakage outside the outer channel.

To figure out the essential minimum velocity in category of laminar flow, temperature of different area under 0.001m/s to 0.015m/s is compared as shown in Figure 9. Due to limitations of the current model, the pressure drop that increases with cable length is not taken into account. Only from the short-distance cooling effect, with the increase of the flow rate, the temperature decreases significantly within a certain speed range.

4. Conclusion

In this paper, a 2D cross-section electromagnetic field model and a fluid-thermal coupling model are established for the three-phase HTS cable. The *T-A* formulation is used in the electromagnetic field model, and the instantaneous AC loss and periodic AC loss of the A, B, and C phases in the HTS cable are solved. The electromagnetic field is analyzed according to the time of the peak value of the instantaneous AC loss of each phase. In the fluid thermal coupling analysis, the state of the fluid at a set flow rate is first determined according to the Reynolds number of the fluid, and then the non-isothermal flow module is used for coupling. Cooling time, temperature distribution, flow velocity distribution and temperature behavior at different flow rates are considered, and the temperature behavior of each layer under ideal laminar flow conditions is analyzed. The constructed models will be used for fault condition analysis and cable design optimization in the future.

References

[1] Nii M, Amemiya N and Nakamura T 2012 *Supercond. Sci. Technol.* **25** 095011
[2] Terao Y, Seta A, Ohsaki H, Oyori H and Morioka N 2019 *IEEE Trans. Appl. Supercond.* **29** 1
[3] Sung B C, Park D K, Park J W and Ko T K 2009 *IEEE Trans. Ind. Electron.* **56** 2412
[4] Ali M H, Wu B and Dougal R A 2010 *IEEE Trans. Sustainable Enery* **1** 38
[5] Zhu Z, Wang Y, Venuturumilli S, Sheng J, Zhang M and Yuan W 2018 *IEEE Trans. Appl. Supercond.* **28** 1
[6] Wang Y, Zhang M, Grilli F, Zhu Z and Yuan W 2019 *Supercond. Sci. Technol.* **32** 025003
[7] Li X, Ren L, Xu Y, Shi J, Chen X, Chen G, Tang Y and Li J 2021 *IEEE Trans. Appl. Supercond.* **31** 1

X-ray Characterizations of Iron-based Superconductor Round Wires with Large Critical Current Density

T Tamegai[1], S Pyon[1], T Sasaki[1], T Ito[1], R Sakagami[1], S Awaji[2] and H Kajitani[3]

[1]Department of Applied Physics, The University of Tokyo

[2]High Field Laboratory for Superconducting Materials, Institute for Materials Research, Tohoku University

[3] Naka Fusion Institute, National Institutes for Quantum and Radiological Science and Technology

tamegai@ap.t.u-tokyo.ac.jp

Abstract. Characterizations of round wires of iron-based superconductors were performed by using two different X-ray techniques. The core of the round wire turns out to be concentrically textured. The texturing parameter (r) defined by $r = I(002)/I(103)$ has a clear correlation with the value of the critical current density (J_c) of the wire. Using ~18 m-long $(Ba,A)Fe_2As_2$ (A: K, Na) wires, small magnets was constructed and successfully generated a magnetic field as large as 3.1 kOe at I = 53 A. The wire for the magnet was carefully cut and local variation of the critical current (I_c) was evaluated. Variation of the core area was also measured by using X-ray tomography for the segments with the largest and smallest I_c. Obtained images demonstrates that the core has a certain degree of sausaging effect and it has good correlation with I_c of the wire. Using the largest value of measured I_c and the minimum core area of that segment, record-high value of J_c = 71 kA/cm^2 at 100 kOe has been achieved.

1. Introduction

Demands for new generation of superconducting wires are growing rapidly. One of good candidates to meet such demands is wires of cuprate superconductors. However, they face several issues such as strong magnetic field dependence of critical current density (J_c) in Bi2223 tapes and delamination in YBCO coated conductors. In addition, due to complicated fabrication processes, fabrication cost of conductors continues to be high. In such a situation, wires and tapes of iron-based superconductors (IBSs) attract much attention, since they have sufficiently high T_c and H_{c2}. In addition, IBSs have sufficiently high J_c in single crystals [1], and it can be enhanced by the introduction of artificial defects [2,3]. Much effort is devoted to the development of tapes and wires of IBSs, in particular to those based on 122-type IBSs. Actually, J_c over the practical level of 1×10^5 A/cm^2 at 4.2 K under the field of 100 kOe has been achieved in 2014 in tapes [4], and that in wires also exceeded a half of the level in 2021 [5].

We have been improving the critical current density (J_c) of round wires of $(Ba,A)Fe_2As_2$ (A: K, Na) prepared under high pressure [5-8]. In the course of this development, we found that the core of the wire was textured to some degree [7]. In addition, a good correlation between the degree of texturing and J_c was also discovered [8]. Since 2020, we started the preparation of long $(Ba,A)Fe_2As_2$ wires over 10 m, and constructed some small coils to generate magnetic fields [9,5,10]. Critical currents (I_c) of these long wires was smaller than that in short segments picked up from the long wire, and vary considerably along the length of the wire. It is important to know the reason for the suppression of I_c and parameters that affect it most. For that purpose, we have employed X-ray tomography for characterization. It turns out that the segment with the large I_c has smaller variation of the core area and large minimum core area. The largest value of J_c calculated from the largest I_c and the minimum core area of the best segment turns out to reach record-high value of 71 kA/cm^2 at 100 kOe.

2. Experiments

2.1. Fabrication of Wires
Superconducting wires of $(Ba,A)Fe_2As_2$ (*A*: K, Na) have been fabricated by powder-in-tube (PIT) method, followed by high-pressure sintering using HIP technique under a pressure of 175 MPa. We chose 6 : 4 ratio between Ba and alkali elements since T_c is the highest at this composition. Polycrystalline powders were prepared by directly mixing starting elements using a planetary ball milling machine as described in Refs. [5,6]. Well-mixed powders were reacted for 30 h at 850°C and 900°C for $(Ba,Na)Fe_2As_2$ and $(Ba,K)Fe_2As_2$, respectively. Prepared powders were packed into Ag tubes with outer and inner diameters (OD and ID) of 4.6 mm and 3.0 mm, respectively. The Ag tube filled with the powder was drawn using wire-drawing dies down to a diameter of 1.9 mm. For seven-core wires, the diameter of the wire was further reduced to 0.86 mm using a rotary swaging machine. Swaged wires were inserted into Cu tubes with (OD = 3.0 mm, ID = 2.0 mm) and with (OD = 4.0 mm, ID = 3.0 mm) for mono-core and seven-core wires, respectively. Finally, Cu/Ag sheathed 122 wires were swaged into a diameter of 1.0 mm. 18 m-long mono-core $(Ba,Na)Fe_2As_2$ and seven-core $(Ba,K)Fe_2As_2$ wires were fabricated in this way.

2.2. Construction of Small Magnets
Using the 18 m-long wires described above, we constructed small magnets with outer diameter 46 mm, inner diameter 19 mm, and height 28 mm following wind & react method. After installing glass fiber sleeve on the wire for insulation, it was wound around a stainless steel bobbin, and the final heat treatment was conducted at 700°C under 175 MPa Ar atmosphere. Fabricated magnets were not impregnated with epoxy resin. After the measurements of *I–V* characterization and generated magnetic field, the wire was carefully removed from the bobbin, and cut into short pieces for the following local characterizations.

2.3. Characterizations of Wires and Magnets
The I_c of the small magnet under the self-field and its generated magnetic field were measured at The University of Tokyo. Generated field at the center of the coil was measured using a Hall probe (THS126, Toshiba). Voltage taps for the evaluation of *I-V* characteristics of the coil were placed at the ends of the coil. The transport I_c measurements under high magnetic fields up to 140 kOe for short segments cut out from the coil were carried out by using the 15T-SM at the High Field Laboratory for Superconducting Materials, IMR, Tohoku University. The magnetization of short pieces of wires was measured to evaluate T_c and magnetic J_c using a commercial SQUID magnetometer (MPMS-XL5, Quantum Design). T_c of these short pieces are slightly lower than the starting powder similar to the previous wires [5,8] and magnetic J_c is consistent with the value of transport measurements. Powder X-ray diffraction measurements with Cu-Kα radiation were performed using Smartlab (Rigaku) for characterizations of polycrystalline powders and the texturing of the core of wires. To evaluate the core area and the sausaging effect, X-ray computed tomography (CT) imaging was applied to mono-core $(Ba,Na)Fe_2As_2$ wire used for the construction of the magnet using CT Lab HX130 (Rigaku) with a setting of 130 kV for X-ray tube voltage.

3. Results and Discussion

3.1. X-ray Characterization of Texturing of Wires

The largest transport J_c in round wires of IBS is lower than that of tapes. More quantitatively, the largest transport J_c at 100 kOe is 150 kA/cm^2 and 54 kA/cm^2 for tapes [4] and wires [5]. It is most probably due to the difference in the *c*-axis texturing of the conductors. Unlike conventional superconductors like NbTi and Nb_3Sn with relatively long coherence length, intergranular current can be suppressed severely, when the grains are misaligned at large angles in superconductors with short coherence lengths. One of representative examples is the demonstration of the misalignment dependence of intergranular critical current at the grain boundary in $YBa_2Cu_3O_{7-\delta}$ [11]. In this case, critical misalignment angle as small as ~5° has been reported. Similar measurements in IBSs proved that the critical misalignment angle is larger than that of cuprates [12,13]. Still, grains in IBS wires needs to be textured. One of the ways to evaluate the degree of texturing is electron backscattering (EBSD) using electron microscope. In the present work, however, we employed ordinary X-ray diffraction for such a purpose. A few mm-long short pieces of wires are cur out and examined by focused X-ray beam using a collimator with multi-layer coating with a diameter of ~0.8 mm. The short piece was polished using lapping films until the top of the core is exposed. Then the wire is set on a diffractometer so that the surface of the polished plane is at the center of X-ray optics. Figures 1(a) and (b) show the profile of (002) and (103) X-ray diffraction peaks for the longitudinal and transverse cross sections. The two figures are displayed so that the intensity of (103) peaks are the same. The intensity of (002) peak is multiplied by a factor of 2.2 to compensate the loss of X-ray intensity due to smaller diffraction angle. Even with this correction, it is obvious that the intensity of (002) peak in the longitudinal cross section is higher. When grains in the core is perfectly textured with their *c*-axis perpendicular to the surface, we should only observe (00*l*) peak as in the case of single crystals. So, the higher intensity of (002) peak on the longitudinal cross section indicates that there are larger proportion of *c*-axis oriented grains on the surface, suggesting that the core is concentrically textured. Even in such a case, the texturing parameter r (=$I(002)/I(103)$) should depend on the measured level of the core defined by the parameter α in Fig. 2(a). Here, $\alpha = 1$ and $\alpha = 0$ correspond to the surface and the center of the core, respectively. When the texturing is weak, r should not depend on α too much as shown in blue broken curve in Fig. 2(b). On the contrary, when the texturing is strong, r depends on α as shown in red broken line in Fig. 2(b). Fig. 2(c) shows examples of the profile of (002) peak, showing the decreasing trend of intensity of (002) peak as α is reduced, consistent with the prediction shown in Fig. 2(b). It also supports our assertion that the core is concentrically textured.

One important observation is that the texturing parameter has a good correlation with J_c of the wire. In table I, J_c values at 100 kOe and texturing parameter r for three different $(Ba,Na)Fe_2As_2$ wires

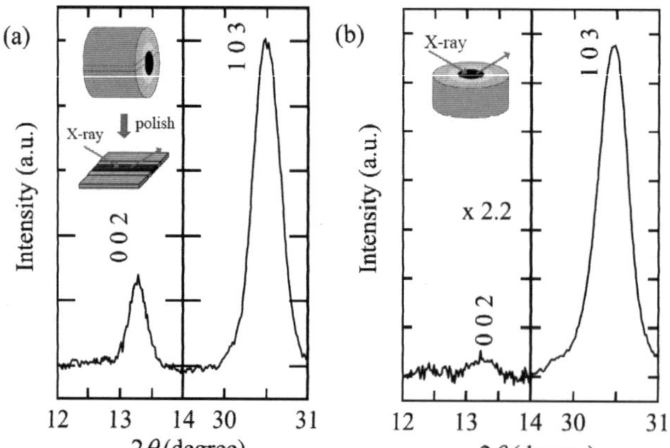

Figure 1. Profiles of (002) and (103) X-ray diffraction peaks of the core of $(Ba,K)Fe_2As_2$ round wire in (a) longitudinal and (b) transverse cross sections, respectively.

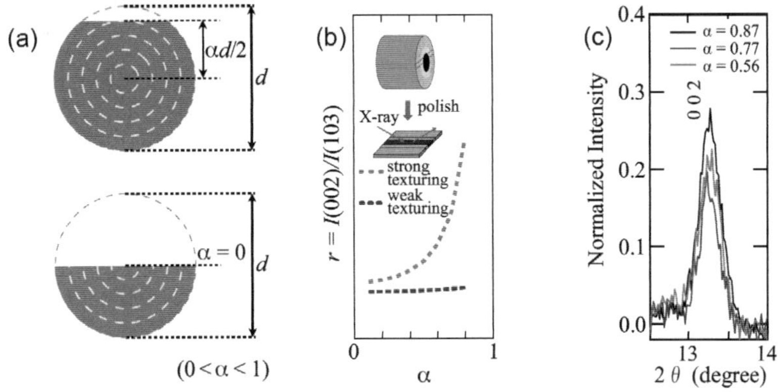

Figure 2. (a) Definition of the location of the polished surface α (0 < α <1) of a wire with an original diameter d. (b) Qualitative change of the texturing parameter $r = I(002)/I(103)$ with α in round wires with strong (red) and weak (blue) texturings. (c) Examples of the normalized profile of (002) peak for different α values in $(Ba,K)Fe_2As_2$ round wire.

processed slightly different conditions are compared. It is clear that as r increases, J_c at 100 kOe increases. So, we can conclude that improving the concentric texturing of the wire is an important strategy to enhance J_c.

Table I Relationship between the texturing parameter r and J_c at high fields.

Material	Name	Reaction T (°C)	$r = I(002)/I(103)$	J_c(100 kOe) (kA/cm^2)
$(Ba,Na)Fe_2As_2$	#1	850	0.27	40
$(Ba,Na)Fe_2As_2$	#2	820	0.20	34
$(Ba,Na)Fe_2As_2$	#3	850	0.33	44

3.2. *Characterizations of Small Magnets*

Small magnets using ~18 m-long mono-core $(Ba,Na)Fe_2As_2$ and seven-core $(Ba,Na)Fe_2As_2$ were fabricated by wind & react method similar to the ones reported in Ref. [5]. Characteristics of the magnets were evaluated by immersing it into liquid He. Figure 1(a) shows I-V characteristics and generated magnetic field of the magnet (photo shown in the inset of Fig. 1(a)) constructed using mono-core $(Ba,Na)Fe_2As_2$ wire. A sharp rise of voltage is seen at ~53 A, and the generated magnetic

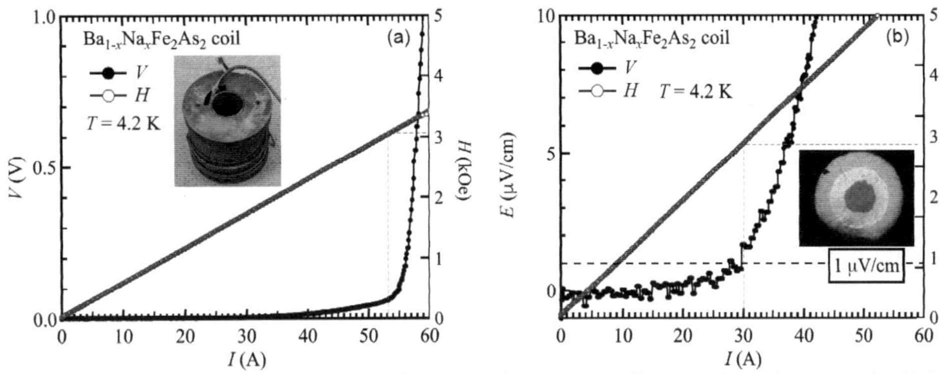

Figure 3. (a) I-V and (b) E-V characteristics and corresponding generated magnetic fields of a small magnet at 4.2 K constructed using $(Ba,Na)Fe_2As_2$ mono-core wires. Insets of (a) and (b) show photos of the small magnet and cross section of the mono-core wire. Red vertical broken lines in (a) and (b) indicate apparent I_c and actual I_c with a criterion of 1 μV/cm. Blue horizontal broken lines in (a) and (b) indicate the corresponding generated magnetic field.

field at this current is 3.1 kOe. However, even below ~53 A, a small voltage is seen. Figure 3(b) shows $E - I$ characteristics of the same magnet and generated magnetic field. When we employ a standard criterion for I_c of 1 μV/cm, the actual I_c of this magnet is 30 A, and the generated magnetic field at this current is 1.75 kOe. Characteristics of the small magnet constructed using the seven-core $(Ba,K)Fe_2As_2$ wire is very similar to that shown in Fig. 3.

3.3. X-ray Tomography of IBS Round Wires

Figure 4(a) shows the I_c variation of the mono-core $(Ba,Na)Fe_2As_2$ wire used for the small magnet at several fields. It is measured by cutting small segments of wires at specific position from the end of the ~18 m-long wire. It is clear that I_c shows an appreciable variation along the length of the wire. Here, segments with smaller distance from the end corresponds to those located close to the outer perimeter of the magnet. The segments at ~5 m and ~12 m display the largest and smallest I_c. The ratio

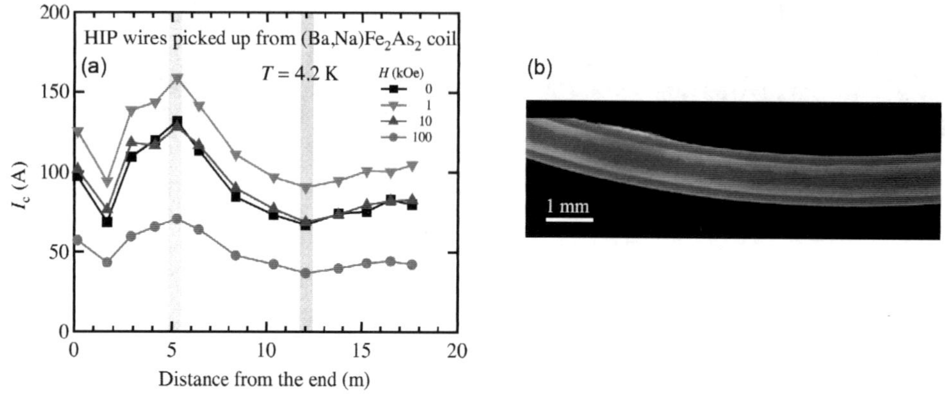

Figure 4. (a) Variation of I_c under several magnetic fields of the short segments of $(Ba,Na)Fe_2As_2$ wires picked up from different locations of the ~18 m-long wire used for the magnet shown in Fig. 3. The distance is measured from the end of the wire. Yellow and light blue bands mark the region with the largest and smallest I_c. (b) X-ray topography image of the segment with the largest I_c.

of the I_c for the "best" segment to the "worst" segment reaches a factor of 1.9. X-ray tomography measurements were performed on the "best" and "worst" segments of the wire. A longitudinal cross section of the "best" segment is shown in Fig. 4(b). It clearly shows that the cross section of the core is changing appreciably along the length of the wire. Figures 5 (a) and (b) show variations of the core area between the voltage contacts for the "best" and "worst" segments. Insets of Fig. 5(a) and (b) are X-ray topography images of the cross sections with the largest (green frame) and smallest (blue frame) core areas. It is observed that the variation of the core area in the "best" segment is much smaller than that in the "worst" segment. It means that the segments with smaller sausaging effect has larger I_c.

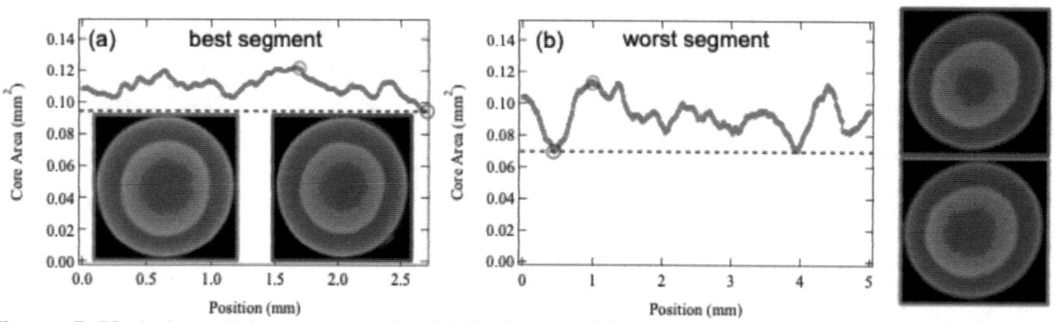

Figure 5. Variation of the core area for (a) the best and (b) the worst short segments of the 18 m-long wire used for the magnet shown in Fig. 3. Inset with green and blue frames in (a) and right side of (b) are X-ray topography images of the area with the largest and smallest areas for each segment.

Using the largest value of measured I_c and the minimum core area of that segment determined by X-ray topography, it turns out that a record-high value of J_c = 71 kA/cm^2 at 100 kOe has been achieved.

4. Summary

Round wires of 122-type iron-based superconductor processed under high pressure were characterized by using two kinds of X-ray measurements. X-ray diffraction measurements of longitudinal and transverse cross sections of the wire demonstrate that the core of the round wire is concentrically textured. The degree of concentric texturing has a good correlation with the J_c values of the wire. Using ~18 m long mono-core (Ba,Na)Fe_2As_2 and seven-core (Ba,K)Fe_2As_2 round wires processed by HIP, small magnets with inner and outer diameters of 19 mm and 46 mm was constructed and evaluated. In both magnets, strong dissipation started at a current of ~50 A, and a magnetic field of ~3.0 kOe was generated. The wire used for the construction of the magnet was disassembled and short segments were evaluated separately. It turns out that local variation of transport I_c of the long wire reaches a factor of 1.9 along the entire length of the wire. Variations of the core area in the wire in much smaller scale were also detected by X-ray tomography measurements, with clear anti-correlation between the value of I_c and variation of the core areas.

Acknowledgements

This work was supported by a Grant-in-Aid for Scientific Research (A) (17H01141) and (C) (22K04904) from the Japan Society for the Promotion of Science (JSPS). A part of this study was performed at the High Field Laboratory for Superconducting Materials, Institute for Materials Research, Tohoku University (Project No. 202112-HMKGE-0044). We thank Division of Creative Activity in The University of Tokyo for technical assistance for X-ray CT equipment.

References

[1] Taen T, Ohtake F, Pyon S, Tamegai T and Kitamura H 2012 *Supercond. Sci. Technol.* **25** 084008
[2] Tamegai T, Taen T, Yagyuda H, Tsuchiya Y, Mohan S, Taniguchi T, Nakajima Y, Okayasu S, Sasase M, Kitamura H, Murakami T, Kambara T and Kanai Y 2012 *Supercond. Sci. Technol.* **25** 084008
[3] Ohtake F, Taen T, Pyon S, Tamegai T, Okayasu S, Kambara T and Kitamura H 2015 P*hysica C* **518** 47
[4] Zhang X P, Yao C, Lin H, Cai Y, Chen Z, Li J, Dong C H, Zhang Q J, Wang D L, Ma Y W, Oguro H, Awaji S and Watanabe K 2014 *Appl. Phys. Lett.* **104** 202601
[5] Pyon S, Mori H, Tamegai T, Awaji S, Kito H, Ishida S, Yoshi Y, KajitaniH and Koizumi N 2021 *Supercond. Sci. Technol.* **34** 105008
[6] Pyon S, Suwa T, Park A, Kajitani H, Koizumi K, Tsuchiya Y, Awaji S, Wanatabe K and Tamegai T 2016 *Supercond. Sci. Technol.* **29** 115002
[7] Pyon S, Suwa T, Tamegai T, Takano K, Kajitani H, Koizumi N, Awaji S, Zhou N and Shi Z Xi 2018 *Supercond. Sci. Technol.* **31** 055016
[8] Tamegai T, Pyon S, Miyawaki D, Kobayashi Y, Awaji S, Kito H, Ishida S, Yoshida Y, Takano K, Kajitani H and Koizumi 2020 *Supercond. Sci. Technol.* **33** 104001
[9] Tamegai T, Miyawaki D, Pyon S, Wu W X, Kajitani H, Koizumi Nand Awaji S 2021 *IEEE Trans. Appl. Supercond.* **31** 7300505
[10] Pyon S, Ito T, Tamegai T, Kajitani H, Koizumi N, Awaji S, Kito H, Ishida S and Yoshida Y 2022 *J. Phys.: Conf. Ser.* **2323** 012020
[11] Dimos D, Chaudri P, Manhart J and LeGoues F K 1988 *Phys. Rev. Lett.* **61** 219
[12] Katase T, Ishimaru Y. Tsukamoto A, Hiramatsu H, Kamiya T, Tanabe and Hosono H 2011 *Nat. Commun.* **2** 409
[13] Iida K, Hanisch J and Yamamoto Y 2020 *Supercond. Sci. Technol.* **33** 043001

Performance evaluation of superconductive-assisted machining (SUAM) with superconducting tape and two permanent magnets

S Iwasaki[1], H Ishii[1], Y Kinoshita[1], E S Otabe[1], K Matsuo[2], T Nakasaki[2], K Suzuki[2]

[1] Department of Computer Science and Electronics, Kyushu Institute of Technology, 680 – 4 Kawazu, Iizuka, Fukuoka 820 – 8502, Japan
[2] Department of Mechanical Information Science and Technology, Kyushu Institute of Technology, 680 – 4 Kawazu, Iizuka, Fukuoka 820 – 8502, Japan

otabe@phys.kyutech.ac.jp

Abstract. In this study, the repulsive force causing magnetic levitation for superconductive-assisted machining (SUAM) is calculated using the finite element method. Although bulk superconductors have been used for SUAM, here, we propose a magnetic levitation tool that uses superconducting tape. To obtain a stronger repulsive force, various models of a SUAM are designed and calculated. The results show that a strong repulsive force can be obtained by using two permanent magnets. In addition, even if the superconducting tape is rotated during stacking due to stacking errors, the repulsive force is not affected, indicating that there is no need to optimize the stacking method. Therefore, we expect that increasing the number of layers will generate a repulsive force that is greater than that obtained with bulk superconductors.

1. Introduction

Progress in industrial product development has led to more complicated product parts. Hollow processing technology can be used to process such complex parts. Hollow processing can be carried out either before or after molding. When it is carried out after molding, it can be difficult to internally process an object with a complicated shape. Superconductor-assisted machining (SUAM), in which a permanent magnet is levitated on bulk superconductors to process an object by utilizing the magnetic flux pinning of superconductors, has thus been proposed [1, 2]. The permanent magnet is a single-sided four-pole magnet and is levitated by using bulk superconductors. The magnet is levitated because the superconducting bulk is cooled in a magnetic field, i.e., it is field-cooled. The permanent magnet rotates when the superconducting part rotates due to the pinning effect. Therefore, a material can be polished from the inside by the surface of the permanent magnet.

The performance of bulk superconductors has improved [3], and high-performance superconducting tape is under development [4]. Superconducting tape can work in the same way as bulk superconductors. By stacking pieces of superconducting tape, each with 130 layers, and then magnetizing them, a magnetic field of 7.9 T can be obtained at 4.2 K [5]. Tape is easy to cut and combine. To design a SUAM using superconducting tape, it is necessary to evaluate the parameters that affect the force required to levitate the permanent magnet. In our previous work [6,7], we showed that a SUAM with superconducting tape has a performance comparable to that of a SUAM with bulk superconductors. For practical applications, it is necessary to improve the levitation performance of the SUAM with superconducting tape.

In this study, an electromagnetic field analysis using the finite element method (FEM) is performed using a simple model that consists of only a permanent magnet and superconducting tape. To obtain a stronger repulsive force, a SUAM that consists of two permanent magnets and superconducting tape is proposed.

2. Calculation Methods

In this study, the FEM was used to analyze the electromagnetic field for the SUAM. JMAG (JSOL Corp.) was utilized for the FEM calculations. For the magnetic field dependence of the critical current density and the electric field-current density (*E-J*) characteristics used in the calculation, the experimental results for $GdBa_2Cu_3O_{7-\delta}$ superconducting tape [8] were used. The critical current for the superconducting tape is 1.7×10^{10} A/m^2 at 77.3 K and 0 T. The critical temperature for the tape is 93 K. For the superconducting layer, the pieces of $GdBa_2Cu_3O_{7-\delta}$ superconducting tape were 20 mm long, 12 mm wide, and 1 μm thick.

As shown in figure 1, the model of the SUAM had six pieces of superconducting tape placed azimuthally at equal intervals from the center. To investigate the variation of the repulsive force with the stacking method used for the superconducting tape, we analyzed the repulsive force for a model in which the pieces of tape were stacked up to four layers by rotating the tape arrangement by 0°, 45°, and 90°, respectively, from the center relative to the bottom layer. The distance between the top and bottom layers of the superconducting tape was 50 μm. The model is shown in figure 2. Here, we introduce two permanent magnets to increase the attractive force for improving the performance of SUAM. As shown in figure 3, the model of the SUAM has six pieces of superconducting tape placed azimuthal at equal intervals from the center. Four layers similar to the upper layer of the superconducting tape were prepared under the superconducting tape with the distances between the layers at intervals of 50 μm. Two permanent magnets were used. One was at a position 10 mm above the superconducting tape, and the other was 1 mm below the superconducting tape. The magnets were oriented to produce an attractive force.

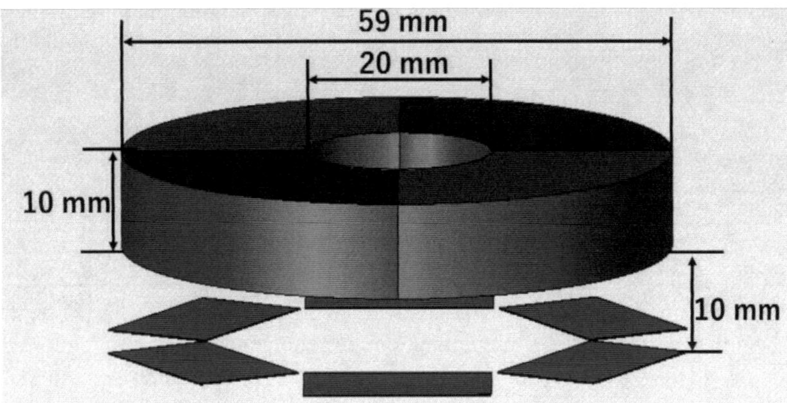

Figure 1. SUAM with four-pole permanent magnet and six pieces of superconducting tape.

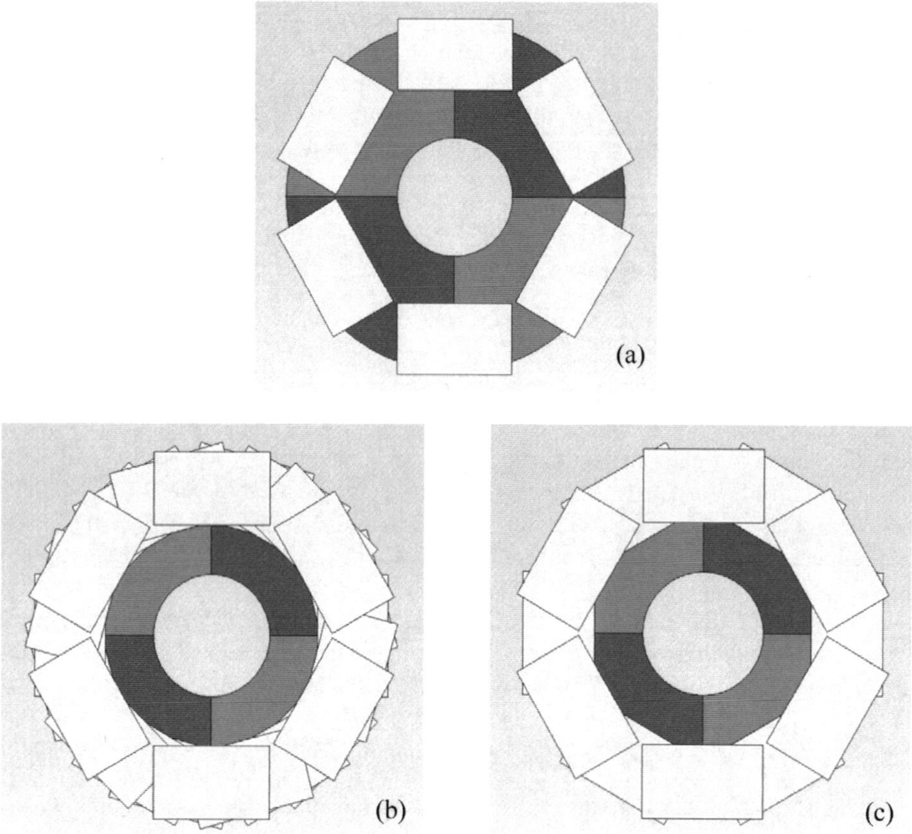

Figure 2. Superconducting Assisted Manufacturing (SUAM), a different method of stacking superconducting tapes. The number of layers is 4. Rotation angle between layers is (a) 0°, (b) 45°, (c) 90°

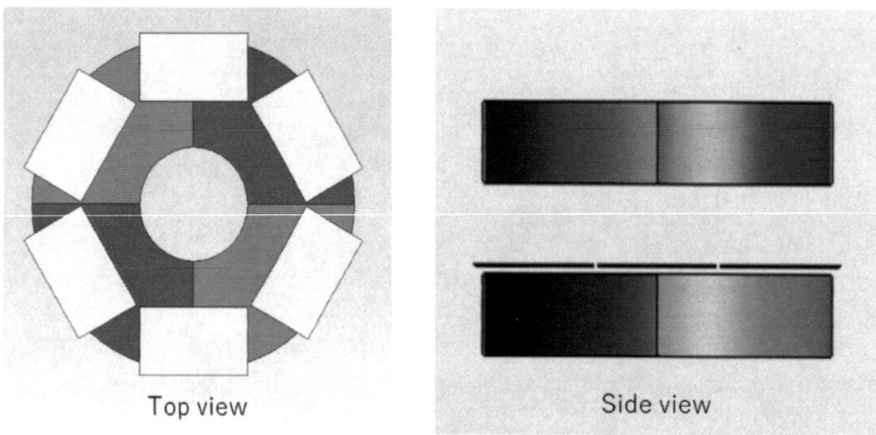

Figure 3. SUAM with two permanent magnets and six pieces of superconducting tape.

The four-pole permanent magnet was placed so that its center coincided with the center of the pieces of tape. We used a permanent magnet with an inner diameter of 20 mm, an outer diameter of 59 mm, a thickness of 10 mm, and a magnetic field of 450 mT at the magnet surface. These values are the same as those for a four-pole permanent magnet used in a previous study [1].

The pieces of superconducting tape were magnetized with field-cooled magnetization at a position 10 mm between the upper part of the tape and the four-pole permanent magnet. The cooling temperature was 77.3 K (liquid nitrogen). The repulsive force on the permanent magnet generated by the superconducting tape was calculated for a permanent magnet moving perpendicularly to the pieces of superconducting tape at a speed of 0.1 mm/s. The distance between the permanent magnet and the pieces of superconducting tape was decreased from 10 mm to 1 mm. In this paper, we focus on the repulsive force since the results for the attractive force and torque are mostly proportional to those for the repulsive force. Therefore, if the repulsive force is improved, the attractive force and torque will also be improved.

3. Results and Discussion

Figure 4 shows the FEM calculation results for a model in which the layers are stacked with the superconducting tape arrangement rotated 0°, 45°, and 90°, from the center with respect to the lower layer. At a distance of 10 mm (i.e., at the magnetization position), the repulsive force is zero. When the permanent magnet is moved closer to the superconducting tape from the magnetization position, a repulsive force is generated. It was confirmed that the repulsive force increases as the distance becomes shorter. However, the stacking method did not affect the repulsive force. Therefore, the layers in the stacking process should be stacked in the same manner as the layers below. Even if the superconducting tape arrangement is rotated due to stacking errors, the repulsive force is not affected. It is believed that the reason for no change between the three cases is related to the gap between the superconducting tape and the permanent magnet being the same for each case.

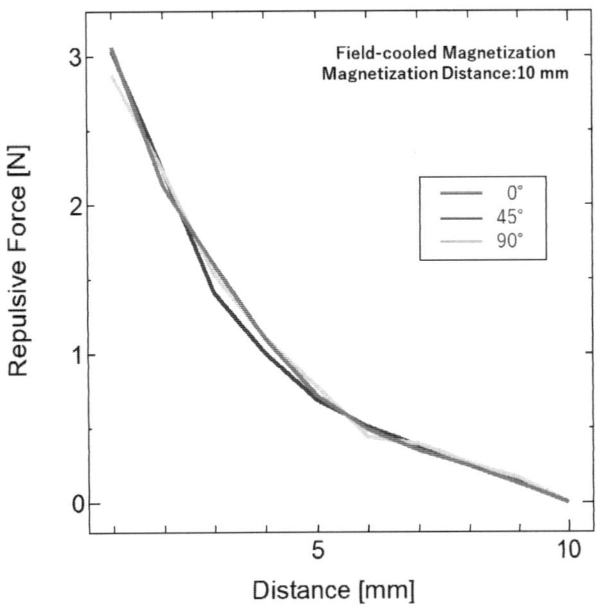

Figure 4. Repulsive force as function of distance between upper permanent magnet and six pieces of superconducting tape for single magnet.

Figure 5 shows the FEM calculation results of (a) repulsive force and (b) attractive force for a SUAM with two permanent magnets and a SUAM with only a single permanent magnet above the superconducting tape. The results show that when there are two permanent magnets, both the repulsive and attractive forces are much greater than that obtained with a single permanent magnet. This is because two permanent magnets attract each other and fix by the superconducting tapes, and the restoring force from the magnetized point (distance = 10 mm) is large compared with a single permanent magnet. Therefore, a SUAM with two permanent magnets provide a better performance.

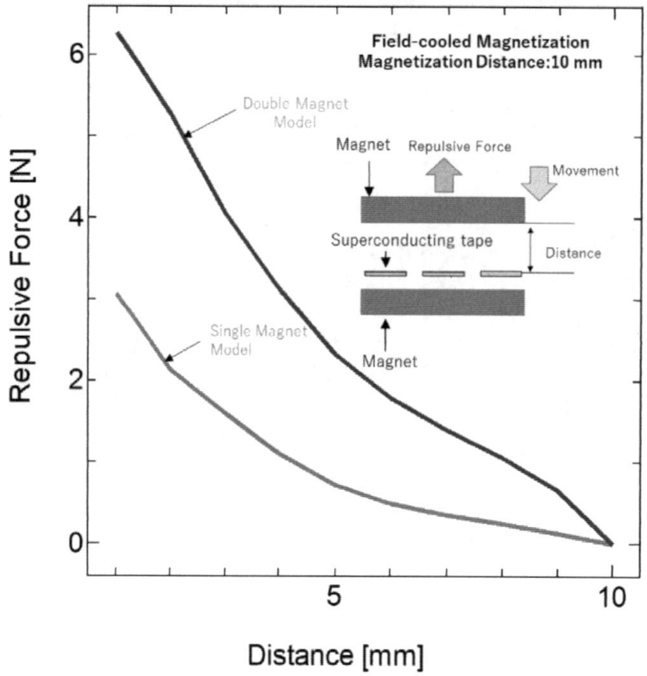

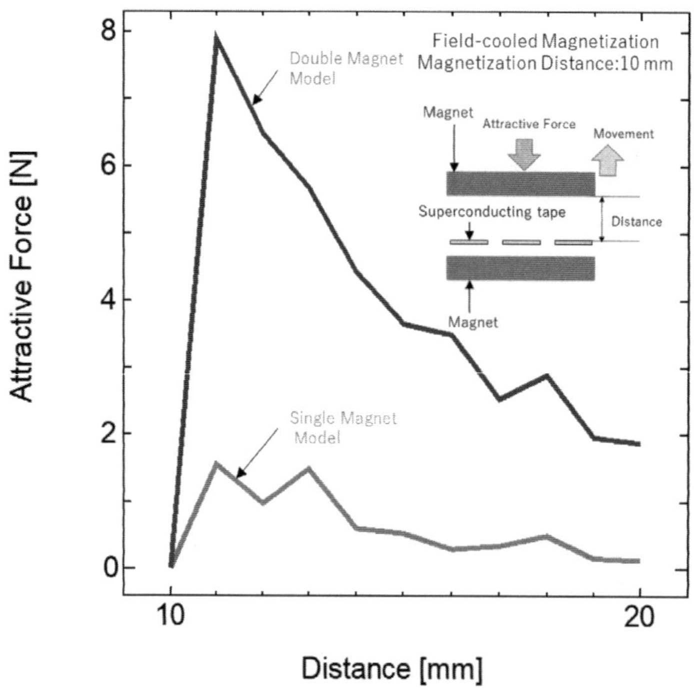

Figure 5. (a) Repulsive force and (b) attractive force as function of distance between upper permanent magnet and superconducting tape for two magnets.

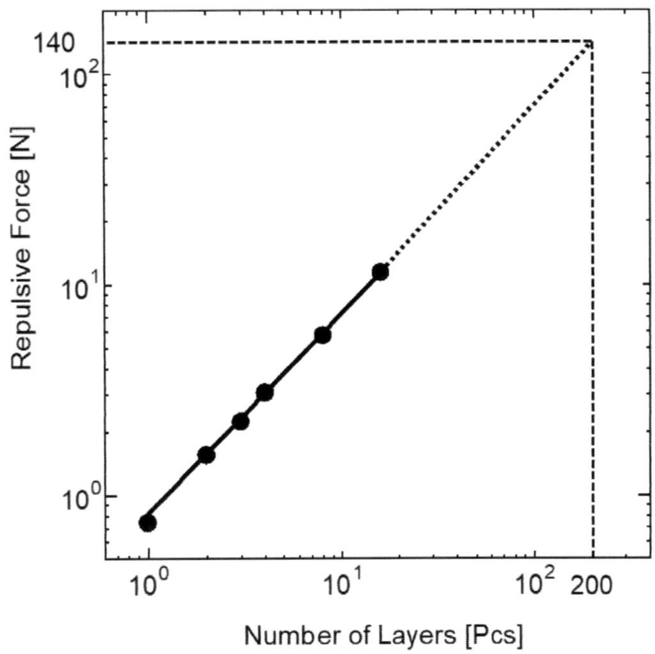

Figure 6. The repulsive force as a function of the number of layers of the superconducting tapes.

In our previous work [6,7], we showed that the repulsive force for a SUAM can be increased by increasing the number of pieces of superconducting tape in the layers, optimizing the geometry, and appropriately arranging the layers. Figure 6 shows the calculation results of the repulsive force with respect to the number of superconducting tape layers. It was found that when the thickness of the superconducting tape stack is the same as the 10 mm thick superconducting bulk, i.e. when the number of layers is 200, the repulsive force is 140 N, which is larger than 85 N of the superconducting bulk. The results in our previous studies and the present study show that it is possible to easily and inexpensively obtain a repulsive force that is greater than that generated by bulk superconductors by using two permanent magnets. Even if the superconducting tape is rotated during stacking due to stacking errors, the repulsive force is not affected. The stacked superconducting tape outperforms a bulk superconductor because it has a higher critical current density and is thinner. Superconducting tape with a width of more than 12 mm should further increase the repulsive force.

4. Conclusion

In this study, various models of a SUAM were fabricated and analyzed to obtain a large repulsive force. The results show that a larger repulsive force can be obtained by fabricating the SUAM with the two permanent magnets. Even if the superconducting tape is rotated during stacking due to stacking errors, the repulsive force is not affected, so there is no need to optimize the stacking method.

It is expected that an increase in the number of layers and/or tape width will generate a repulsive force that is larger than that obtained with bulk superconductors.

Acknowledgements
The present study was supported in part by the Japan Society for the Promotion of Science (JSPS) KAKENHI (grant number 19H00771).

References

[1] Nakashima H, Nakasaki T, Tanaka T, Kinoshita Y, Khajornrungruang P, Otabe E S and Suzuki K 2021 *Int. J. of Automation Technology* **15** 234 – 42
[2] Hiramatsu Y, Takahashi Y, Otabe E S, Suzuki K, Tanaka Y and Kiuchi M 2017 *J. Phys.: Conf. Ser.* **871** 012048
[3] Murakami M 2007 *Int. J. Appl. Ceram. Technol.* **4** 225 – 241
[4] Lao M, Willa R, Meledin A, Rijckaert H, Chepikov V, Lee S, Petrykin V, Van Driessche I, Molodyk A, Holzapfel B 2019 *Supercond. Sci. Technol.* **32** 094003
[5] Tamegai T, Hirai T, Sun Y and Pyon S 2016 *Physica C* **530** 20 – 23
[6] Kinoshita Y, Zhang R, Otabe E S, Suzuki K, Tanaka Y, Nakashima H, Nakasaki T 2020 *J. Phys.: Conf. Ser.* **1590** 012023
[7] Iwasaki S, Kinoshita Y, Ishii H, Otabe E S, Suzuki K, Nakasaki T 2022 J. Phys.: Conf. Ser. **2323** 012025
[8] Inoue M, Tanaka K, Imamura K, Higashikawa K, Awaji S, Watanabe K, Taneda T, Yoshizumi M, Izumi T and Kisu T 2015 *Physics Procedia* **67** 903 – 907

Journal of Physics: Conference Series **2545** (2023) 012013 doi:10.1088/1742-6596/2545/1/012013

High-Speed Spin Testing of Bi-2212 Superconducting Wires for High-Field NMR Magnet Qualification: Final Progress Report

J Voccio

[1]School of Engineering, Wentworth Institute of Technology, Boston, MA 02115

Abstract. This paper summarizes an NIH-funded research project designed to study the strength of high temperature superconductors under high circumferential hoop stress, in order to qualify these materials for high-field (> 1 GHz-class) NMR magnets. The unique approach presented here is to spin test coils at high rotational speeds, approaching 50,000 rpm, in order to induce mechanical hoop stresses which are equivalent to the magnetic Lorentz stresses experienced by a high-field solenoid insert coil. Bi-2212 wires and coils were tested in sub-cooled nitrogen to determine if any mechanical damage was imparted by each spin test. These preliminary experimental results demonstrate that, with further development, this test could prove to be a simple method for qualifying superconducting wires for high-field insert magnets.

1. Introduction

High-temperature superconducting (HTS) materials are an enabling technology for a new class (> 1 GHz) of NMR magnets due to their ability to transport large electrical currents in the presence of high magnetic fields beyond the limitations of low temperature superconductors (LTS). [1-7] As described in previous papers [8, 9], only the HTS types known as REBCO, Bi-2212 and Bi-2223 can break the LTS high-field (> 22 T) Je barrier. Here, a new spin test method has been proposed and developed to simulate the magnetic Lorentz stresses by imposing spin-induced circumferential hoop stresses.

The circumferential hoop stress, σ_h, generated in a superconducting magnet winding, assuming a thin radial build, may be expressed in terms of the conductor current density, J, magnetic field, B, and radius, R, as:

$$\sigma_h = J\,B\,R \tag{1}$$

From (1), the circumferential hoop stress for a 500 A/mm^2, 30 T at a winding radius of 0.03 m is 450 MPa. This winding radius represents the smallest cold winding bore (i.e., 60-mm diameter) for a high-field insert coil.

For the proposed spin test, this same circumferential hoop stress, σ_h, may be generated mechanically in terms of the density, ρ, angular velocity, ω, and radius, R, as:

$$\sigma_h = \rho\,\omega^2\,R^2 \tag{2}$$

From (2), the speed required to simulate this magnetic condition for a 7.5 cm diameter winding is ~6300 rad/s, or ~60,000 rpm. Since the critical temperature of Bi-2212 wires is close to the 77 K boiling point of liquid nitrogen, a subcooled nitrogen test station was designed and built, allowing for testing down to 63 K, which is the temperature of solid nitrogen. With this setup (shown in figure 1), we could more quickly and easily test the superconducting winding after each subsequent spin test. Some of the key feature of this set-up include a vacuum roughing pump and Bell jar with a 7.5-cm high standoff to accommodate the increased length of the Inconel spindle which is attached to the rotor.

Figure 1. Subcooled nitrogen critical current set-up.

2. Low Speed Rotor Design (50,000 rpm)

Initially, a low-speed, 2-part rotor was fabricated for this study. The low speed rotor was made with an aluminum arbor, which connects the Inconel spindle to the 304 stainless mandrel. The choice of aluminum for the arbor simplified the machining, while the mandrel was formed from a tube of stainless steel. This rotor was designed for a maximum speed of 50,000 rpm.

As shown in figure 2a, the aluminum arbor was shrunk fit into the stainless mandrel thereby eliminating the need for welding which could lead to rotor imbalance. A plot of the von Mises stresses for the 50,000 rpm operating condition is provided in figure 2b, showing a nominal stress of ~270 MPa at the outermost diameter of the rotor. With this initial design, the machined groove was sized to accommodate the Bi-2212 winding, although the method of bonding and attaching current leads and voltage taps would end up requiring further development.

(a) (b)

Figure 2. Low speed spin test rotor: (a) low-speed, 2-part rotor; (b) stress analysis @ 50,000 rpm.

3. Spin Testing Experiments

3.1 Initial Low Speed Testing

Due to pandemic-related delays in the Bi-2212 coil delivery, initial coil testing was conducted with SuperPower 2G YBCO tape. Since this tape is thinner and wider than the corresponding Bi-2212 winding, the 2G winding was bonded to the upper, non-grooved section of the rotor as shown in figure 3a.

In order to bond the coil to the rotor, small copper mounting pads (5-cm long, 2.5-cm wide and 0.13-mm thick) were epoxy bonded to the rotor. Then, the ends of 2G YBCO tape were soldered to these pads using lead tin solder, and the tape was also epoxy bonded to the rotor along its length. Indium-based solder was used to connect both the resistive copper braided current leads to the pads and voltage taps to the wires during critical current testing.

While this method of coil attachment was sufficient to allow initial low speed balancing, during testing at ~10,000 rpm, there was enough motion of the epoxy-bonded pads to throw off the rotor balance, thereby tripping the test.

Next, we tried to make current mounting pads from continuous bands of the 2.5-cm wide copper material. These bands were soldered to themselves under tension and epoxy mounted to the rotor. In addition, fiber-reinforced tape was used to further secure the coil to the rotor, as shown in figure 3a. This technique was successful achieving a 15,000 rpm test which showed no degradation in the 2G tape as depicted in the graph of figure 3b. However, a subsequent test at 25,000 rpm failed due to the copper bands debonding from the rotor and falling apart.

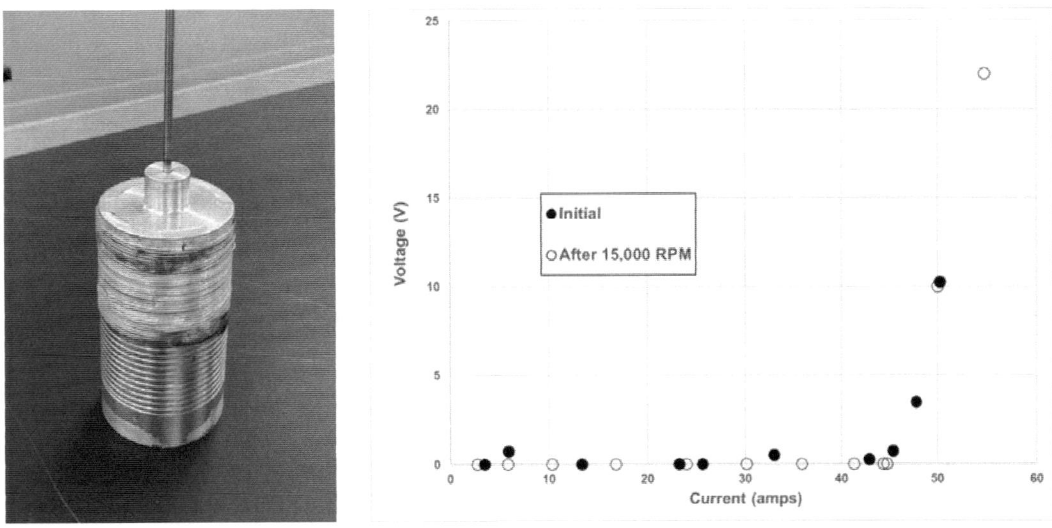

(a) (b)

Figure 3. Low speed spin test rotor: (a) 2G YBCO coil mounted on rotor; (b) 15,000 rpm data.

3.2 Final High Speed Testing

In order to attempt higher speeds, we had designed and fabricated a new monolithic aluminum rotor. By using a one-piece construction, as opposed to the two-piece, shrink-fit assembly of the initial low speed rotor, we would ensure that the rotor would not fall apart due to the centrifugal displacement at high speeds.

Since this new rotor was made from aluminum and since epoxy bonding attempts on the first rotor had failed, we were left to determine a new method of bonding to this rotor. After much consideration of various methods, we decided to nickel plate this aluminum rotor. Then, we tried to spot weld nickel strips to the nickel plating; however, the nickel plating bond was not sufficient to withstand the stresses imposed by spinning. While this new method did not work, it led us to the final successful solution which was to spot-weld nickel strips to the stainless steel rotor shown previously in figure 3a. While

Journal of Physics: Conference Series **2545** (2023) 012013 doi:10.1088/1742-6596/2545/1/012013

we would not be able to exceed 50,000 rpm with this rotor, we could at least attempt to test a Bi-2212 coil at higher speeds.

As shown in figure 4a, a Bi-2212 coil was mounted onto the spin test rotor and secured along most of its length with spot-welded nickel strapping. Some sections were left open to allow for soldering of current leads and voltage taps after spinning. High-speed balancing was performed by preferentially spot-welding additional pieces of nickel straps on the rotor. Successful spin tests were conducted at both 30,000 rpm and 40,000 rpm. A schematic drawing for this coil is provided in figure 5, showing the 6.3-mm long unsupported, cantilevered ends, which were used for the initial current lead connections, and the 12.5-mm long open sections used to attach the initial voltage taps to the 150-mm long coil winding.

While the critical current tests proved useful in isolating the damage on the Bi-2212 coil winding, the coil performance was possibly affected by a few factors. First of all, in order to mount the coil onto this rotor, the coil was slightly unwound by twisting to enlarge its diameter. This mechanical strain may have slightly damaged the wire. Secondly, the rotor material is slightly magnetic, so this affects the critical current at these higher temperatures. Lastly, the extended height of the Inconel spindle necessitated using an additional vacuum Bell jar standoff, which limited the operating temperature to only ~65 K.

Despite these factors, we were able to get a useful baseline critical current measurement after the 30,000 rpm spin test as shown in figure 4b. Subsequently, the 40,000 rpm test showed some degradation. Voltage taps were then added to a reduced section which did not include the suspect unsupported regions, and this test showed good comparison to the previous 30,000 rpm test.

(a)

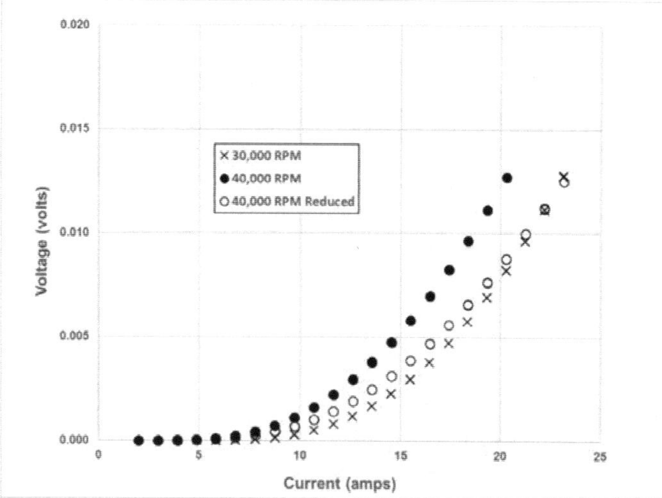

(b)

Figure 4. Bi-2212 spin testing: (a) photograph of rotor mounted on spin test flange; (b) critical current testing after 30,000 rpm and 40,000 rpm spin tests.

Upon inspection of the rotor after the 40,000 rpm spin test, there was visible damage to the Bi-2212 winding. As depicted in figure 5, the 6.3-mm long, cantilevered, unsupported end sections which were used for current lead attachment during critical current testing, had broken off during spinning. The calculated stress to cause this failure was ~320 MPa, which is beyond the yield strength of the reinforced wire. We were able to still attach current leads at those locations; however, the critical current test showed damage in the voltage tap region, which is ~12.5-mm long. The calculated stress in these voltage tap regions was ~270 MPa, which is close to the measured yield strength of this wire. In a subsequent test, the current leads were relocated to these damaged sections, and new voltage taps were installed in a reduced section showing little or no damage as compared to the previous 30,000 rpm spin test result.

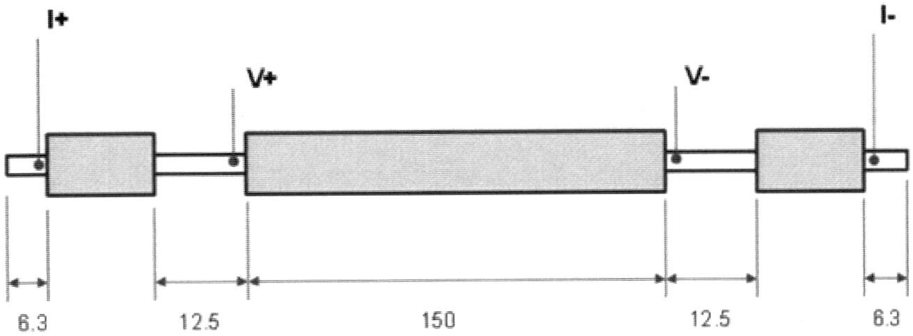

Figure 5. Schematic of Bi-2212 coil winding, showing 6.3-mm long unsupported ends used for soldering current leads and 12.5-mm long open sections for connecting voltage taps after testing. The Bi-2212 wire is shown in white, while the nickel strapping is depicted in grey.

4. Summary and Conclusions

This project involved the development of a new mechanical spin test designed to impose stresses equivalent to those experienced by a high-field insert coil. In this manner, stresses can be imposed on all features of a superconducting magnet, including end connections, crossover joints, etc. This technique is simpler to implement in that the assembly into a high field (i.e., ~30 T) superconducting magnet is not necessary. Also, after mechanically imposing the stresses with a given spin test, critical current testing in either liquid or sub-cooled nitrogen is simple to perform. The results of this research project demonstrate that this is a viable technique which could be useful to qualify various superconductors for use in high-field magnets.

Acknowledgments

This work was supported by the National Institute of Biomedical Imaging and Biology of the National Institutes of Health under Grant No. R15EB0260224. The author would like to acknowledge Dr. Alex Otto of Solid Materials Solutions for providing the reinforced Bi-2212 superconducting material for this project.

References

[1] Iwasa Y 2006 HTS and NMR/MRI magnets: unique features, opportunities, and challenges *Physica C: Superconductivity and its Applications* **445-446 (1)** 1088-1094.

[2] Markiewicz W D 2006 Perspective on a Superconducting 30 T/1.3 GHz NMR Spectrometer Magnet *IEEE Transactions on Applied Superconductivity* **16 (2)** 1523-1526.

[3] Kiyoshi T et. al. 1999 Development of 1 GHz superconducting NMR magnet at TML/NRIM *IEEE Transactions on Applied Superconductivity* **9** (2) 559-562.

[4] Hahn S, Park D K, Voccio J, Bascunan J and Iwasa Y 2011 No-Insulation (NI) HTS Inserts for >1 GHz LTS/HTS NMR Magnets *IEEE Transactions on Applied Superconductivity* **22 (3)** 4302405-4302405.

[5] Fukushima K et. al. 2001 Fabrication and transport properties of Bi-2212/Ag prototype magnets for 1 GHz-NMR magnet system *Physica C: Superconductivity*, **357-360 (2)** 1297-1301.

[6] Weijers H 2010 High field magnets with HTS conductors *IEEE Transactions on Applied Superconductivity*, **20 (3)** 576-582.

[7] Otsuka A 2010 A 1.3 GHz NMR magnet design under high hoop stress condition *IEEE Transactions on Applied Superconductivity* **20(3)** 596-599.

[8] Voccio J et. al 2019 High speed spin testing of reinforced 2212 coils for high-field NMR magnets *IEEE Transactions in Applied Superconductivity* **29 (5)**.

[9] Voccio et al 2020 Progress in High-Speed Spin Testing of Superconducting Wire and Tapes for High-Field NMR Magnet Qualification *J. Phys. Conf. Ser.* **1590** 012021.

Journal of Physics: Conference Series **2545** (2023) 012014 doi:10.1088/1742-6596/2545/1/012014

Effect of nano ZrO_2 addition on the properties of $Ca_{0.86}Sr_{0.14}CuO_2$ added Bi 2223 composites

Pawan K Verma[1], T. Rajasekharan[2], Shamili C Das[3], K P Surendran[3], and V Seshu Bai [1]

[1] School of Physics, University of Hyderabad, Hyderabad-500046
[2] Hymod Advanced Products, TIDE, University of Hyderabad, Hyderabad - 500046
[3] Material science and technology division, NIIST, Industrial Estate PO, Thiruvananthapuram, Kerala 695019
E- mail id- seshubai@uohyd.ac.in

Abstract: Effect of introducing nano ZrO_2 particles into $(Ca,Sr)CuO_2$ added Bi 2223 superconductor composites is studied. Microstructures revealed that 20 mol. % $(Ca,Sr)CuO_2$ phase added to Bi 2223 formed as micron sized particles among the Bi 2223 platelet shaped grains. Distribution of $(Ca,Sr)CuO_2$ in Bi 2223 has enhanced the critical current density (J_c) and flux pinning force density (F_p) substantially to fields up to 9 T at 20 K. Introduction of nano ZrO_2 into the composite reacted with $(Ca,Sr)CuO_2$ particles and formed additional secondary phases of increasing amount with a rise in ZrO_2 content. This resulted in lowering of $J_c(0)$ due to a gradual reduction in the fraction of superconducting phases, but retained the enhanced field range in which flux pinning was achieved. At 10 wt.% addition of ZrO_2, Bi 2223 phase was suppressed, and Bi 2212 phase was promoted. Scaling behaviour of pinning force density has shown the dominant mechanism at low fields to be normal surface pinning due to interfacial defects. The fact that $(Ca,Sr)CuO_2$ distributes itself as fine spherical particles in Bi 2223, without causing degradation of the superconducting matrix material, is of significance and opens up a scope to optimize the microstructures further for enhancement of flux pinning.

1. Introduction

Cuprate superconductors have layered perovskite lattice structure. Bismuth Strontium Calcium Copper Oxide (BSCCO) system are a class of cuprate superconductors having three superconductors of different transition temperatures (T_c) based on the number of CuO_2 planes in the unit cell. The general formula for BSCCO superconductors is $Bi_2Sr_2Ca_{n-1}Cu_nO_{10+\delta}$. For n = 1, the compound is Bi 2201, having T_c =20 K and one CuO_2 plane in the unit cell whereas, for n =2 and 3, the compounds are Bi 2212 and Bi 2223, having T_c = 85 K and 110 K, respectively. The number of CuO_2 planes in Bi 2212 and Bi 2223 are two and three, respectively [1]. (Bi,Pb)SrCaCuO system is one of the multi-component systems having a very complex phase diagram [2]. A narrow temperature window for phase formation and high reactivity rate make it challenging to optimise the processing conditions. High flux creep, large anisotropy, and poor grain connectivity are the common problems of high T_c superconductors, that limit their potential for practical applications close to 77 K [3].

Partial substitution of Pb for Bi helps in Bi 2223 phase formation [4]. Far below T_c, the flux lines move from their position by the action of Lorentz force causing deterioration in the critical current density (J_c), much below the irreversibility field (B_{irr}), due to flux creep. To suppress the flux creep, defects of the size of coherence length are introduced into the superconducting matrix. Secondary phase addition helps in generating defects that act as centres to pin the flux lines and lead to enhancement of J_c. Choice of the second phases should be such that they do not react with matrix phase or deteriorate its superconducting properties [5]. Secondary phase addition or elemental substitution into BSCCO superconductors were reported by various groups. Studies on Sm, Pr, Y, and Nd substitution into Bi 2223 were reported and it was found that even at low concentrations, Bi 2223 phase decomposes to form

Bi 2212 and other phases, and also the T_c decreases with an increase in the dopant concentration [6–9]. MnO and TiO_2 addition to Bi 2223 was reported to cause a lowering of T_c and to decease the Bi 2223 phase fraction, thus affecting the superconducting properties at concentrations above 1 wt.% [10,11]. Secondary phase additions in nano form to Bi 2223 was also studied with an aim to control the flux creep. For instance, Al_2O_3 nano particles were added to Bi 2223 and were found to improve J_c by 30 % when the Al_2O_3 concentration was 0.2 wt.%. The increase in J_c was attributed to the pinning effect of Al_2O_3 nano particles [12]. However, higher concentration leads to agglomeration and causes a reduction in J_c. SnO_2 nano particles added to Bi 2223 was reported to cause grain size to decrease, while the extent of voids and porosity increased. Higher concentration of SnO_2 addition affects the Bi 2223 phase formation and T_c [13]. Gua et al. reported that nano SiC addition reduces the Critical current (I_c). Bi 2223 phase formation was not affected till 1 wt.% nano SiC addition, however concentrations above 2 wt. % resulted in incomplete Bi 2223 phase formation. The 0.15 wt.% nano SiC added Bi 2223 sample showed improvement in J_c due to grain alignment, enhanced grain connectivity, and effective pinning [14].

Nano ZrO_2 addition to Bi 2223 was reported by various groups. Jia et al. reported that T_c (Onset) is not much affected even at 50 mol.% nano ZrO_2 addition. J_c improved at 10 and 20 mol.% whereas at 50 mol.% nano ZrO_2 addition, the J_c decreased [15]. Zouaoui et al. reported that 0.1 wt. % nano ZrO_2 addition to Bi 2223 improves the J_c in low field regime due to the formation of Zr- rich phases, embedded between the Bi 2223 grains and act as effective pinning centers [16].

In the present work, we bring in an anology to the enhancement of pinning in $YBa_2Cu_3O_{7-\delta}$ (Y 123) system, which is another cuprate superconductor with T_c around 92 K. Y_2BaCuO_5 (Y 211) is a non superconducting phase that occurs in the YBCO pseudo-binary phase diagram [17]. It is established that unifrom distribution of about 30% of Y 211 particles of size ~ 1 μm in Y 123 matrix shows dramatically improved superconducting properties as compared to pure Y 123 [18–20]. In the case of Sm_2BaCuO_5 (Sm 211) addition to YBCO, highest J_c was observed at 20 mol% of Sm 211 to Y 123 due to flux pinning at higher fields [21]. It was also shown that defects of different size ranges would result in effective pinning at different field regimes [22]. So far there are no reports on the effect of subtantial amounts of such second phases added to BSCCO superconductors, on the lines similar to the role of Y 211 in Y 123, or RE 211 in RE 123 [23],(RE is Rare Earth element like Gd, Sm Nd etc.), to enhance their superconducting properties.

In this direction, we have synthesized composites of Bi 2223 with 20 mol.% addition of $Ca_{0.86}Sr_{0.14}CuO_2$ into which x wt.% of nano ZrO_2, where x= 0, 0.5, 1 and 10 wt.%, is added. The composition for our Bi 2223 phase is $Bi_{1.2}Pb_{0.3}Sr_{1.54}Ca_{2.06}Cu_3O_{10+\delta}$, derived from the general formula $(Bi,Pb)_2Sr_2(Ca_{0.86}Sr_{0.14})_{n-1}Cu_nO_y$ for n= 5 and normalized to 3 Cu atoms per unit cell [24]. $Ca_{0.86}Sr_{0.14}CuO_2$ phase [25], can also be derived from the general formula when n goes to infinity [24]. $Ca_{0.86}Sr_{0.14}CuO_2$ phase can coexist with Bi 2223 as a second phase without interacting with it. Nano ZrO_2 was then added to mixed powders of Bi 2223 and 20 mol. % of $Ca_{0.86}Sr_{0.14}CuO_2$, using a sol-casting method [26] that enables uniform distribution of particles without agglomeration and is different from adding it directly to the precursor powders of the Bi 2223.

2. Experimental details

The precursor powders of Bi 2223 were prepared using the metal nitrate decomposition method reported in the literature [24,27]. The $Ca_{0.86}Sr_{0.14}CuO_2$ powder which was also synthesized using nitrate method, was calcined at 900 °C before mixing with Bi 2223 precursor powder. The precursor powder mixed with $Ca_{0.86}Sr_{0.14}CuO_2$ powder was tumbled for several hours before being pelletised. The pellets were then processed at 845 °C for 48 hours and then ground to powder form. A sol-casting method was employed, for uniform distribution of ZrO_2 nano particles to the ground powder. This method is discussed elsewhere [26] in detail. The resultant gel was dried, debindered, and ground to obtain them in powder form. The powder thus obtained was again pelletized and sintered twice at 845 °C with one

Journal of Physics: Conference Series **2545** (2023) 012014 doi:10.1088/1742-6596/2545/1/012014

intermediate pressing to enhance the density. The pellet, after final sintering, was cut into requisite sizes for characterization using various techniques.The samples synthesized in the present work are coded as follows: Pure Bi 2223 as BZ 0, 20 mol.% $Ca_{0.86}Sr_{0.14}CuO_2$ added Bi 2223 as BZ 1 and 20 mol.% $Ca_{0.86}Sr_{0.14}CuO_2$ added Bi 2223 + 0.5, 1, and 10 wt.% nano ZrO_2 as BZ 2, BZ 3 and BZ 4, respectively.

3. Results and discussion

Fig. 1(a) shows the powder XRD pattern indexed to $(Ca,Sr)CuO_2$ as the main phase with minor amounts of CuO and other oxide phases present. Micrograph in Fig. 1(b) shows the size of nano ZrO_2 particles to be in 50 -200 nm range. Before adding to Bi 2223 powder by sol-casting method, these particles were dispersed uniformly in water using high power ultrasonicator.

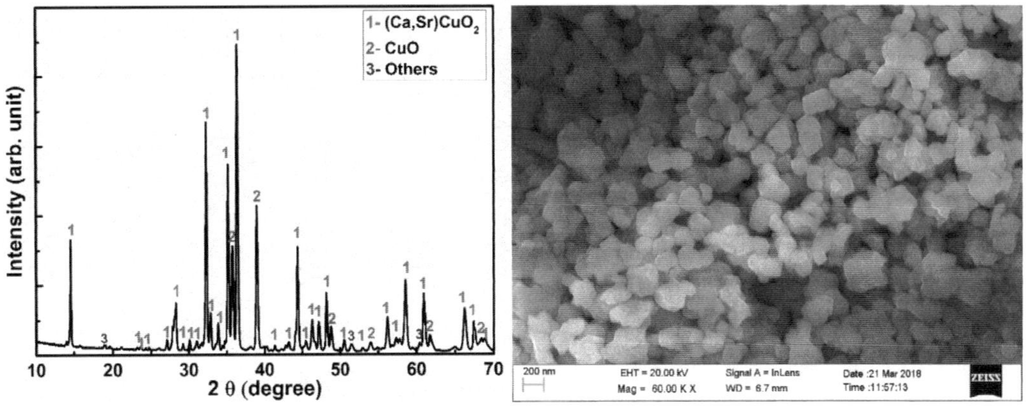

Figure 1: (a) XRD pattern indexed to $(Ca,Sr)CuO_2$ as major phase (b) FESEM micrograph showing ZrO_2 nano particles in 50- 200 nm range

3.1 XRD analysis

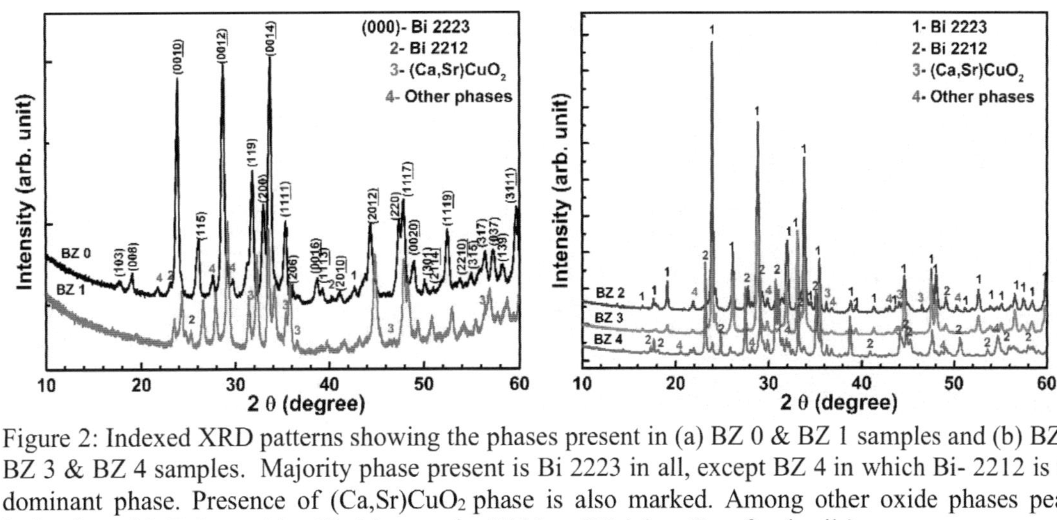

Figure 2: Indexed XRD patterns showing the phases present in (a) BZ 0 & BZ 1 samples and (b) BZ 2, BZ 3 & BZ 4 samples. Majority phase present is Bi 2223 in all, except BZ 4 in which Bi- 2212 is the dominant phase. Presence of $(Ca,Sr)CuO_2$ phase is also marked. Among other oxide phases peaks indicative of $SrZrO_3$ are identified in samples BZ 2 to BZ 4 (see Text for details).

Indexed XRD patterns for all the samples in BZ series are given in Fig. 2 (a & b). Analysis of XRD patterns confirms that all the samples are crystalline and the majority phase present is Bi 2223, except in sample BZ 4. The Bragg peaks from $(Ca,Sr)CuO_2$ phase are also marked in Fig. 2. High concentration

(10 wt.%) of nano ZrO_2 addition appears to have suppressed the Bi 2223 phase and promoted the Bi 2212 phase formation in sample BZ 4. Traces of oxide phases other than Bi 2223, Bi 2212 and $(Ca,Sr)CuO_2$ are indicated as Others in Fig. 2 and are present in minute amounts. Among the other oxide phases (marked as Others in Fig. 2), in the ZrO_2 added samples BZ 2 to BZ 4, increasing amounts of Strontium Zirconate ($SrZrO_3$) phase is noticed as indicated by the peaks at 2θ of 30.78°, 44.03° and 54.74°. Lattice parameters of samples BZ 0 to BZ 3 are given in Table 1 which are comparable to those of Bi 2223 phase reported.

Table 1: Lattice parameters for samples BZ 0 to BZ 3 for Bi 2223 phase.

Sample	Lattice parameters of Bi 2223 phase		
	a (Å)	b (Å)	c (Å)
BZ 0	5.413(7)	5.381(3)	37.076(8)
BZ 1	5.409(7)	5.398(9)	37.078(2)
BZ 2	5.404(2)	5.403(2)	37.046(2)
BZ 3	5.403(9)	5.407(4)	37.051(4)

Table 2: T_c (Onset) of all samples in BZ series

Sample	Starting composition	*T_c (Onset), K*
BZ 0	Pure Bi 2223	104.5
BZ 1	Bi 2223+ 20 mol.% $Ca_{0.86}Sr_{0.14}CuO_2$	105
BZ 2	Bi 2223+ 20 mol.% $Ca_{0.86}Sr_{0.14}CuO_2$ with 0.5 wt.% nano ZrO_2	104.2
BZ 3	Bi 2223+ 20 mol.% $Ca_{0.86}Sr_{0.14}CuO_2$ with 1 wt.% nano ZrO_2	104.5
BZ 4	Bi 2223+ 20 mol.% $Ca_{0.86}Sr_{0.14}CuO_2$ with 10 wt.% nano ZrO_2	77

3.2 Microstructure analysis

FESEM micrographs of cracked surfaces for all samples of BZ series are shown in Fig. 3(a-e). Large platelet like grains corresponding to Bi 2223 phase (except for sample BZ 4) can be seen. The grains are dense, stacked and tightly packed. Platelet morphology of BSCCO superconductors are reported by various groups [9,12–14]. Micron sized spherical particles of $(Ca,Sr)CuO_2$ phase are clearly seen in Fig. 3 (b) for sample BZ 1, distributed among the platelets of Bi 2223 without any reaction. In samples BZ 2 to BZ 4 with ZrO_2 addition, $(Ca,Sr)CuO_2$ phase particles have reacted with ZrO_2, as can be seen from a change from their spherical morphology in Fig. 3(c - e). As the concentration of ZrO_2 increased, the amount of the reacted phase also increased consistently. From an examination of XRD patterns for samples BZ 2 to BZ 4, we identify the reacted phase to be $SrZrO_3$. In BZ 2 and BZ 3 samples containing 0.5 and 1 wt.% ZrO_2, there was no visible reaction with the platelets of Bi 2223. For BZ 4, where the reacted phase is more and widespread, the reaction can be seen to have extended also to the edges of the grains rendering them smaller. Additionally, the platelets are found to be of Bi 2212 composition. In other words, at higher contents of ZrO_2, Bi 2223 phase is suppressed and formation of Bi 2212 phase is promoted, as confirmed by EDAX and XRD analysis in sample BZ 4. Presence of secondary phases formed due to reaction of ZrO_2 are indicated in the region R of Fig. 3. Region R is found to contain Strontium Zirconate ($SrZrO_3$) and Ca-Sr-Cu-O phases from an analysis of XRD and EDAX in Samples BZ 2 to BZ 4.

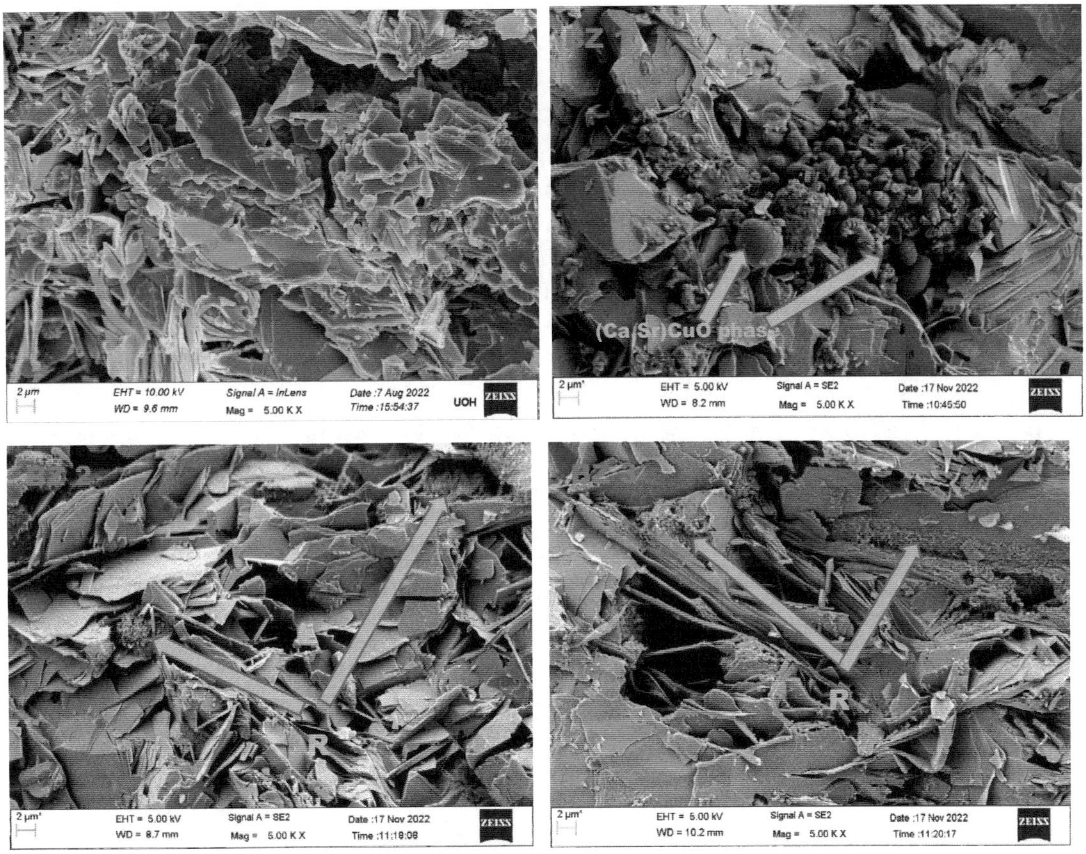

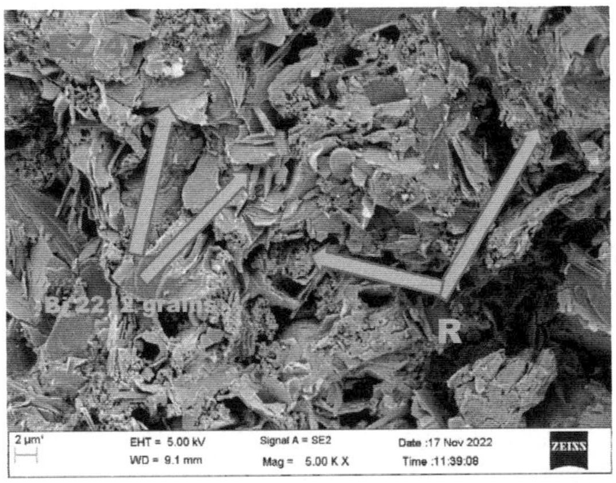

Figure 3 (a-e): FESEM microstructures of fractured surfaces of the samples show large platelet like grains of Bi 2223 phase (except in BZ 4). BZ 1 shows spherical particles of $(Ca,Sr)CuO_2$ present among Bi 2223 platelets. Nano ZrO_2 is found to react with $(Ca,Sr)CuO_2$ particles as shown in region marked as R. Among the resultant products of reaction, $SrZrO_3$ is identified from XRD of samples BZ 2 to BZ 4. Platelets seen from EDAX in BZ 4 are of Bi 2212, confirming that 10 wt. % ZrO_2 addition suppressed Bi 2223 phase.

3.3 Temperature and magnetic field dependence of magnetization

In order to observe the superconducting transition, the temperature (T) dependence of magnetization (M) was measured in an applied field ($B = \mu_0 H$) of 5 mT on cooling from 120 K down to 20 K. The M-T transition curves are shown in Fig. 4(a).

The T_c (Onset) for samples BZ 0 to BZ 3 is nearly 105 K, suggesting that the main phase is Bi 2223 and that there is no substantial reaction of Bi 2223 phase with ZrO_2, at lower concentrations. Sample BZ 4 has T_c (Onset) around 77 K indicative of Bi 2212 phase as the main phase, in agreement with its XRD pattern and microstructural features. The diamagnetic drop for sample BZ 4 is the lowest among all the samples, suggesting the superconducting fraction to be low, in agreement with extensive formation of secondary phases seen in Fig. 3e. The values of T_c (onset) are given in Table 2 for all the samples.

3.4 Field dependence of critical current density (J_c) and Flux pinning force density (F_p)

The field dependence of magnetization was studied by recording MH loop at 20 K. The MH loops for all samples at 20 K is shown in Fig. 4 (b). The samples BZ 0 to BZ 3 exhibit open MH loops up to 9 T applied field. As revealed by XRD and M-T curve in Fig. 4(a), the sample BZ 4 has Bi 2212 as the main phase and hence it is irrelevant to compare its magnetic behaviour with other samples that are Bi 2223 composites and hence is not included here. The critical current density(J_c) was determined from the irreversibility in M, using the relation from Bean's critical state model [28]

$$J_c = 20\ \Delta M/d,$$

where $d= a(1-a/3b)$ and ΔM is the difference in magnetization at a field and has units in emu/cc, and a, b (a<b) are the dimensions of the cross-section of the sample having units in cm. Field dependence of critical current density of samples BZ 0 to BZ 3 are shown in Fig. 5 (a).

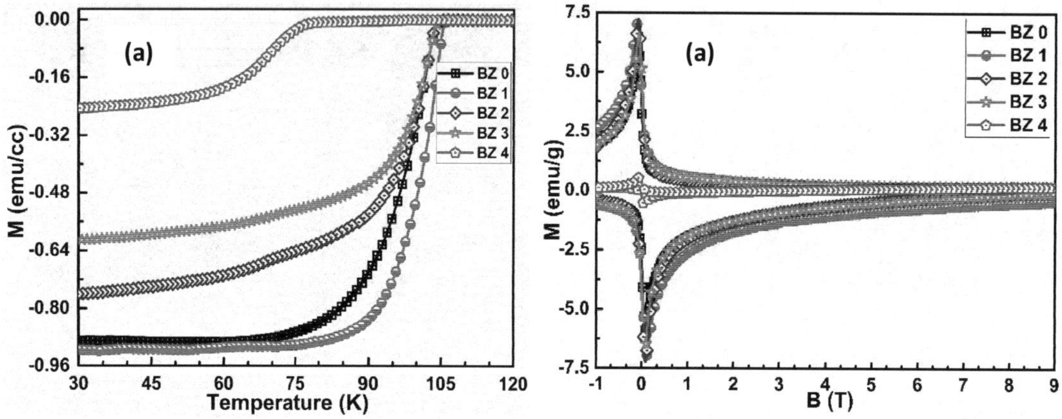

Figure 4: (a) Diamagnetic transitions showing T_c (Onset), (b) Magnetic hysteresis at 20 K in different samples of BZ composites

Samples BZ 0 to BZ 3 have finite J_c values till 9 T applied field at 20 K. The $J_c(0)$ values (i.e. J_c at zero field) for samples BZ 1 to BZ 3 are better than sample BZ 0. This reveals that distribution of secondary phases in Bi 2223 has been effective in improving $J_c(0)$. Sample BZ 1 has exhibited highest $J_c(B)$ values up to 9 T, compared to other samples. In order to gain an insight into the flux pinning behaviour of the composites, the flux pinning force density (F_p) is calculated using the relation $F_p=J_c \times B$.

Field dependence of F_p shown in Fig. 5 (b) for all the samples reveals that Sample BZ 1 exhibits the largest flux pinning force density in the entire field range at 20 K. It is established [22,29,30] that presence of non-superconducting Y(RE) 211 particles in Y(RE) 123 superconducting matrix would give rise to fine defects like stacking faults at the interface of the particles with the matrix phase; these defects are of size comparable to the coherence lengths and can aid in flux pinning. This suggests that enhanced J_c and F_p observed in the entire field range in BZ 1 has origin in the effective flux pinning provided by the secondary defects generated by the spherical particles of $(Ca,Sr)CuO_2$ present in the Bi 2223 matrix. Addition of nano ZrO_2 to the composite of Bi 2223 with 20 mol.% $(Ca,Sr)CuO_2$ phase has decreased the overall pinning. However, Fig. 5(b) shows that Sample BZ 3 with 1 wt.% nano ZrO_2 shows higher flux pinning compared to BZ 2 with 0.5 wt.% ZrO_2 and BZ 0, at fields above 3 T.

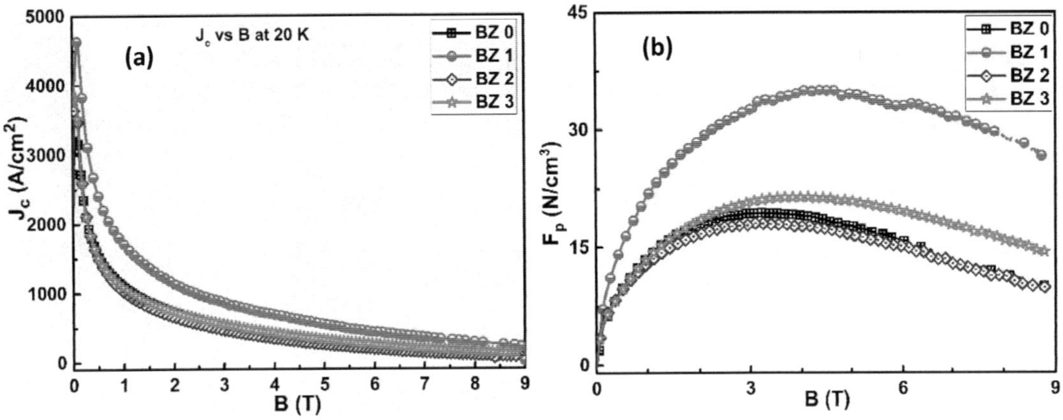

Figure 5: (a) Field dependence of J_c in the field range 0-9 T, (b) field dependence of F_p for samples BZ 0- BZ 3 at 20 K.

3.5 Scaling behaviour of Flux pinning force density

In the case of high T_c superconductors, multitude of pinning mechanisms can be operative in different field regimes [16,29]. In order to understand the nature of pinning and the effective field range in which pinning is effective, scaling laws are used based on the models discussed in the literature [30,31]. B_p is the field at which pinning force curve shows a peak and $F_{p\ max}$ is the maximum pinning force density observed. Normalized pinning force density $f = F_p/F_{p\ max}$ vs reduced field ($h = B/B_p$) are plotted in Fig. 6 (a) along with the curves defined [30,31] for theoretical models for ΔK, normal surface and normal point pinning, as given below

For ΔK pinning, $f(h) = 3h^2(1-2h/3)$

For normal point pinning, $f(h) = 9/4h(1-h/3)^2$

For normal surface pinning, $f(h) = 25/16h^{1/2}(1-h/5)^2$

From Fig. 6(a), one can see close agreement of experimental data to the curve-representing surface pinning, suggesting inter-granular /interfacial defects to be the dominant mechanism at low fields.

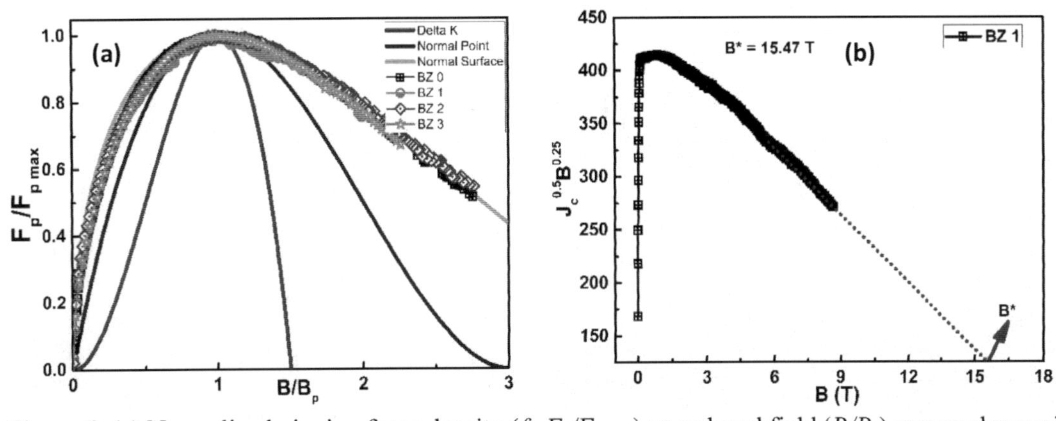

Figure 6: (a) Normalized pinning force density ($f = F_p/F_{p\ max}$) vs reduced field (B/B_p) curves along with theoretical curves (solid lines), showing the nature of pinning to be surface pinning (due to defects

generated at the inter-grain/ interfacial boundaries), (b) determination of B^*, the characteristic field from Kramer's model.

Dew Hughes model [32] suggests that normalized F_p can have a functional form $\boldsymbol{f(b)= F_p / F_{p\,max} = A * b^p (1-b)^q}$ using a scaling field B^*such that $b= B/B^*$, where B* is the field at which the pinning force due to a given mechanism vanishes. *A* is the numerical constant, *p* and *q* are the fit parameters, the values of *p* and *q* lead to identifying the dominating pinning mechanism present in the sample.

Table 3: $J_c(0)$, $B_P(T)$, $F_{p\,max}$ (N/cm³), $B^*(T)$, *A*, *p* and *q* parameters for samples BZ 0 to BZ 3

Sample	$J_c(0)$ (A/cm^2)	$B_p(T)$	$F_{p\,max}$ (N/cm^3)	$B^*(T)$	A	p	q
BZ 0	3265	3.19	19.43	11.37	2.90	0.54	1.21
BZ 1	4633	4.39	35.35	15.47	3.08	0.56	1.31
BZ 2	3779	3.18	18.1	11.43	2.56	0.48	1.07
BZ 3	3923	3.89	21.46	11	2.34	0.5	0.77

In order to have an estimate of the characteristic field B^*, beyond which surface pinning observed at low fields is effectively zero, field dependence of $J_c^{0.5}B^{0.25}$ is examined as per Kramer's model [33] as shown in Fig. 6(b), typically for BZ 1. Ideal values for surface pinning are $p= 0.5$ and $q= 2$, if only one mechanism contributes to pinning. The fit parameters support the view that dominant pinning is from surface pinning; the deviation from the ideal value suggests that more than one pinning mechanism exists in the present samples. Similar deviation from ideal values for fit parameters *p* and *q* were also reported [34,35].

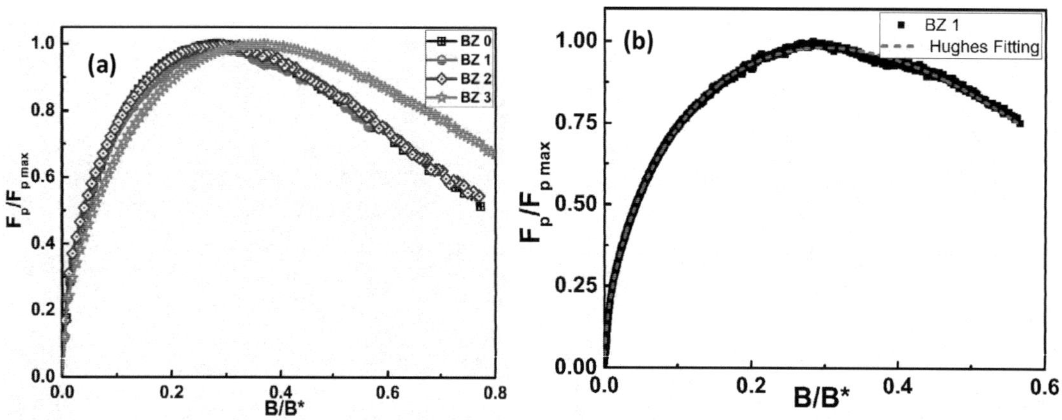

Figure 7: (a) Experimental data analysed by Hughes model (b) fit of the data to Hughes' model for sample BZ 1.

Fig. 7 shows the fit of experimental data to the Hughes' equation. It can be seen that the curve for sample BZ 3 deviates from the other curves at high fields. The higher pinning in sample BZ 3 at higher fields can be associated with the effective pinning caused by larger amounts of second phase formed due to reaction of 1 wt.% nano ZrO_2 addition. Table 3 presents the B* values calculated using Kramer's model along with the fit parameters of Hughes' model [32].

4. Summary

$(Ca,Sr)CuO_2$ added Bi 2223 composites with nano ZrO_2 addition were synthesized and characterized using various techniques. Analysis on XRD patterns showed that the main phase for samples BZ 0 to BZ 3 was Bi 2223. Larger amounts of nano ZrO_2 in Sample BZ 4 had promoted the formation of Bi 2212 phase. Microstructural studies showed large platelet like grains of Bi 2223 in samples BZ 0 to BZ 3. Addition of 20 mol. % of $(Ca,Sr)CuO_2$ enhanced J_c up to 9 T at 20 K, in sample BZ 1. It is interesting that $(Ca,Sr)CuO_2$ formed as spherical non-reacted particles in Bi 2223 matrix and caused substantial flux pinning to high fields. Distribution of non-reacted $(Ca,Sr)CuO_2$ particles in Bi 2223 matrix in BZ 1 createed a two phase microstructure, like that of Y 123 with Y 211 precipitates. High density of fine structural defects that get generated at the interface of second phase particles with the matrix was successful in providing effective pinning and enhancement of J_c. Addition of nano ZrO_2 to the above composite is found to react with the $(Ca,Sr)CuO_2$ particles forming $SrZrO_3$ as an impurity phase. This had decreased the superconducting phase fraction and lowered the $J_c(0)$ compared to BZ 1 but enhanced the field range in which pinning was effective, in samples with ZrO_2 addition. Pinning force density was analysed in terms of scaling models and it was found that surface pinning is the dominant mechanism at low fields. Examination of the data, fit to Hughes model, suggests that sample BZ 3 has higher pinning properties at fields close to B^* compared to samples with lower ZrO_2 addition. This is attributed to flux pinning caused by the enhanced defect density created by the reaction products after nano ZrO_2 addition. Pinning from different mechanisms is thus observed to be operative at different field regimes. It will be of interest to optimize the microstructures in textured tapes on similar lines, with larger superconducting fraction retained, for achieving enhanced performance.

Acknowledgment:

DST- SERB is greatly acknowledged for funding of the project EMR/2016/006004. PKV is grateful to SERB for fellowship.

References:

[1] Uglietti D 2019 *Supercond. Sci. Technol.* **32** 053001

[2] Majewski P 2000 *J. Mater. Res.* **15** 854–70

[3] Gurevich A 2014 C *Annu. Rev. Condens. Matter Phys.* **5** 35–56

[4] Knauf N, Harnischmacher J, Müller R, Borowski R, Roden B and Wohlleben D 1991 *Phys. C Supercond. its Appl.* **173** 414–24

[5] Glatz A, Sadovskyy I A, Welp U, Kwok W K and Crabtree G W 2020 T *J. Supercond. Nov. Magn.* **33** 127–41

[6] Nanda Kishore K, Muralidhar M, Hari Babu V, Peña O, Sergent M and Bénière F 1993 *Phys. C Supercond. its Appl.* **204** 299–304

[7] Singh R, Gupta A, Agarwal S K, Singh D P and Narlikar A V 1998 *Supercond. Sci. Technol.* **11** 311–4

[8] Sedky A 2009 *J. Phys. Chem. Solids* **70** 483–8

[9] Aloysius R P, Guruswamy P and Syamaprasad U 2005 *Supercond. Sci. Technol.* **18** 427–31

[10] Grivel J C, Jeremie A and Flukiger R 1995 *Supercond. Sci. Technol.* **8** 41–7

[11] Grivel J C and Flükiger R 1995 *Supercond. Sci. Technol.* **8** 751–4

[12] Annabi M, M'chirgui A, Ben Azzouz F, Zouaoui M and Ben Salem M 2004 *Phys. C Supercond. its Appl.* **405** 25–33

[13] Yavuz Ş, Bilgili Ö and Kocabaş K 2016 *J. Mater. Sci. Mater. Electron.* **27** 4526–33

[14] Guo Y C, Tanaka Y, Kuroda T, Dou S X and Yang Z Q 1999 *Phys. C Supercond. its Appl.* **311** 65–74

[15] Jia Z ., Tang H, Yang Z ., Xing Y ., Wang Y . and Qiao G . 2000 *Phys. C Supercond.* **337** 130–2

[16] Zouaoui M, Bessais L and Ben Salem M 2010 T *Supercond. Sci. Technol.* **23** 095013

[17] Murakami M 1992 Processing of bulk YBaCuO *Supercond. Sci. Technol.* **5** 185–203

[18] Thoma M, Shi Y, Dennis T, Durrell J and Cardwell D 2015 *J. Cryst. Growth* **412** 31–9

[19] Pavan Kumar Naik S, Devendra Kumar N, Missak Swarup Raju P, Rajasekharan T and Seshubai V 2013 *Phys. C Supercond. its Appl.* **487** 72–6

[20] Devendra Kumar N, Missak Swarup Raju P, Pavan Kumar Naik S, Rajasekharan T and Seshubai V 2014 *J. Low Temp. Phys.* **174** 113–27

[21] Naik S P K, Raju P M S, Rajasekharan T and Seshu Bai V 2016 superconductors *Supercond. Sci. Technol.* **29** 115001

[22] Devendra Kumar N, Rajasekharan T, Gundakaram R C and Seshubai V 2011 *IEEE Trans. Appl. Supercond.* **21** 3612–20

[23] Reddy E S and Rajasekharan T 1998 *J. Mater. Res.* **13** 2472–5

[24] Bai V S, Ravi S, Rajasekharan T and Gopalan R 1991 *J. Appl. Phys.* **70** 4378–82

[25] Siegrist T, Zahurak S M, Murphy D W and Roth R S 1988 T *Nature* **334** 231–2

[26] Raju P M S, Seshubai V and Rajasekharan T 2015 *Mater. Chem. Phys.* **161** 59–64

[27] Chen D -X., Hernando A, Conde F, Ramírez J, González-Calbet J M and Vallet M 1994 *J. Appl. Phys.* **75** 2578–83

[28] Bean C P 1962 *Phys. Rev. Lett.* **8** 250–3

[29] Devendra Kumar N, Rajasekharan T, Muraleedharan K, Banerjee A and Seshubai V 2010 *Supercond. Sci. Technol.* **23** 105020

[30] Civale L, McElfresh M W, Marwick A D, Holtzberg F, Feild C, Thompson J R and Christen D K 1991 *Phys. Rev. B* **43** 13732–5

[31] Goto T, Inagaki K and Watanabe K 2000 *Phys. C Supercond. its Appl.* **330** 51–7

[32] Dew-Hughes D 1974 *Philos. Mag.* **30** 293–305

[33] Kramer E J 1973 *J. Appl. Phys.* **44** 1360–70

[34] Zhang S, Li C, Hao Q, Lu T and Zhang P 2015 *Phys. C Supercond. its Appl.* **511** 26–32

[35] Pu M H, Song W H, Zhao B, Wu X C, Sun Y P and Du J J 2001 *Supercond. Sci. Technol.* **14** 299–304

Yoke geometry study using pulsed field magnetization on superconducting joined bulk Gd-Ba-Cu-O

K Takemura[1], K Sudo[1], R Dorget[2], J L Dadiel[1], M Sakafuji[1], K Yokoyama[3], T Oka[1] and N Sakai[1]

[1] Shibaura Institute of Technology, 3-7-5 Toyosu Koto-ku Tokyo, Japan
[2] Université de Lorraine, GREEN, F-54000 Nancy, France
[3] Ashikaga University, 268-1 Omae-cho Ashikaga-shi Tochigi, Japan

Email: na21104@shibaura-it.ac.jp

Abstract. Conditions for obtaining high-quality superconducting joint in Gd-Ba-Cu-O bulk superconductors were investigated, and the effects of PFM properties of the joined bulk and the change of yoke geometry in magnetization were examined. For a good superconducting joint fabricated by the local melting method using sintered Er-Ba-Cu-O, the J_c-B property and multiple Hall sensors indicates that the joined part is the preferential field penetration path. PFM was performed using a changed geometry of yoke in which concentrated long pulse to the joined part, but the result was equivalent to the trapped field before the changing yoke because the volume of the yoke was reduced. This is the first attempt to approach an effective PFM method for the joined bulk from the viewpoint of a changed geometry of yoke.

1. Introduction

The RE-Ba-Cu-O bulk superconductor (abbreviated as REBCO, RE : rare earth) has a high critical current density and is applied to superconducting magnets. This is more than 17 T at 29 K [1], which is much higher than the highest value of 0.5 T for Nd-Fe-B permanent magnets, making it a compact and powerful source of magnetic field generation. Applications include flywheels, which are devices that use a magnetic field to rotate a disk and store electric power as rotational energy, magnetic separation, which improves water quality and Ni recovery by separating ferromagnetic materials from aqueous solutions, and nuclear magnetic resonance (NMR) [2]-[7]. Increasing the trapped field properties of RE-Ba-Cu-O requires increasing the bulk radius. Although the melt-growth method is very widely used to produce RE-Ba-Cu-O, the crystal growth rate is very slow, making difficult to produce a bulk with a diameter of 10 cm or larger, and the J_c value worsens in the periphery [8], [9]. One solution is the multi-seeding method [10], [11]. In this method, multiple seed crystals are placed on a large-diameter precursor, and crystals are grown from each seed crystal to fabricate a large bulk. However, since the crystal growth edges are weakly bonded. The superconducting current is divided, and the trapped field distribution is not a single peak. Therefore, we chose a local melting method, which is a kind of superconducting joint [12]-[15]. By selecting RE'-Ba-Cu-O, which has a lower melting point than the matrix RE-Ba-Cu-O, for the joined part, only the joined part can be melted and grown in the furnace. In addition, by making the joined surface of the matrix (110)/(110), the segregation can be reduced and the trapped field properties will not be degraded [13].

In addition, the magnetization method must be considered when using the bulk superconductor. Field cooling magnetization (FCM) is a method in which the bulk superconductor is magnetized by applying a magnetic field to it at room temperature and cooling it below the critical temperature (T_c). On the other hand, pulse field magnetization (PFM) applies a pulsed field to a bulk superconductor cooled below T_c using a solenoid coil etc. This method allows magnetization in the equipment due to the small size of the magnetizing equipment [16]-[18]. However, rapid magnetic flux motion occurs in the bulk while the PFM magnetic field is applied, due to the heat generated. As a result, it is difficult to obtain high trapped field properties. In addition, with the previous method, higher trapped fields are obtained at the periphery than at the center of the bulk [17]. Therefore, methods to magnetize multiple times are being explored [19]. There is also the use of split coils and the long-pulse method, which extend the pulse width of the PFM applied to the bulk, thereby improving the trapped field. However, this requires modifications to the inductance and capacitor bank, and is expensive because it modifies the device itself. Similarly, a method to extend the pulse width is to use a yoke, usually made of soft iron or permendur. This method can be introduced at low cost. Normally, the larger volume of the yoke, the longer pulse width, but this is not practical in terms of mass increase. Therefore, by appropriately changing the yoke geometry, a long pulse can be obtained with reduced mass, resulting in a high trapped field [20]. There is potential to improve trapped field by concentrating long pulse in regions of different J_c-B characteristics within the bulk superconductor.

In this study, a good superconducting joint was made using Er-Ba-Cu-O sintered to a Gd-Ba-Cu-O bulk superconductor. The joined sample showed a single peak in the trapped field properties, confirming that a good superconducting joint was achieved. In the previous study, the joined sample was magnetized using PFM, and several Hall sensors confirmed that the joined part is the preferential field penetration path because the J_c-B characteristics of each region are different [21]. The yoke geometry was changed to concentrate the long pulse in the joined part, but the trapped field was equivalent to that of the yoke before processing because the volume of the yoke was reduced.

2. Experimental procedure

2.1. Joined Sample Preparation

Melt-grown Gd-Ba-Cu-O was selected as the matrix. This was manufactured by Nippon Steel Co., Ltd. and is a mixture of Gd123 ($GdBa_2Cu_3O_{7-\delta}$) and Gd211 (Gd_2BaCuO_5) in the ratio of 5:2, with 9.4 wt% Ag_2O and 0.5 wt% Pt added. The diameter is 60 mm and the thickness is 15 mm. The joining material is a mixture of Er123 ($ErBa_2Cu_3O_{7-\delta}$) and Er211 (Er_2BaCuO_5) in the ratio of 5:2, with 10 wt% Ag_2O and 1 wt% CeO_2. The mixed powders were sintered at 900 °C for 10 hours using an electric furnace. Their melting points were measured by DTA and were 1012 °C and 945 °C, respectively. To begin, the Gd-Ba-Cu-O bulk was cut into 25 mm square pieces, cut again in half so that the cut surface was on the (110) plane, and mirror polished to 0.3 μm with lapping film manufactured by 3M. The joining material was then polished to a thickness of 0.3 mm and placed between the Gd-Ba-Cu-O cut surfaces. The temperature program was adjusted so that the maximum temperature exceeded 945 °C and did not exceed 1012 °C so that only the joining material would melt and grow in the electric furnace. The temperature program was heated to 980 °C, held for 3 hours, cooled to 940 °C in 0.5 hours, and then slowly cooled to 920 °C at -1.6 °C/h for melt growth. The joined samples were annealed at 400 °C for 300 hours in an oxygen atmosphere to make the joined part superconducting. The joined samples were formed into Φ30 mm using a low-melting-point alloy (U-Alloy made from Osaka Asahi Co., Ltd), for installation in a refrigerator.

2.2. Evaluation Method

The joined sample was placed in a GM refrigerator and the sample was cooled to 50 K. For magnetization, a solenoid coil cooled by liquid nitrogen was used to apply a pulsed magnetic field of 1.0 to 5.0 T. Figure 1 (a) shows a view of the area around the cryostat where the joined sample was placed. The right side of the figure shows the pulse current generator using a capacitor, and the left side shows the GM refrigerator and magnetizing coil. The joined sample is placed on the cold stage, and its temperature is controlled using a thermometer and heater. Hall sensors were placed on the centre of the joined sample surface, as shown in the figure 1 (b). In addition, a cylindrical yoke and a machined yoke called "bar yoke" is installed. The base height of the bar yoke is 5 mm, and two bar thicknesses, 5 mm, and 2.5 mm, are prepared. As shown in the figure 1 (c), the bar yoke was placed parallel to the joined part of the sample so that the long pulse was concentrated on the joined part. After magnetised the joined sample was measured for the distribution of the trapped field using a Hall sensor and an x-y stage. The measurement range was 50 mm × 50 mm centered on the joined sample, 2 mm above the cryostat. The trapped field was defined as the magnetic field obtained by the Hall sensor placed on the sample surface at 800 ms after magnetization.

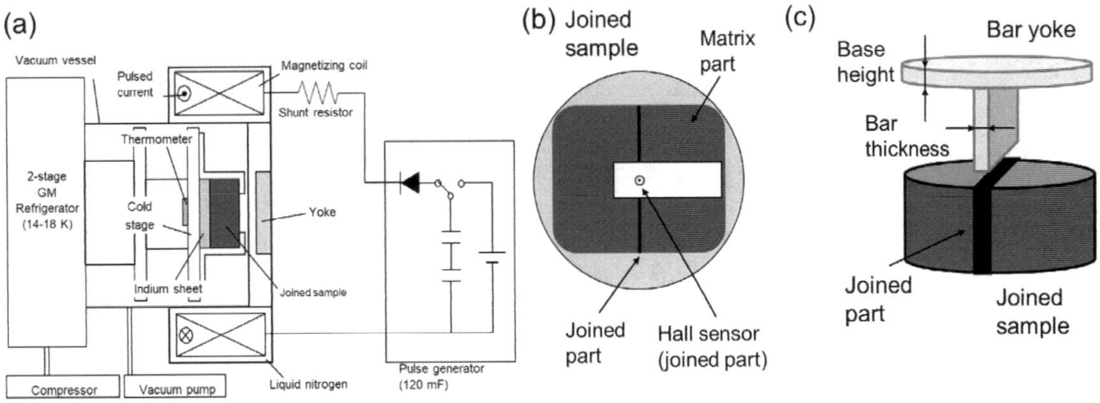

Figure 1. (a) External view of the area around the pulse generator and refrigerator (b) Enlarged view of the area around the surface of the joined sample and the installed Hall sensors (c) Schematic diagram of bar yoke installation location.

3. Result and discussion

3.1. Trapped field distribution of joined sample

Crystal orientation is an important aspect of superconducting joint. When the (100) plane is used as the joined plane, crystal growth proceeds parallel to the joined plane, and particles such as RE211 (RE_2BaCuO_5) are pushed out of the crystal growth region. Finally, at the center of the joined plane, particles such as RE211 that inhibit the superconducting current segregate. When the (110) plane is used as the joined plane, crystal growth proceeds in the form of a mountain range inclined at 45° from the joined plane. Therefore, more particles such as RE211 are incorporated into the crystal growth region, and segregation does not occur in the central part, resulting in a good superconducting joint. This technique makes it possible to fabricate larger bulk sizes and complex shapes.

The figure 2 shows the trapped field distribution of a joined sample to which 5 T is applied by PFM. A perfect single peak is seen, and the maximum trapped field is 0.52 T. Since there is no reduction in the trapped field at the joined part, it can be said that a good superconducting joined sample was obtained.

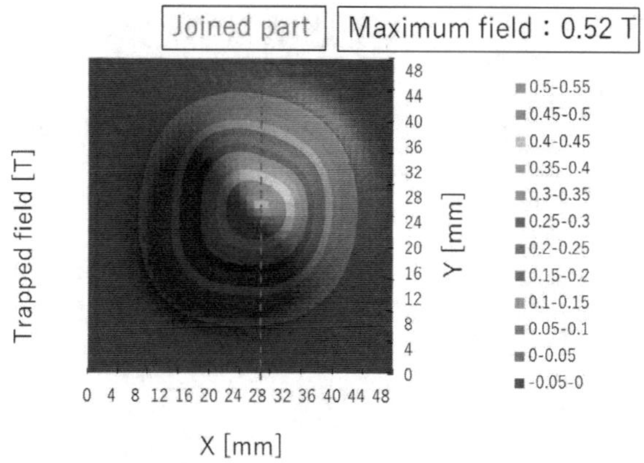

Figure 2. Trapped field distribution above 2 mm of joined sample at 50 K 5 T.

3.2. Trapped field using cylindrical and bar yoke

To check the difference in the penetrated field due to the change in yoke geometry, a magnetic field was applied with the refrigerator stopped. In other words, the experiment was conducted with the joined sample in a non-superconducting state. Figure 3 shows the waveforms of the penetrated field when the yoke geometry was changed. The cylindrical yoke decays more slowly than bar yoke. The reason for the cylindrical yoke is the larger volume.

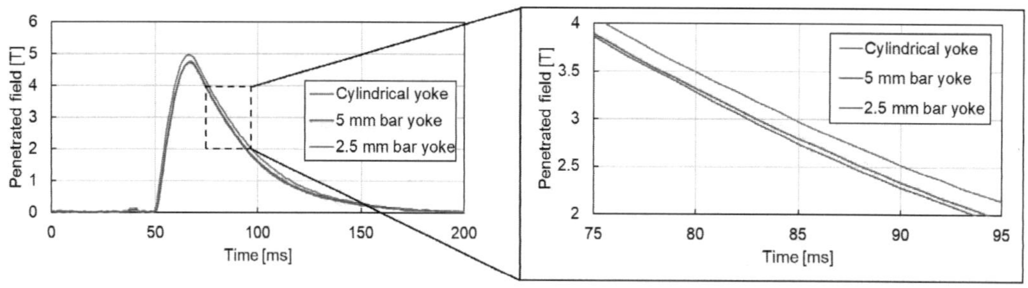

Figure 3. Difference in the penetrated field between a cylindrical yoke and a bar yoke at room temperature, and a magnified view of the damping region.

The figure 4 shows the results of trapped field at each applied field in a PFM using a cylindrical yoke and a bar yoke. The trapped field increases linearly when the applied field is 4 T or lower, but when 5 T is applied, the trapped field increases sharply, and a high trapped field of 1.2 T is obtained. The difference between the cylindrical and bar yokes was almost the same result, although it was slightly smaller when a 2.5 mm bar yoke was used. The applied field of 5 T resulted in a high trapped field. The reason for this is that the high magnetic field entering the bulk causes a rapid flux movement, which lowers J_c-B due to heat generation in the bulk and promotes the penetration of the magnetic field. With the bar yoke, a concentration of long pulse occurred near the joined part, but because of the smaller volume as shown in figure 3, the long pulse was not as long as those of the cylindrical yoke, and this change in yoke geometry did not change the trapped field. In addition, Figure 5 (a) shows the J_c-B characteristics measured from a superconducting quantum interference device (SQUID) sample fabricated under the same conditions as the joined sample. Figure 5 (b) shows a region cut from the sample. The J_c-B characteristic of the joined part showed a lower value than that of the matrix part at high magnetic fields of 50-70K. This is one of the pieces of evidence that the joined part is the preferential field penetration path. This discussion is done in [21], and the effect is tremendous. Therefore, it can be concluded that the effect of the preferential field penetration path is larger than the change in yoke geometry.

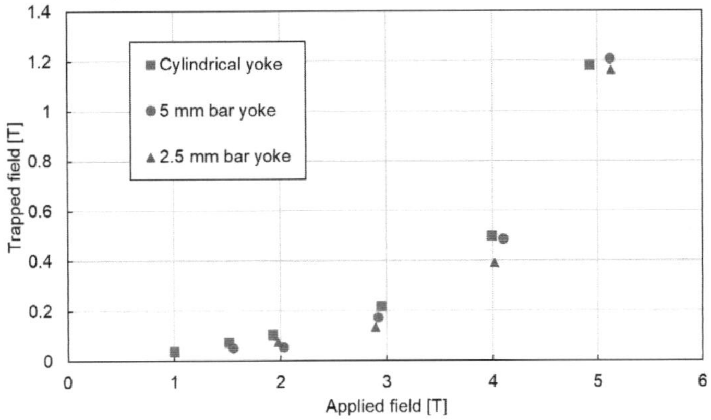

Figure 4. Difference in the trapped field using cylindrical yoke, 5 mm bar yoke and 2.5 mm bar yoke.

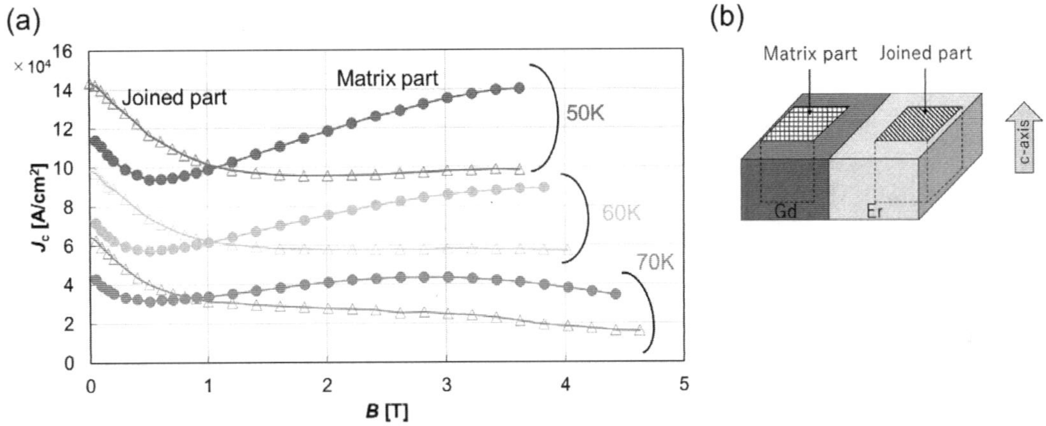

Figure 5. (a) J_c-B curves for each region of the matrix and joined part from joined sample (b) SQUID sample cut-out position from joined sample.

4. Conclusion

Using Gd-Ba-Cu-O melt-grown bulk and sintered Er-Ba-Cu-O plates, we obtain a good joined sample using the local melting method. The difference in J_c-B characteristics between the matrix and the joined part confirms that the joined part is the preferential field penetration path when PFM is performed. When 5T was applied, a high trapped field of 1.2T was obtained. The yoke geometry was modified to improve the trapped field of the entire bulk by concentrating long pulse at the joined part. However, the result was equivalent to the trapped field before processing. This is thought to be because the yoke volume was reduced despite the concentration of long pulse at the joined part due to the change in yoke geometry, so there was no difference. The effect of the preferential field penetration path at the joined part is considered to have been larger than the change in the yoke geometry. This experiment has enabled us to approach the effective PFM method for the joined bulk from the viewpoint of yoke fabrication, and we believe that it will contribute to the improvement of magnetizing properties.

References

[1] Tomita M and Murakami M 2003 *Nature* **421** 517–20
[2] Durrell J H, Ainslie M D, Zhou D, Vanderbemden P, Bradshaw T, Speller S, Filipenko M, and Cardwell D A 2018 *Supercond. Sci. Technol.* **31** 103501
[3] Murakami M 2007 *Int. J. Appl. Ceram. Technol.* **4** 225–41
[4] Zhou D, Izumi M, Miki M, Felder B, Ida T, and Kitano M 2012 *Supercond. Sci. Technol.* **25** 103001
[5] Nagashima K, Seino H, Miyazaki Y, Arai Y, Sakai N, and Murakami M 2008 *QR of RTRI* **49** 127-32
[6] Hayashi H, Tsutsumi K, Saho N, Nishizima N, and Asano K 2003 *Physica C: Supercond.* **392–396** 745-8
[7] Nakamura T, Tamada D, Yanagi Y, Itoh Y, Nemoto T, Utumi H, and Kose K 2015 *J. Magn. Reson.* **259** 68–75
[8] Sakai N, Kita M, Nariki S, Muralidhar M, Inoue K, Hirabayashi I, and Murakami M 2006 *Physica C* **445–448** 339-42
[9] Shi Y, Gough M, Dennis A R, Durrell J H, and Cardwell D A 2020 *Supercond. Sci. Technol.* **33** 044009
[10] Werfel F N, Floegel-Delor U, Riedeld T, Goebel B, Rothfeld R, Schirrmeister P and Wippich D 2013 *Physica C: Supercond.* **484** 6-11
[11] Kim C-J, Kim H-J, Jee Y A, Hong G-W, Joo J-H, Han S-C, Han Y-H, Sung T-H and Kim S-J 2000 *Physica C: Supercond.* **338** 205-12
[12] Mukhopadhyay S M, Mahadev N, and Sengupta S 2000 *Physica C: Supercond.* **329** 95-101
[13] Iida K, Yoshioka J, Sakai N, and Murakami M 2002 *Physica C: Supercond.* **370** 53-8
[14] Iida K, Yoshioka J, Negichi T, Noto K, Sakai N, and Murakami M 2002 *Physica C: Supercond.* **378–381** 622-6
[15] Harnois C, Desgardin G, Laffez I, Chaud X, and Bourgault D 2002 *Physica C: Supercond.* **383** 269-78
[16] Fujishiro H, Tateiwa T, Fujiwara A, Oka T, and Hayashi H 2006 *Physica C Supercond.* **445–448** 334-8
[17] Ainslie M D, Fujishiro H, Mochizuki H, Takahashi K, Shi Y-H, Namburi D K, Zou J, Zhou D, Dennis A R, and Cardwell D A 2016 *Supercond. Sci. Technol.* **29** 074003
[18] Zhou D, Ainslie M D, Shi Y, Dennis A R, Huang K, Hull J R, Cardwell D A, and Durrell J H 2017 *Appl. Phys. Lett.* **110** 062601
[19] Mizutani U, Oka T, Itoh Y, Yanagi Y, Yoshikawa M, and Ikuta H 1998 *Appl. Supercond.* **6** 235-46
[20] Yokoyama K and Oka T 2020 *IEEE Trans. Appl. Supercond.* **30** 6800105
[21] Takemura K, Sudo K, Dorget R, Dadiel J L, Sakafuji M, Yokoyama K, Oka T, Murakami M and Sakai N *Supercond. Sci. Technol.* **35** 04005

Orientation degrees in Dy123 powders by using Linear-drive type Modulated Rotating Magnetic Field system

W B Ali[1], S Adachi[1], F Kimura[1], and S Horii[1]

[1]Kyoto University of Advanced Science, 18 Gotanda-cho, Yamanouchi, Ukyo-ku, Kyoto 615- 8577, Japan.

Email: 2021md01@kuas.ac.jp

Abstract: Bi-axial magnetic alignment of $DyBa_2Cu_3O_y$ (y ~ 7) grains with high in-plane orientation degrees has been achieved using only permanent magnets as magnetic field sources. When compared to electromagnets such as a superconducting magnet, arrayed magnets offer a lower cost and greater versatility as magnetic field sources for magnetic alignment. In this study, permanent magnets were arranged in such a way that they were able to create a rotating magnetic field with a strength of 0.8 T-class. This helps with the magnetic alignment of functional materials like cuprate superconductors.

Keywords: $DyBa_2Cu_3O_y$, linear-drive type, modulated rotating magnetic field (MRF), magnetic alignment

1. Introduction:

The $REBa_2Cu_3O_y$ (RE: rare-earth elements, RE123) superconductor has a critical temperature (T_c) higher than the boiling point of liquid nitrogen and a high critical current density (J_c) in a magnetic field, which requires bi-axial crystal orientation control for practical use. These cuprate superconductors are intriguing contenders for superconducting applications such as superconducting wires and high-field magnets. The anisotropic crystal structure of RE123 is widely known to comprise alternate 2-dimensional superconducting CuO_2 layers and a 1-dimensional Cu-O chain, which increases anisotropic critical current densities (J_{cs}), $J_c//c < J_c//ab$ [1]. The intergrain J_c is drastically reduced when the misorientation angle between bicrystals in cuprate superconductors increases, even for *c*-axis oriented materials [2]. Bi- or tri-axial orientation and consolidation of RE123 grains are indispensable to increase the J_c in RE123 microstructures to a level suitable for practical use. Until now, researchers have used epitaxial growth methodologies such as melt-solidification [3] and thin film [4] processes to develop bi- or tri-axially oriented RE123 superconductors.

Recently, to achieve bi- or tri-axial grain orientation, a modulated rotating magnetic field (MRF) [5][6][7][8] has been used at room temperature. This proceeds to a fabrication technique to produce a single crystalline microstructure and an epitaxial growth method. Tri-axial grain orientation via MRF is achieved by magnetically aligning two magnetization axes: the first easy axis of magnetization is parallel to the static magnetic field (SF), and the hard axis of magnetization is perpendicular to the rotating magnetic field (RF). In the MRF process, the grains with tri-axial magnetic anisotropy and enough magnetic orientation energy may achieve bi- or tri-axial alignment. When using a magnetic field for alignment purposes, the orientation energy used is proportional to a square of the magnetic field strength (B^2). Since the cost of a superconducting solenoidal magnet is much higher than that of a permanent magnet, the permanent magnets are the favored option for producing a magnetic field in the RE123 magnetic alignment method. In practice, RE123 powders cured in epoxy-resin were bi-axially aligned under the MRF of 10 T using a SC-magnet with a sample rotation system at room temperature [9]. Taking this into account, linear drive-type modulated rotating magnetic field (LDT-MRF) equipment was recently established [10]. In this equipment, the array of permanent magnets reciprocates over the sample. This creates the static and the rotating magnetic field at the same time, even though the sample is not moving in real space. Through this LDT-MRF, bi- or tri-axial orientation can be achieved, and magnetic alignment using LDT-MRF is a low-cost continuous production process. The use of this equipment in conjunction with the colloidal method is intended to allow the fabrication of bi-axially oriented thick film superconductors. Also, LDT-MRF not only attains bi- or tri-axially superconducting materials but is also thought to be useful in orienting other anisotropic functional materials.

In this study, we performed bi-axial magnetic alignment experiments on the grains of $DyBa_2Cu_3O_y$ (Dy123) cuprate superconductor using an LDT-MRF apparatus with enhanced rotating magnetic field strength compared to our previous study [10]. The increment of the rotating magnetic field is expected to support the magnetic alignment of the hard axis and achieve a higher in-plane orientation degree. In terms of sample selection, Dy123 (y ~ 7) was chosen because it has the highest magnetic anisotropy among RE123. In other words, the magnetic alignment behaviour of Dy123 (y ~ 7) was expected to be easy to observe even at the magnetic field strength of the permanent magnets. For comparison with the samples made with LDT-MRF, other samples were also prepared by a SC-magnet. The in-plane orientation degrees of the samples obtained by LDT-MRF, and SC-magnet were evaluated, and the location dependence of the in-plane orientation degrees was also investigated by cutting the Dy123 sample prepared by LDT-MRF.

2.Experimental details

The Dy123 polycrystals were fabricated by using a standard solid-state reaction. The mixture was prepared from powders of Dy_2O_3, $BaCO_3$, and CuO (3N or higher) mixed in the desired ratios. The nominal cation ratio of Dy: Ba: Cu = 1:2:3 was chosen in the mixture. The mixture was heat treated thrice in air at 880, 900, and 920 °C with intermediate grindings. Then, the mixture was pelletized and sintered at 950 °C for 24 h. By annealing the sintered polycrystal, the amount of oxygen in $DyBa_2Cu_3O_y$ was adjusted to y ~ 7. The details of the annealing conditions for oxygen content adjustment are described in our previous papers [11][12]. Furthermore, annealed polycrystalline pellets were ground thoroughly in an agate motor to obtain the powders of Dy123 (y ~ 7).

The obtained Dy123 (y ~ 7) powder were gently mixed with epoxy-resin at a weight ratio of 1: 10 for LDT-MRF equipment. After that, the Dy123 (y ~ 7) and the epoxy-resin mixture was transferred to an acrylic mold and placed in LDT equipment for 12 h at room temperature. Figure 1(a) shows a schematic diagram of the LDT-MRF equipment. By linearly reciprocating the arranged magnets, (SF) and (RF) were alternatively applied to the sample. Figure 1(b) displays the dimension of the rectangular shaped epoxy cured sample as labeled the as-aligned, center-part, and side-part. The as-aligned sample fabricated by LDT-MRF was cut to investigate the location dependence of the in-plane orientation degree, as described later. The α and β planes represent the first and second easy axes of magnetization, while γ plane represents the hard axis of magnetization, and these planes are normal to each other. The magnetization axes χ_1, χ_2, and χ_3 correspond the first easy axis, the secondary easy axis, and the hard axis of magnetization, respectively. Ideally, the first easy, second easy, and hard axes will be normal to the α, β, and γ planes, respectively. To compare with magnetically aligned samples using LDT-MRF, a sample consisting of Dy123 powders and epoxy-resin with dimensions of $15mm^W$x $10mm^D$x $10mm^H$ was fabricated for intermittent MRF [7][8] with 0.7 T and 1 T by a SC-magnet. The in-plane orientation degrees of the magnetically aligned Dy123 grains in epoxy-resin were examined using the (103) pole figure of the α-plane. In details, the in-plane orientation degrees of them were evaluated from the full widths at half maximum (FWHMs) of 4-fold symmetric spots along the rotation angle direction in the (103) pole figure measurements.

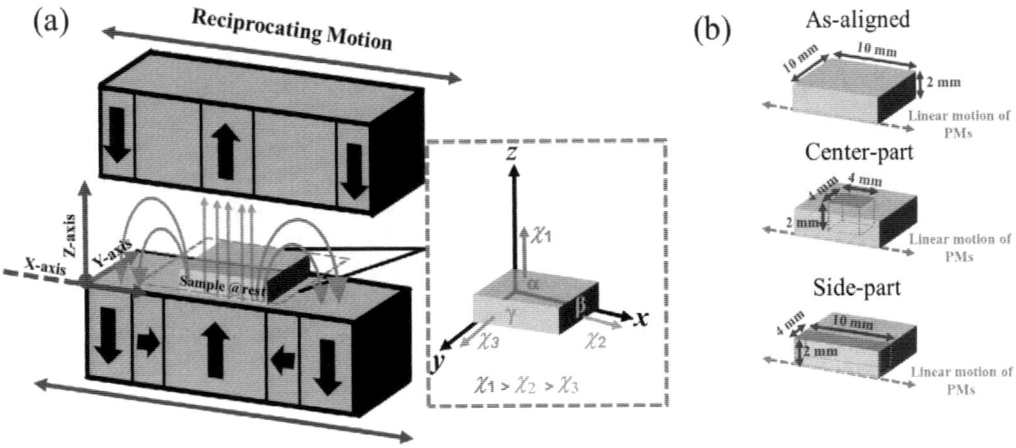

Figure 1. (a) Schematic diagram of the linear drive type MRF apparatus. A rectangular shaped sample (mixture of Dy123 powders and epoxy resin) does not move in real space. The permanent magnet array is reciprocated in X-axis direction during the magnetic alignment of the sample. The Y-axis is parallel to the width of the magnets. The X-Y plane is parallel to the horizontal plane and corresponds to the flat top surface of the lower part of the magnet array. The Z-axis is perpendicular to the X-Y plane. The magnetization axes with $\chi_1 > \chi_2 > \chi_3$ mean the first easy axis, second easy axis, and hard axis, respectively. (b) The position and size of each part in the aligned sample were schematically illustrated. The sections in the aligned sample were labeled the as-aligned, center-part, and side-part.

3.Results and Discussion

Figure 2 displays (103) pole figure of the as-aligned Dy123 sample magnetically aligned in LDT-MRF equipment. In Figure 1(b), the dimension of the as-aligned sample is schematically illustrated. When an orthorhombic crystal structure with twin microstructure is aligned magnetically, four-fold symmetric spots are observed in the (103) pole figure. These results were qualitatively similar to those obtained from bi-axially aligned Dy123 grains with 10 T MRF measured at the α-plane using an SC-magnet [9]. The angle between the (103) plane and the α-plane in orthorhombic Dy123 with twin microstructures was reflected by these 4-fold symmetric spots, which appeared around a tilted angle Ψ = 45°. The FWHM values in the rotational angle (ϕ) direction were determined using these spots to quantitatively investigate the in-plane orientation degrees of the Dy123 grains in epoxy-resin. In this case, increasing/decreasing FWHM corresponds to decreasing/increasing in-plane orientation degrees. After alignment, the (103) pole figure showed an FWHM of ~ 12.5° for the as-aligned sample.

Table 1 elaborates the magnetic field strength ($|B|$) of the two components of MRF and FWHMs of the Dy123 samples aligned in LDT-MRF and SC-magnet. Using SC-magnet, we observed the (103) pole figures for the aligned Dy123 (y ~ 7) samples and discovered that the FWHM values are approximately 9.1° for 0.7 T MRF and 8.0° for 1 T MRF. From this, one can easily understand that FWHM values decrease with higher magnetic fields. In other words, a higher in-plane orientation degree was achieved with an increase in the magnetic field. In contrast, the FWHM value for the as-aligned sample under LDT-MRF showed 12.5°. Compared to the FWHM values for LDT-MRF and SC-magnet, the sample aligned in the SC-magnet has a lower FWHM value than the LDT-MRF. That is, the results in Table 1 suggested the presence of a difference in the in-plane orientation degrees due to the difference in the method of generating the magnetic fields for the magnetic alignment.

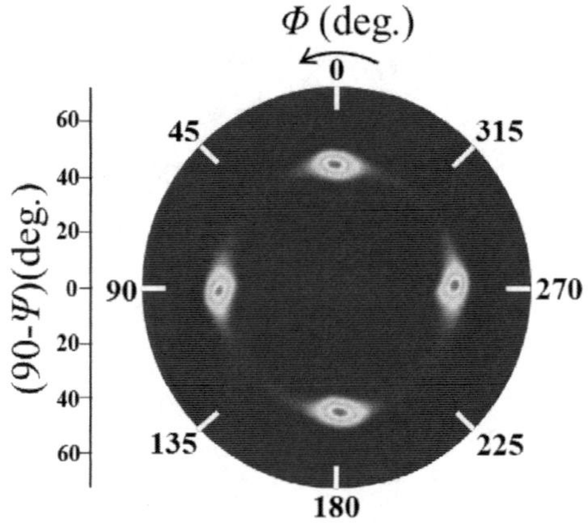

Figure 2. For the as-aligned sample, a (103) pole figure at α-plane (normal to the z-axis in Figure 1(a)). The Dy123 (y ~ 7) grains in epoxy-resin were magnetically aligned under the LDT-MRF equipment.

Table 1. The magnetic field strength ($|B|$) generated in SC-magnet and LDT-MRFs, as well as the average FWHM along the rotation angle direction at each spots in the (103) pole figure measurements. $|B|$ values of SC-magnet are the actual value. $|B|$ (at z = 2mm in Figure 1(a)) values of LDT-MRFs were calculated by the finite element method.

Types of MRF	Component of magnetic field strength ($|B|$) (T)		FWHM (deg.)
	Static field (T)	Rotating field (T)	
Linear drive type	0.9	0.8	12.5
SC-magnet	0.7	0.7	9.1
SC-magnet	1	1	8.0

To investigate the origin of the difference in the in-plane orientation degrees between LDT-MRF and MRF of SC-magnet, the as-aligned sample by LDT-MRF was cut into three parts parallel to the direction of reciprocating motion: the center-part and both side-parts. In the present study, we discuss only one side of the as-aligned sample due to the symmetric condition of the magnetic array. In Figure 1(b), the dimensions of the center-part and the side-part are schematically illustrated. Figures 3(a) and 3(b) displays the (103) pole figure of the center-part and the side-part, respectively. When comparing the (103) pole figures of Figure 2 (for the as-aligned sample) and Figure 3(a) (for the center-part), the 4-fold symmetric spots of the center-part sample were more sharpened remarkably. The symmetric spots in Figure 3(a) are less widened in the rotational angle direction, and the FWHM for the center-part was approximately 8.5°. As a result of the side-part in Figure 3(b), the 4-fold symmetric spots are also sharper in comparison to the as-aligned sample. The FWHM value for the side-part was approximately 9.1°. Moreover, the centroid of the four spots in pole figure of the side-part was parallelly shifted by ~ 6° to the right from the center of the spots of Figure 3(a). The FWHM value for the side-part was also lower than that for 0.7 T-MRF sample. From the above results, we found that various in-plane orientation states were realized in the as-aligned samples. As a representative example, the in-plane orientation degree of the center-part is higher/lower than that of 0.7 T/1 T-MRF sample by SC-magnet. In other words, these experimental results suggest that a rotating magnetic field strength of 0.7 T or more and less than 1 T is generated for magnetic alignment in the spatial region of the center-part under LDT-MRF. Such rotating magnetic field strengths are in good agreement with the values expected from the finite element method (see Table 1). In addition, considering the tilting angle direction of the Dy123 grains as clarified from the pole figure in Figures 3(a) and 3(b), increasing the width of the magnet in the Y-axis direction in real space may be able to lead to a homogeneous in-plane orientation degree in the sample. This research almost proved that a 0.8 T-class rotating magnetic field strength $|B|$ can be generated only by permanent magnets and can be used for magnetic alignment. The rotating magnetic field strength $|B|$ of the magnet array in the previous work [10] was ~ 0.5 T. Thus, we have achieved magnetic alignment using magnetic orientation energy about 2.5 times (~ $(0.8\ T)^2 / (0.5\ T)^2$) as the rotating field component compared to the previous magnet array [10]. The above results of this research expand the possibilities of magnetic alignment in both basic and applied aspects.

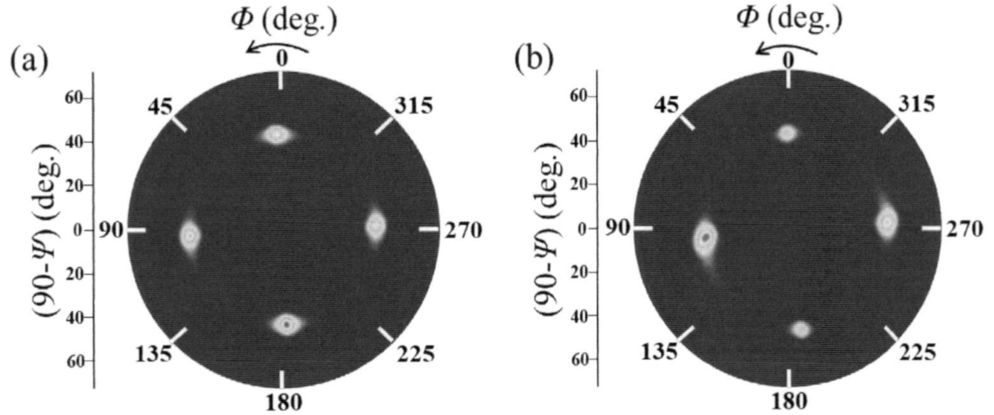

Figure 3. The (103) pole figure for (a) center-part, (b) side-part (after cutting the as-aligned sample by using the LDT-MRF equipment).

4. Conclusion:

We used LDT-MRF with enhanced rotating magnetic field strength (0.8 T at 2 mm from the top of the lower magnet) to achieve bi-axial alignment of Dy123 (y ~ 7) grains with a high in-plane orientation degree. Our experimental results raised the possibility of magnetically aligning bi- or tri-axially oriented high-T_c cuprate superconducting materials and products using only permanent magnets as magnetic field sources. Naturally, the enhancement of the rotating magnetic field strength in LDT-MRF contributes not only to the bi- or tri-axial magnetic alignment of superconducting materials but also to the magnetic alignment of other functional materials. This study demonstrated that the application range of magnetic alignment is extended by controlling the magnetic circuit in the LDT-MRF.

Reference:

[1] Yamaki M, Furuta M, Doi T, Shimoyama J and Horii S 2014 *Physics Procedia* **58** 62.
[2] Dimos D, Chaudhari P, Mannhart J and Legoues F K 1988 *Physical Review Letters* **11** 61.
[3] Wu M K, Ashburn J R, Torng C J, Hor P H, Meng R L, Gao L, Huang Z J, Wang Y Q and Chu C W 1987 *Physical Review Letters* **58** 908.
[4] Sawano K, Morita M, Tanaka M, Sasaki T, Kimura K, Takebayashi S, Kimura M and Miyamoto K 1991 *Japanese Journal of Applied Physics* **30** L1157.
[5] Kimura T and Yoshino M 2005 *Langmuir* **21** 4805.
[6] Fukushima T, Horii S, Ogino H, Uchikoshi T, Suzuki T S, Sakka Y, Ishihara A, Shimoyama J I and Kishio K 2008 *Applied Physics Express* **1** 1117011.
[7] Horii S, Yamaki M, Ogino H, Maeda T and Shimoyama J 2010 *Physica C: Superconductivity and its Applications* **470** 1056.
[8] Horii S, Okuhira S, Yamaki M, Kishio K, Shimoyama J I and Doi T 2014 *Journal of Applied Physics* **115** 113908.
[9] Horii S, Nishioka T, Arimoto, Fujioka S, and Doi T 2016 *Superconductor Science and Technology* **29** 125007.
[10] Horii S, Arimoto I and Doi T 2018 *Journal of the Ceramic Society of Japan* **126** 885.

[11] Adachi S, Ichinose A, Kimura F and Horii S 2022 *IEEE Transactions on Applied Superconductivity* **32** 7200105.
[12] Kimura F, Adachi S, Horii S and Kimura T 2022 *CrystEngComm* **24** 3807.

Trapping Magnetic Field in Bulk Iron-based Superconductor Sintered under High Pressure

Wenxi Wu[1], Sunseng Pyon[1], and Tsuyoshi Tamegai[1]

[1]Department of Applied Physics, The University of Tokyo, 7-3-1 Hongo, Bunkyo-ku, Tokyo 113-8656, Japan

E-mail: wu-wenxi856@g.ecc.u-tokyo.ac.jp

Abstract. We fabricated two bulk superconductors of $(Ba,Na)Fe_2As_2$ using a cubic press and hot press sintering technique. For bulk #1 sintered by cubic press, despite its small dimensions, with diameter 4.1 mm and height 3.64 mm, a trapped field value of ~9.97 kG has been achieved on its surface, which is only 230 G lower than the value reported in a bulk with a diameter of 10 mm before. Critical current density (J_c) is estimated to be 81.2 kA cm^{-2} at 4.2 K under self-field, which is larger than the value reported before. It suggests that the improvement of J_c achieved by high pressure could make up for the negative influence from the smaller dimensions and achieve a large trapped field. In the second larger bulk #2 sintered by hot pressing technique, with diameter 9.45 mm and height 4.58 mm, the value of J_c is ~ 33.9 kA cm^{-2} and the trapped field is ~8.81 kG on the surface. The expected better texturing has not been observed. Possible origins for inferior properties of bulk #2 are discussed.

1. Introduction

After their discovery in 2008, iron-based superconductors (IBSs) have attracted much attention [1]. With large critical density (J_c), high upper critical field (H_{c2}), and high critical temperature (T_c), IBSs have great potential for high-field applications [2-9]. Among IBSs, $(Ba,A)\ Fe_2As_2$ (A=K, Na) (Ba122) compounds in polycrystalline forms are most anticipated for the fabrication of superconducting wires, tapes, and bulks. In particular, the fact that Ba122 bulks could be fabricated in a much larger dimensions than that of fragile $RBa_2Cu_3O_{7-\delta}$ (R: rare earths) bulks together with their H_{c2}, which is higher than that of MgB_2, make them candidates to be applied for powerful magnets by trapping field [10].

The J_c value, which is crucial for the application of IBSs, could be improved via high-pressure sintering. A magnetic J_c value of 327 kA cm^{-2} at 4.2 K under the self-field condition has been achieved in a $(Ba,Na)Fe_2As_2$ round wire sintered by hot-isostatic pressing (HIP) [11], and the transport J_c reached

54 kA cm^{-2} and 47 kA cm^{-2} at 4.2 K under 100 kOe in (Ba,Na)Fe_2As_2 and (Ba,K)Fe_2As_2 round wires [12]. Better texturing of grains in the sample is another key factor to improve J_c value. In the well-textured (Ba,K)Fe_2As_2 tapes, the transport J_c value has reached 150 kA cm^{-2} at 4.2 K under 100 kOe [13], which is much larger than that in round wires. For the bulks used as trapped-field magnets, when the dimensions are fixed, the maximum trapped fields in them are proportional to their J_c value. Thus, a high-pressure sintering and a better texturing are demanded.

In this study, we fabricated two (Ba,Na)Fe_2As_2 bulks via different sintering methods. First, a bulk was sintered by a cubic press which could provide an isotropic pressure, although its dimension is limited. Second, a larger bulk was sintered via hot pressing with pressure only along the *c*-axis. The cubic press could provide a pressure higher than 2 GPa, while it is suitable only for small samples. The high pressure helps to prepare highly dense bulk material leading to a large J_c, although the isotropic pressure could not help to improve the texturing. On the other hand, hot pressing is expected to enhance the texturing and allows fabrication of larger bulks, although it is not safe to apply excessive pressure only in one direction. The magnetic J_c and trapped field in these two bulks are evaluated.

2. Experiments

Polycrystalline powder of (Ba,Na)Fe_2As_2 was synthesized by a mechanochemical reaction. Ba, Na, Fe, and As were mixed with a ratio of 0.6:0.48:2:2.1 in a ball-milling machine for ten hours. The mixture was finally heated up to 850 °C and kept for 30 hours in a Nb tube. In the case of bulk #1, powder was pressed into a Ag tube with an outer diameter (OD) of 5 mm and an inner diameter (ID) of 4 mm in a glovebox filled with Ar gas. A Ag disc was finally pressed into the tube for sealing the material to avoid possible reaction between the superconductor core and the air during the sintering. A metallic mold with a circular hole was used to avoid the deformation during the preliminary pressing. The sealed tube was put into a carbon heater by isolating it with a boron nitride sleeve and further inserted into a pyrophyllite block and sintered under ~ 2.2 GPa, 850 K for 1 hour in the cubic press. On the other hand, in the case of bulk #2, the powder was put in a double-sheathed tube with a Cu tube (OD: 11 mm and ID: 10 mm) and a Ag tube (OD: 10 mm and ID: 9.2 mm) with the bottom made of Cu and Ag discs in the glovebox. The tube was then put into a metallic mold with a circular hole and sealed by Ag and Cu discs. Finally, the sample tube together with the metallic mold was set in a small vertical furnace and pressed by a hydraulic unit under ~250 MPa, ~700 °C for 1 hour. To avoid oxidation of metallic parts during the heating process, nitrogen gas was flown.

After sintering, the core was exposed by mechanically removing the sealing discs, and it was polished by fine lapping films to check the quality of the surface of the bulk. Dimensions of the bulks are summarized in Table 1. The cross sections of these bulks are shown in figures. 1 (a) and (b). Figure 1 (c) shows the surface of bulk #1 under a microscope. With only small and isolated voids on the surface, it is concluded that bulk #1 experienced a successful sintering process. However, for bulk #2, there is a crack on its surface, suggesting that the sintering process shall still be improved.

The core of the bulk was observed by scanning electron microscope (SEM, S-4300, Hitachi High Technologies). The quality of the sample was evaluated by X-ray diffractions (XRD) and magnetization measurements using a superconducting quantum interference device magnetometer (MPMS-5XL, Quantum Design). The texturing of the sample was also evaluated by XRD. Measurements of local magnetic induction were performed using a GaAs Hall probe (THS126, Toshiba) by physical property measurement system (PPMS, Quantum Design).

Figure 1. Optical photos of the cross sections of (a) bulk #1 and (b) bulk #2. The red circle in (b) marks the crack on the surface of bulk #2. (c) Optical micrograph of the surface of bulk #1.

Table 1. Dimensions of our 2 bulks.

Parameters	Bulk #1	Bulk #2
Diameter of the core	4.1 mm	9.45 mm
Thickness	3.64 mm	4.58 mm

3. Results and discussion

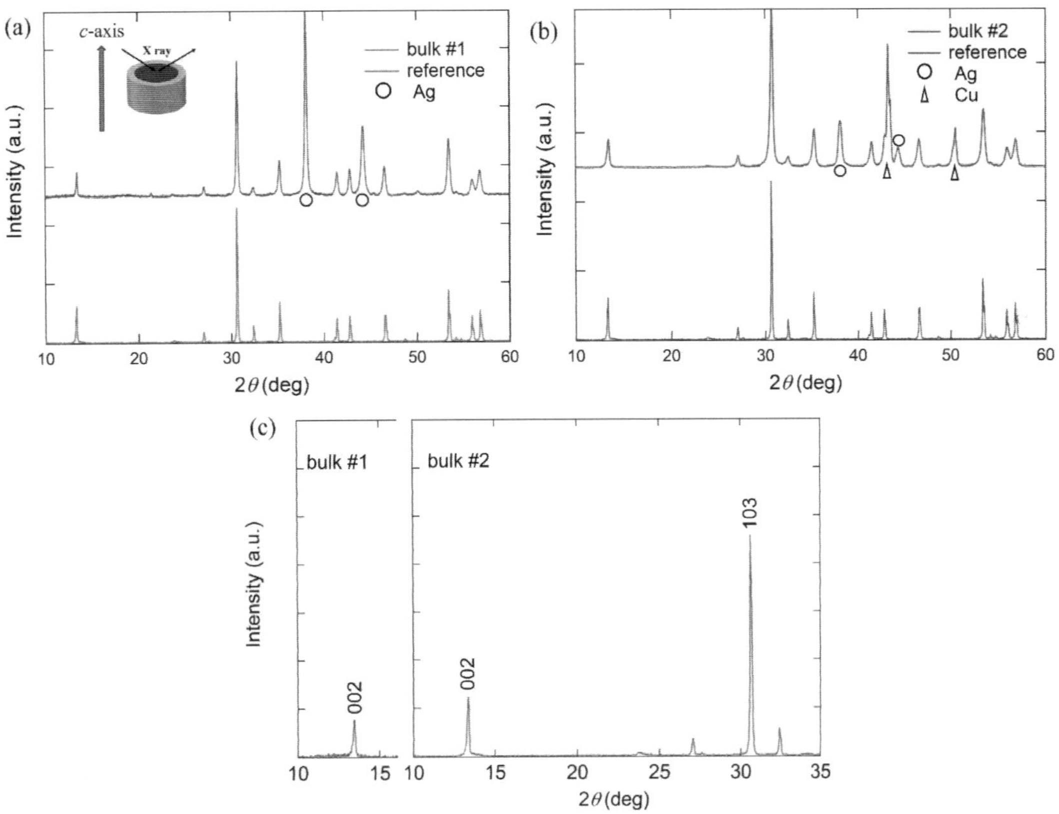

Figure 2. XRD patterns for (a) bulk #1 and (b) bulk #2. (c) XRD patterns of (002) and (103) peaks for the cross sections of bulk #1 and #2 normalized by the intensity of the intensity of (103) peaks.

Figures 2 (a) and (b) show the XRD patterns of bulk #1 and #2, respectively, measured as shown in the inset of figure 2 (a). Except for peaks from Cu and Ag sheaths, there are no obvious peaks from impurities (e.g., FeAs), which suggests that both cubic press and cold pressing could successfully seal the tubes and there was no severe decomposition during the sintering processes. Figure 2 (c) shows the XRD patterns of (002) and (103) peaks for the cross sections of bulk #1 and #2 normalized by the intensity of the (103) peaks. A better c-axis texturing, which results in larger intensity of (002) peak, was expected in bulk #2. However, figure 2 (c) shows that degrees of texturing in bulk #1 and #2 are comparable. It indicates that the texturing in bulk #2 is not along *c*-axis as expected and the hot pressing was not conducted as we anticipated. We will discuss this point later.

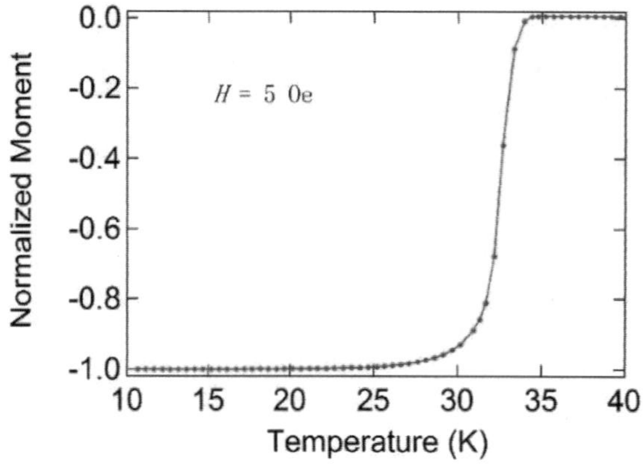

Figure 3. Temperature dependence of normalized magnetization at 5 Oe for bulk #1.

T_c is another important value to evaluate the quality of the sample and the temperature dependence of the normalized magnetization at 5 Oe for bulk #1 is shown in figure 3. Because bulk #2 is too large to put into the MPMS, its T_c was not measured. The magnetization curve shows a sharp transition at around $T_c \sim 34$ K, which is slightly higher than that in the round wires and could match the value in powders [11]. Combining the results of XRD and magnetization measurements, it is concluded that for both bulks there was no decomposition during the sintering and the qualities of these bulks are good.

Figures 4 (a) and (b) show the measured trapped field by a Hall probe set at a position 0.5 mm away from the bottom of bulk #1 and #2 at 4.2 K after zero-field cooling, respectively. After temperature was reduced to 4.2 K, an external field was ramped up to H_{max} (H_{max} = 50 kOe for bulk #1 and 80 kOe for bulk #2) and then decreased back to zero at a rate of 10 kOe min^{-1}. No flux jumps were observed at this ramp rate. The measured trapped fields are 4.67 kG and 5.85 kG for bulk #1 and #2, respectively. Note that there is a gap between the Hall probe and the bottom of the bulk. The magnetic induction generated by a cylindrical bulk superconductor at a finite position on the central axis can be calculated by the following formula;

$$B = \frac{\mu_0 J_c}{2}\left[(z+t)\ln\frac{\frac{d}{2}+\sqrt{\left(\frac{d}{2}\right)^2+(z+t)^2}}{z+t} - z\ln\frac{\frac{d}{2}+\sqrt{\left(\frac{d}{2}\right)^2+z^2}}{z}\right], \qquad (1)$$

where d and t are the diameter and thickness of the bulk and z is the distance between the probe position and the bottom of the sample [14]. The J_c values of the bulks can be calculated by setting the value of B in the formula as the measured induction and z = 0.5 mm. Thus estimated J_c values are 81.2 kA cm^{-2} and 33.9 kA cm^{-2} in bulk #1 and #2, respectively. The value of trapped field on the surface of the bulk is evaluated by introducing the calculated J_c value into the formula and by setting $z = 0$. Tapped field values on the surface are 9.97 kG and 8.81 kG for bulk #1 and #2, respectively. These values are listed in Table 2 together with the value from the previous study [10] for comparison. The bulk #1 sintered under a high pressure of 2.2 GPa shows a large J_c and traps comparable magnetic field to that reported in the previous study on (Ba,K)Fe_2As_2 bulk despite being nearly a half the diameter. This fact suggests that the improvement of J_c could make up for the negative influence of smaller dimensions.

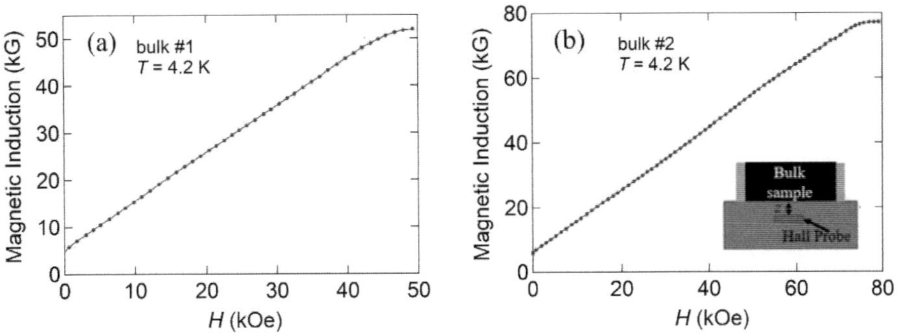

Figure 4. The measured magnetic induction by a Hall probe set at a position 0.5 mm away from the bottom of (a) bulk #1 and (b) bulk #2 at 4.2 K.

On the other hand, the J_c value of bulk #2 is lower than expected partly due to insufficient texturing. Application of uniaxial pressure at the final powder-filling process may not be enough to achieve sufficient texturing as seen in tapes [15]. Application of mechanical vibration during the powder packing into metallic sheath using ultrasonic may help to enhance the texturing. The fact that cracks were found on the surface of bulk #2 may suggest that pressure has not been applied properly along the intended direction. It may be necessary to use a proper tool to ensure that the force from the pressing piston is applied parallel to the axis of cylindrical bulk sample. Another possibility is that the temperature of the bulk sample may have been lower than the furnace temperature measured by the thermocouple placed close to the sample. Due to large thermal conduction of the base and pressing piston, the temperature of the bulk can be lower than the set temperature for the furnace. Higher set temperature for the sintering as well as higher pressure may be helpful to improve J_c and trapped field.

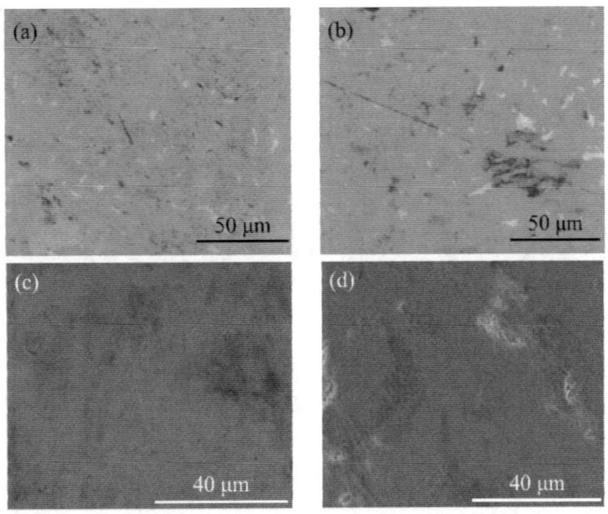

Figure 5. The photos taken with an optical microscope of the microstructure of the surface of (a) bulk #1 and (b) bulk #2 and the SEM images of (c) bulk #1 and (d) bulk #2.

Figures 5 (a)-(d) show the photos of the microstructure of bulk #1 and #2 taken with an optical microscope and SEM. It is obvious that the volume of impurities phases, which show up with brighter contrasts in the optical photos, is more in bulk #2. There are more voids in bulk #2 as seen in figure 5(d) compared with bulk #1. All these differences make intergranular connections in bulk #2 worse than that in bulk #1.

Table 2. Values of J_c and trapped field of bulk #1, bulk #2, and previous work.

Bulk	J_c	Trapped field
Bulk #1 (d = 4.1 mm) (T = 4.2 K)	81.2 kA cm^{-2}	9.97 kG
Bulk #2 (d = 9.45 mm) (T = 4.2 K)	33.9 kA cm^{-2}	8.81 kG
Previous work (d = 10 mm) (T = 5 K)	50 kA cm^{-2}	10.2 kG

4. Summary

Two bulk samples of (Ba,Na)Fe_2As_2 have been fabricated using a cubic press and hot-pressing technique. In bulk #1 with diameter of 4.2 mm sintered by a cubic press at the pressure of 2.2 GPa, a J_c value of 81.2 kA cm^{-2} and a trapped field value on the surface of 9.97 kG have been achieved. The value of the trapped field is comparable to that reported in the previous study despite the much smaller diameter. This fact suggests that the improvement of J_c could make up for the negative influence from the smaller dimensions. In bulk #2 sintered by hot pressing technique, both the J_c value and texturing are worse than expected, resulting in lower trapped field despite its larger diameter. Higher pressure, higher furnace temperature, proper application of pressure during the sintering as well as application of mechanical vibration during the powder packing may help to improve the trapped magnetic field.

References

[1] Kamihara Y, Watanabe T, Hirano M and Hosono H 2008 *J. Am. Chem. Soc.* **130** 3296-7
[2] Hosono H, Yamamoto A, Hiramatsu H and Ma Y 2018 *Mater. Today* **21** 278
[3] Ma Y 2012 *Supercond. Sci. Technol.* **25** 113001
[4] Shimoyama J 2014 *Supercond. Sci. Technol.* **27** 044002
[5] Ma Y 2015 *Physica C* **516** 17
[6] Pallecchi I, Eisterer M, Malagoli A and Putti M 2015 *Supercond. Sci. Technol.* **28** 114005
[7] Hosono H, Tanabe K, Takayam-Muromachi E, Kageyama H, Yamanaka S, Kumakura H, Nohara M, Hiramatsu H and Fujitsu S 2015 *Sci. Technol. Adv. Mater.* **16** 033503
[8] Yao C and Ma Y 2019 *Supercond. Sci. Technol.* **32** 023002
[9] Pyon S, Tamegai T, Takano K, Kajitani H, Koizumi N and Awaji S 2019 *J. Phys.: Conf. Ser.* **1293** 012042
[10] Weiss J D, Yamamoto A, Polyanskii A A, Richardson R B, Larbalestier D C and Hellstrom E E 2015 *Supercond. Sci. Technol.* **28** 112001
[11] Pyon S, Miyawaki D, Tamegai T, Awaji S, Kito H, Ishida S and Yoshida Y 2020 *Supercond. Sci. Technol.* **33** 065001
[12] Pyon S, Mori H, Tamegai T, Awaji S, Kito H, Ishida S, Yoshida Y, Kajitani H, Koizumi N, 2021 *Supercond. Sci. Technol.* **34** 105008
[13] Huang H, Yao C, Dong C, Zhang X, Wang D, Cheng Z, Li J, Awaji S, Wen H and Ma Y 2018 *Supercond. Sci. Technol.* **31** 015017
[14] Naito T, Sasaki T and Fujishiro H 2012 *Supercond. Sci. Technol.* **25** 095012
[15] Gao Z, Togano K, Matsumoto A and Kumakura H 2014 *Sci. Rep.* **4** 4065

Current–voltage characteristics of 100-pixel 200-μm-square superconducting tunnel junction X-ray detector array

***Shigetomo Shiki[1] and Go Fujii[1]**

1 Device Technology Research Institute, National Institute of Advanced Industrial Science and Technology, 1-1 Umezono, Tsukuba, Ibaraki 305-8568, Japan
s-shiki@aist.go.jp

Abstract. The superconducting tunnel junction (STJ) X-ray detector is an important tool for materials analysis because of its high energy resolution and high counting rate. In this work, the current–voltage (*I*–*V*) characteristics were investigated for a 100-pixel STJ array with a pixel size of 200 μm having a layer structure of Nb–Al/AlO_x/Al–Nb and working at 0.3 K. The fabrication yield of the 200-μm STJ reached approximately 90%. A series of *I*–*V* curves for a single pixel in different cooling runs showed a discrete excess current, possibly due to the trapping of single fluxoids. With a reduction of the ambient magnetic field during cooling, larger STJ pixels can be used.

1. Introduction

X-ray absorption spectroscopy (XAS) is a materials analysis tool that can characterize the electronic state and bond distance of a specific element; it is indispensable in the research and development of functional materials. Superconducting tunnel junctions (STJs) are promising for XAS because of their high energy resolution and high counting-rate capability [1-3]. We have developed an XAS apparatus that utilizes a 100-pixel array of 100-μm-square Nb-STJs; the array has a total effective area of 1 mm^2 working at 0.3 K [4]. Table 1 shows examples of energy-dispersive X-ray detector systems in XAS apparatus at various synchrotron sources. The advantage of using an STJ detector is that it can achieve both high resolution and a high count rate in the soft X-ray range. However, STJ detectors have a shortcoming of a small sensitive area. An STJ detector with a larger sensitive area is required for the realization of trace analysis in the soft X-ray region.

Table 1. Performance of energy-dispersive detectors in XAS apparatus at synchrotron sources.

	X-ray energy [keV]	count rate [count per sec]	energy resolution [FWHM eV]	area [mm^2]
STJ [4]	0.1–2	0.5 M	10 at 0.5 keV	1
TES* [5]	5–37	2 k	4	2.4
SDD** [6]	3–30	3.2 M	150 at 5.9 keV	9

* TES: Transition Edge Sensor microcalorimeter
** SDD: Silicon Drift Detector

Several efforts have been made to develop a large STJ detector. In a single-element study of a 20–500-μm square, the detector showed a subgap current due to trapping of fluxoids in a 500-μm STJ, although detailed magnetic shielding was used during cooling [7]. In mass spectrometer studies using 100-element 200-μm-square STJs, 85% of pixels showed a typical subgap current less than 1 μA [8].

Although the performance was sufficient for mass spectrometry, it was insufficient for X-ray spectroscopy, which requires high energy resolution and precise readout noise. To realize high-resolution X-ray spectrometry with an STJ array, it is necessary to reduce the size of an STJ pixel to 100-μm square [9].

The sensitive area of an STJ array X-ray detector can be improved by increasing the pixel size. However, no such attempts have thus far been reported. Therefore, in the present study, we investigated the current voltage (*I–V*) characteristics for a 200-μm 100-pixel STJ detector array. Because fluxoid trapping occurs stochastically, multiple measurements were performed to obtain *I–V* curves without random fluxoids trapping effects.

2. Experiment

A 100-pixel STJ array was fabricated with a pixel size of 200 μm and a layer structure of Nb (300 nm) – Al (70 nm) / AlO_x / Al (70 nm) – Nb (100 nm), as shown in figure 1(a), in the clean room for analog-digital superconductivity (CRAVITY) at the National Institute of Advanced Industrial Science and Technology (AIST) in 2015. The details of the fabrication and performance of a 100-μm-square STJ with a similar layer structure have been reported elsewhere [9].

The cryogenic setup is shown in figure 1(b). A ^{3}He cryostat was used to cool the STJ chip to 340 mK. The STJ was covered by a magnetic shield made of Permalloy PC placed at a position corresponding to 30 K. The shield reduces the geomagnetism perpendicular to the tunnel barrier to ~200 nT, which corresponds to ~4 Φ_0 in a 200-μm square, where Φ_0 is a single magnetic flux quantum. The magnetic shield was fixed using nylon posts to suppress the decrease in permeability of the

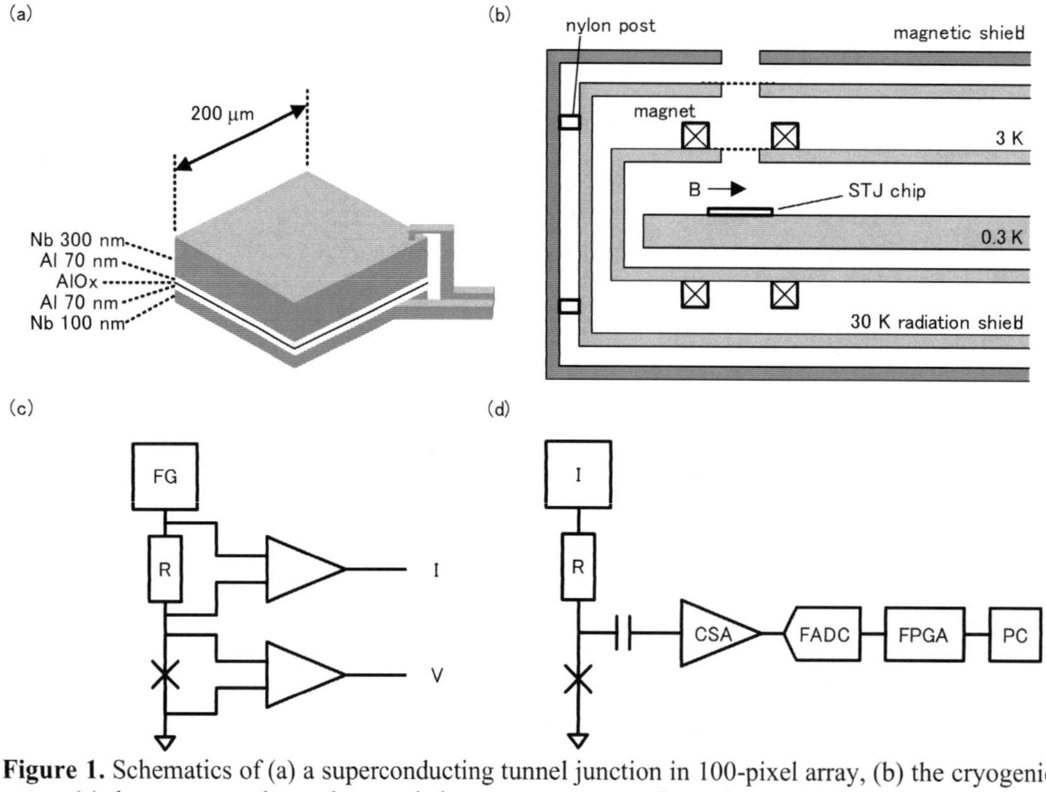

Figure 1. Schematics of (a) a superconducting tunnel junction in 100-pixel array, (b) the cryogenic setup, (c) the current–voltage characteristic measurement configuration, and (d) the X-ray response measurement configuration. The abbreviations FG, CSA, FADC, and FPGA are Function Generator, Charge Sense Amplifier, Flash Analog to Digital Converter, and Field Programmable Gate Array, respectively.

permalloy due to a temperature drop [10]. A superconducting magnet was placed in the magnetic shield to apply a magnetic field of ~10 mT in parallel with the tunnel barrier to reduce supercurrent and Fiske modes.

Figure 1(c) shows the setup used to measure the *I–V* characteristics. The subgap current is defined as the current at a voltage of 0.3 mV in the *I–V* curve. When the measurement of the *I–V* curve for 100 STJ pixels was completed, the applied magnetic field was removed and the temperature was raised above 10 K, which is higher than the critical temperature (T_c) of Nb (9.2 K), to release the trapped fluxoids. The cooldown and measurement were then repeated. A pixel with a sufficiently low subgap current value was selected, and its X-ray spectrum was recorded. Figure 1(d) shows the setup used for the X-ray performance measurement of the STJ pixel. The response of the STJ pixel was amplified using a charge-sensitive amplifier and acquired using a digital multichannel analyzer, which consisted of a 200-MHz analog-to-digital converter and a field-programmable gate array.

3. Results and Discussion

Figure 2(a) shows the *I–V* curves of an STJ pixel in 12 different cooling runs. There are two groups of *I-V* curves. One group exhibits a low subgap current. The *I–V* curves of the low-current group are similar to that of a previously reported 100-μm STJ that showed excellent X-ray detection properties [8]. In the other group, the curves show an excess current at a voltage greater than 0.1 mV. This discrete behavior suggests that the excess current in the *I–V* curve might be due to the trapping of a single fluxoid.

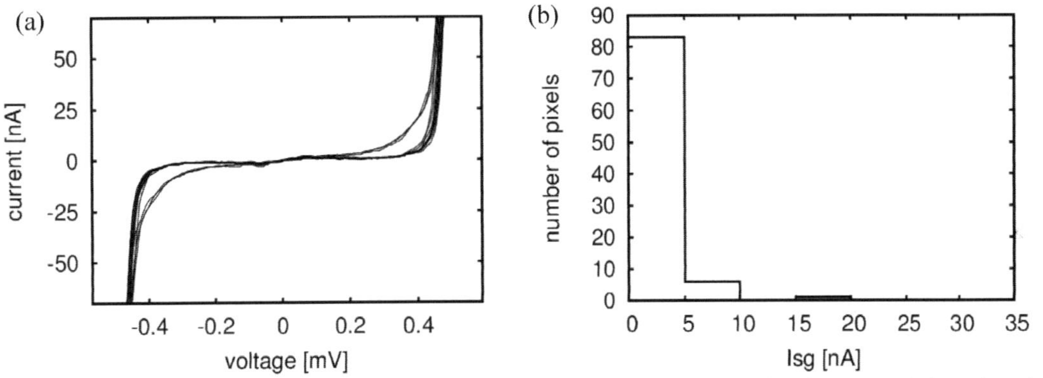

Figure 2. (a) Current–voltage curves of a 200-μm-square superconducting tunnel junction in 12 different cooling runs. (b) A histogram of the subgap current (Isg) at 0.3 mV in the 100-pixel 200-μm-square superconducting tunnel junction array.

The value of the subgap current was recorded for each curve and each pixel. A histogram of the lowest subgap current value in each pixel is shown in figure 2(b). The yield of the 200-μm STJ, which is defined by the portion of pixels having a subgap current less than 10 nA, is approximately 90%. The typical subgap current values in the present study are two orders of magnitude lower than those in a previous study [8].

Figure 3 shows the X-ray spectrum for the 1.5-keV Al-K line, as recorded using an STJ pixel. The energy resolution and readout noise are 25 eV and 10 eV in the full-width at half-

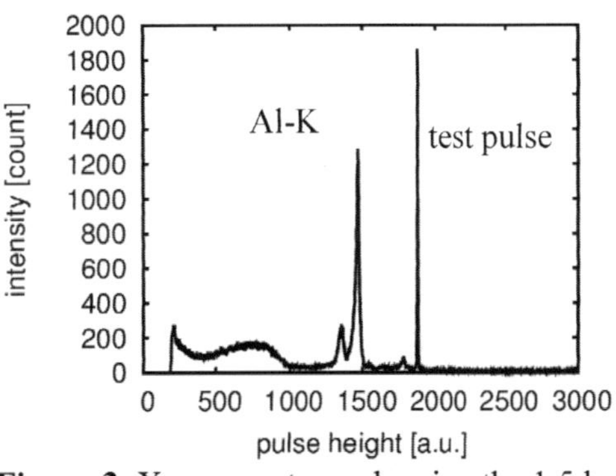

Figure 3. X-ray spectrum showing the 1.5 keV Al-K line measured using the 200-μm-square superconducting tunnel junction.

maximum, respectively. This energy resolution is comparable to that of a typical 100-μm STJ [4]. Further research is needed to evaluate the performance across the entire array. The entire array can be operated as an X-ray detector having a sensitive area of 4 mm^2, which is four times larger than the sensitive area of the detector reported in a previous study [4]. The sensitive area could be expanded if hybrid permalloy-based magnetic shields for use at low temperatures, such as Cryoperm10 or A4K, and a superconductor were used to reduce the ambient magnetic field during cooling [10-13].

4. Conclusions

The *I–V* characteristics of a 100-pixel STJ array with pixel size of 200 μm and a layer structure of Nb–Al/AlO_x/Al–Nb were investigated at 0.3 K. The fabrication yield of the 200-μm STJ reached 90%. STJ array detectors with a sensitive area of 4 mm^2 are expected to be available. The size of the sensitive area might be further improved by enhancing the magnetic shielding for STJs at cryogenic temperatures.

Acknowledgement

The devices were fabricated in the clean room for analog-digital superconductivity (CRAVITY) at the National Institute of Advanced Industrial Science and Technology (AIST).

References

[1] Kurakado M and Mazaki H 1981 *Nucl. Instr. and Meth.* **185**, 141-147
[2] Kurakado M, Takahashi T and Matsumura A 1993 *J Low Temp Phys* **93**, 567-572
[3] Friedrich S, Drury O B, Cramer S P and Green P G 2006 *Nucl. Instr. and Meth.* A **559,** 776–778
[4] Shiki S, Ukibe M, Matsubayasi N, Zen N, Koike M, Kitajima Y and Ohkubo M 2014 *J Low Temp Phys* **176**, 604–609
[5] Yamada S et al. 2021 *Rev. Sci. Instrum.* **92**, 013103
[6] Rachevski A et al. 2019 *Nucl. Inst. and Meth. Phys. Res.* A **936**, 719-721
[7] Sato H et al. 2000 *Jpn. J. Appl. Phys.* **39**, 5090-5094
[8] Ukibe M, Kushino A, Chen Y and Ohkubo M 2007 *X-ray spectrometry* **36**, 260
[9] Ukibe M, Shiki S, Kitajima Y and Ohkubo M 2012 *Jpn. J. Appl. Phys.* **51** 010115
[10] Tsuchiya K et. al 2006 *Proc. 2006 European Particle Accelerator Conference*, 505-507
[11] *ColdFacts* 2010 **26-4**, 30
[12] Cabrera B 1975 PhD thesis Stanford University
[13] Antoine R et. Al 2019 *J. Low Temp. Phys.* **194** 433–442

Superconducting neutron transmission imaging for investigating a sequential change in phase separations of low-melting Wood's metal

The Dang Vu[1,2,*], Hiroaki Shishido[3], Kazuya Aizawa[2], Takayuki Oku[2], Kenichi Oikawa[2], Masahide Harada[2], Kenji M. Kojima[4], Shigeyuki Miyajima[5], Kazuhiko Soyama[2], Tomio Koyama[1], Mutsuo Hidaka[6], Soh Y. Suzuki[7], Manobu M. Tanaka[8], Masahiko Machida[9], Shuichi Kawamata[1] and Takekazu Ishida[1]

[1]Division of Quantum and Radiation Engineering, Osaka Metropolitan University, Sakai, Osaka 599-8570, Japan

[2]Materials and Life Science Division, J-PARC Center, Japan Atomic Energy Agency, Tokai, Ibaraki 319-1195, Japan

[3]Department of Physics and Electronics, Graduate School of Engineering, Osaka Metropolitan University, Sakai, Osaka 599-8531, Japan

[4]Centre for Molecular and Materials Science, TRIUMF, 4004 Wesbrook Mall, Vancouver, BC V6T 2A3, Canada

[5]Advanced ICT Research Institute, National Institute of Information and Communications Technology, 588-2 Iwaoka, Nishi-ku, Kobe, Hyogo 651-2492, Japan

[6]Advanced Industrial Science and Technology, Tsukuba, Ibaraki 305-8568, Japan

[7]Computing Research Center, Applied Research Laboratory, High Energy Accelerator Research Organization (KEK), Tsukuba, Ibaraki 305-0801, Japan

[8]Institute of Particle and Nuclear Studies, High Energy Accelerator Research Organization (KEK), Tsukuba, Ibaraki 305-0801, Japan

[9]Center for Computational Science & e-Systems, Japan Atomic Energy Agency, 178-4-4 Wakashiba, Kashiwa, Chiba 277-0871, Japan.

*vuthedang@omu.ac.jp

Abstract. We built a new cryostat system for a current-biased kinetic inductance detector (CB-KID) to be able to place a sample at room temperature for conducting the neutron transmission imaging experiments with pulsed neutrons. A Wood's metal alloy sample (Bi 50wt.%, Pb 25wt.%, Sn 12.5wt.%, Cd 12.5wt.%) of melting temperature 75.2°C was used to check a practical test by observing fine microstructures under beam power of 812 kW at beamline BL10 of J-PARC (MLF). The Wood's metal is composed of four phases, of which one is a Cd-rich needle-like phase of an average width of 25 μm and length of 5 mm. Since Cd is a strong neutron absorber, it is suitable for observing the fine mosaic structure contrastingly in neutron transmission imaging. After neutron-transmission imaging with the Wood's metal in the initial state, we subsequently melted the Wood's metal sample and solidified it again by slow cooling during the same beam time. We were successful in observing an impressive change in morphology of the phases by neutron-transmission imaging. Room-temperature sample imaging is very convenient for possible users, and is expected to be very useful for applying our CB-KID system to versatile different materials of interests.

1. Introduction

As opposed to the X-rays, neutrons directly interact to nuclei [1]. Therefore, it is useful for revealing structure and dynamics at different-length and different-time scales to elucidate deeper insight on microscopic nature of condensed-matter systems. The neutron imaging technique has intensively been developed as a non-destructive and non-invasive technique [2] in combination with the X-ray imaging [3]. It became a powerful tool to clarify the characteristics of any interesting materials. The advance of scattering techniques, with the aid of the development of intense neutron sources and sophisticated instruments [4], made an energy-dispersive analysis more attractive in investigating various material properties over different fields, such as physics, biology, chemistry, metallurgy, geology, and engineering. In recent decades, a neutron-imaging technology has been developed rather rapidly to achieve a high-resolution imaging, an energy-dispersive imaging, and a high-speed readout system.

Some of typical neutron imaging systems can be itemized as follows: (1) A gaseous detector is one of the most popular detectors and is known to have a very high detection efficiency up to 54% and a large sensitive area. However, the detection efficiency would be deteriorated down to 1% when a resolution of 0.1 mm is pursued by downsizing to a small gaseous detector [5]. (2) A scintillation detector is one of the most versatile-purpose systems, of which the best spatial resolution was reported as 2 μm while a detection efficiency remains to be lower than 1% [6]. In addition, it is not perfectly suitable for the instrumentation under pulsed neutrons. The high detection efficiency was recorded as 43.5% with the ZnS:Ag/6LiF powder scintillator [7]. (3) A semiconductor detector or a solid-state detector has an advantage in its small size and a low operating voltage, and hence is profitable when it is used for time-of-flight measurements under pulsed neutrons. A high spatial resolution of 15 μm was achieved by using a CMOS imaging sensor covered with $NaGdF_4$ nanoparticle scintillators [8]. However, their detection efficiency is known as low as 2% [9]. (4) One of the outstanding detectors is known as a ^{10}B-doped multichannel plate (MCP) because it has a high spatial resolution of 15 μm at low count rates and a moderate resolution of 55 μm at high count rates. Nonetheless, it also has a high detection efficiency of 43% for cold neutrons and a moderate efficiency of 16% for thermal neutrons [10].

Our group proposed a current-biased kinetic inductance detector (CB-KID) as an unconventional neutron detector [11]. Regarding the operating principle, it is based on a rapid reduction in the local Cooper pair density n_S at a spot of the nanowire stripline with a very short length $\Delta\ell$ ($\ll$ total length ℓ) when a neutron reacts with a ^{10}B nucleus in the conversion layer. Under a small DC bias current I_b, a pair of pulsed-voltage signals is produced at the hot spot in the stripline of the detector by a sudden change in kinetic inductance. Those pulsed-voltage signals start to propagate toward both end-electrodes with opposite polarity as an electromagnetic-wave pulse [12,13]. The nanowire meanderline-stripline works for the two signals as not only a low-loss impedance-matched waveguide but also as an indispensable part of a delay-line detection system for signals. Characteristics of our superconducting neutron detector have been systematically investigated to improve the spatial resolution and the neutron detection efficiency of CB-KID in our preceding works [14,15]. A spatial resolution of neutron imaging was achieved down to 16 μm and a detection efficiency was 1% at 25 meV [16,17]. We already demonstrated that the most-visible Bragg edge of Fe sample can be discerned even with a very narrow sampling area of 3.1× 2.3 μm^2 in the neutron transmission spectrum with wavelength resolution of 0.0007 nm [18]. We confirmed that the energy resolution is good enough to obtain the resonance absorption dips to reveal mosaic crystalline structure appeared in Sm-Sn mixed metal [19]. In order to check the practical usefulness of the CB-KID imaging, we demonstrated a good linearity between the Gd-island sizes obtained from neutron transmission images and those obtained from SEM images over a wide length range from 20 μm to 130 μm [20].

In the present work, we intend to improve the effective easiness of using our superconducting neutron detector. For that purpose, we built a new cryostat system so as to be able to mount a room-temperature sample for conducting the transmission experiments under pulsed neutrons. This is in marked contrast with a constraint that the former cryostat required to mount a sample at the vicinity of the cryogenic detector in vacuum. It is noteworthy to mention that one can perform a very high-resolution imaging when a sample is placed at the close vicinity of the CB-KID detector at a cryogenic temperature even when one uses the room-temperature cryostat.

We chose the Wood's metal to test the new-cryostat system by placing the sample at room temperature. The Wood's metal is one of the best practical materials for testing because one of the four phases in Wood's metal is a fine needle-like Cd-rich phase, which is easily observed by the neutron transmission imaging. It is also easy to mount a sample or to remove a sample from the cryostat even in the same beam time. We heated the sample to melt and to re-solidify the low-melting-temperature alloy during the beam time to change sample morphology. Details of our observations are reported in this paper. A flexible planning of installing sample for conducting imaging would open wider application fields with our high spatial resolution CB-KID system.

2. Experimental details

2.1. *Superconducting Neutron Sensor*

We designed a superconducting neutron sensor using a computer-aided design software. The structure of the superconducting neutron sensor consists of a pair of meanderlines with an orthogonal stacking of X and Y sensors [11,21]. A ^{10}B layer deposited on top of the X and Y meanderlines is to convert neutrons into charged particles for making it easier to sense an electroneutral particle. The nanowire width of the X (or Y) Nb meanderline was 900 nm, and the space between two neighbouring nanowire segments was 600 nm. These values gave a 1.5-μm repetition period with 10,000 repetitions to cover an effective sensing area of 15×15 mm^2 and a total length ℓ of 151 m of the meanderline. In summary, the CB-KID sensor was prepared on a thermally oxidized Si substrate as eight-layered stackings from the bottom: (1) a 300-nm-thick SiO_2, (2) a 300-nm-thick Nb ground plane, (3) a 350-nm thick SiO_2 layer, (4) a 50-nm-thick Nb Y meanderline, (5) a 150-nm-thick SiO_2 layer, (6) a 50-nm-thick Nb X meanderline, (7) a 150-nm-thick SiO_2 layer, and (8) a 220-nm ^{10}B neutron conversion layer at the top.

We requested the clean room for analog-digital superconductivity (CRAVITY) at the National Institute of Advanced Industrial Science and Technology (AIST) to fabricate our superconducting devices. The characteristics of the sensor were systematically investigated to optimize operation conditions, to improve the spatial resolution and the detection efficiency, and to conduct check test of sensors in previous studies [14,15].

2.2. *Measurement System*

Figure 1 presents the block diagram for the CB-KID measurement system. We built the new cryostat with two concaved Al-alloy windows at the upstream-side of neutron beams while flat flanges were used in the present study in the downstream side (see figure 1(c)). This is to reduce a distance between a sample and a CB-KID sensor to minimize the blurring due to the beam divergence. Figure 1(b) is a three-leg sample supporting tool, of which an Al-alloy (A2017) sample substrate with the Wood's metal (Wood#1) on the sample substrate (see figure 1(a)). We chose Al-alloy in fabricating the window because it affects almost no effects in conducting the neutron transmission imaging over a wide range of wavelength. The inner concaved window has an inner diameter of 54 mm, which is large enough to accommodate the three-leg tool at a distance between the sample and the detector as close as 20 mm.

The superconducting CB-KID sensor was cooled down to an operating temperature 5.0 K. The neutron sensor was installed on the sensor stage attached to the bottom side of the anti-vibration second stage of a Gifford–McMahon (GM) refrigerator. The temperature of the anti-vibration second stage was controlled at 4.7 K while the superconducting sensor stage was controlled at 5.0 K by using a high precision AC bridge temperature controller (Cryocon Model 54). A temperature difference of 0.3 K is in accordance with moderate thermal link between the second stage and the detector stage. This instrument has two-channel PID controllers with two electric resistance heaters and two Cernox thermometers at the neutron-sensor stage and the low-vibration cold stage (see figure 1(d)). Two-step control of the temperature is to suppress the unavoidable effect of temperature breathing in the GM cryocooler. We performed long-time transmission measurements with CB-KID, of which the temperature was controlled at the extremely stable temperature as $T = 5.00000 \pm 0.00008$ K by optimizing PID parameters of the two-control loops. This is crucial to stabilize a propagation velocity of a signal, for example, along the X meanderline of with the error of ±160 m/s (0.0002%) in the CB-KID sensor because it is relevant to realize a high spatial resolution as far as we employ a delay-line technique in conducting imaging. Semirigid cables (50 Ω) with microminiature coaxial (MMCX) connectors and subminiature version A (SMA) connectors were used to transmit a high-frequency signal pulse from the cryogenic temperature to the room temperature in the inside of the cryostat to reduce noises.

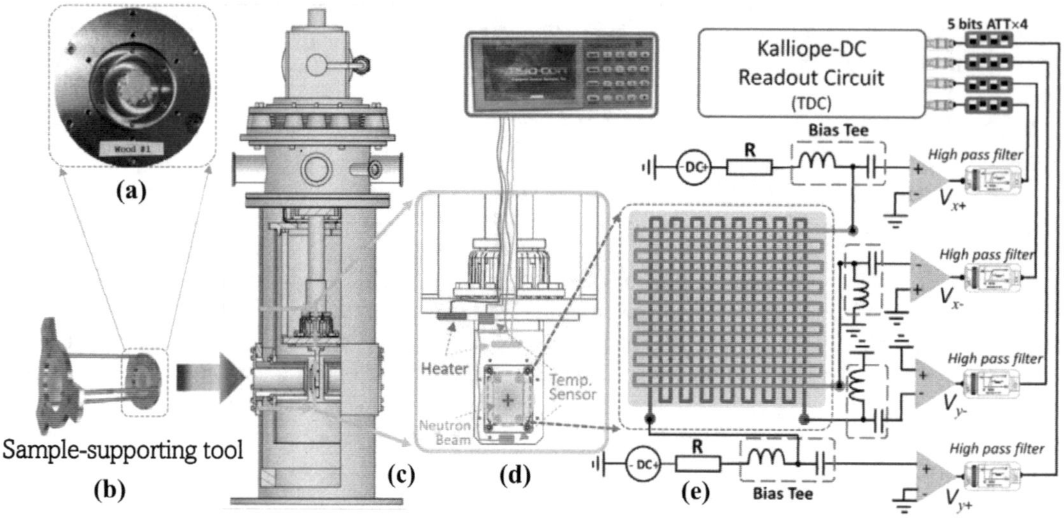

Figure 1. The measurement system of the superconducting neutron imaging system: **(a)** A sample-holder plate made of Al alloy (A2017). **(b)** A sample-supporting tool with three legs and the sample-holder plate. **(c)** The cryostat with a concaved window, in which the sample-supporting tool can be inserted to place samples at a 20-mm distance from the CB-KID sensor at the upstream side of pulsed neutrons (see arrow). The detector was cooled down to a cryogenic temperature using a Gifford–McMahon cryocooler. **(d)** The temperature sensors and the heaters are installed to control a temperature very precisely. The anti-vibration second-cold stage and the detector stage were controlled at the two different temperatures by a temperature controller (Cryocon 54). **(e)** Measurement diagram: The neutron detector consists of the X and Y-meanderlines and a ^{10}B neutron conversion layer. Neutron signals arising from both ends of the two meanderlines were amplified by ultra-low noise amplifiers to input to a Kalliope-DC readout circuit via high-pass filters (HPF) and adjustable 5-bit attenuators (ATT). The system was controlled by two computers.

The cryostat was placed in the inside of the irradiation room. The X/Y meanderlines of the superconducting neutron detector need to be biased by two direct current (DC) bias currents, which were supplied from two DC adjustable stable voltage sources via two 10-kΩ bias resistors. Four ultra-low-noise amplifiers (NF SA-430F5) were used to amplify voltage signals caused by passage of a charged particle produced by a neutron nuclear reaction in the conversion layer across the meanderline

The amplified signals were transmitted from the irradiation room to the outside measurement room through 20-m long 50 Ω coaxial cables, where we used four high pass filters (Thorlabs EF505 with a cut-off frequency 130 kHz) in series connection to suppress noticeable low-frequency noises, which are typical in the large-scale accelerator facility. Four 50 Ω splitters were used to input signals both to a 2.5-GHz digital oscilloscope (Teledyne LeCroy HDO8108A) and to a 32-channel time-to-digital converter (TDC) module of the 1-GHz Kalliope-DC readout circuit [22]. We also used two power oscillators (NF EC1000SA) to provide the AC 100 V powers of 60 Hz to our measurement instruments instead of commertial AC 100 V power supply of 50 Hz from the facility.

It is an easy-to-use merit of our CB-KID system so that the amplitude of the CB-KID signal produced by neutron irradiation can easily be tuned by changing the amplitudes of the bias currents in feeding to the X or Y meanderline. Thus, we achieved the best measurement conditions for our readout circuit under irradiation of neutron beams to the superconducting neutron sensor at a chosen operating temperature. The signal amplitudes of the four channels were adjusted so as to be suitable for a threshold setting of reading timestamps by TDC of the Kalliope-DC readout circuit. Since a setting resolution of the threshold value to read a timestamp is not so high enough, we also used 5-bit variable attenuators (ATT, Hoshin Electronics N032) of 50 Ω impedance for fine tuning of the input-signal amplitudes.

Neutron imaging experiment was performed under proton beam power of 812 kW at BL10 of MLF (J-PARC) to obtain the transmission neutron images of the test samples. A beam collimator ratio value *L/D* was appropriately chosen by considering the purpose of each measurement.

2.3. Sample preparation

We used a commercial Wood's metal alloy in shots (FUJIFILM Wako Pure Chemical Corporation, Bi 50wt.%, Pb 25wt.%, Sn 12.5wt.%, Cd 12.5wt.%, a melting temperature of 75.2°C). The Wood's metal with the same composition was investigated by Kamioka and Kataoka [23]. Multi-component eutectic system tends to have different solidification temperatures. They reported the appearance of four-different solid-state phases [23]. Phase I (solidifying at 74.5°C) represents Cd-rich needles or dendrites of fraction of 6.2%. Phase I is supposed to have the width 22 μm and the length 5 mm in average. Phase II (solidifying at 72.8°C) represents Sn-rich particles of fraction of 10.3%. Phase II has a diameter 25.2 μm in average. Phase III (solidifying at 71.8°C) Bi-rich particles of fraction of 22.1%. Phase III has a diameter 39.1μm in average. Phase IV (solidifying at 71.4°C) is supposed to be matrix with eutectic structure of fraction 61.8%. Phase IV has a small diameter smaller than 0.01μm. Upon cooling, phase I appears at the highest temperature, and is very suitable for neutron transmission imaging because the Cd is a good neutron absorber to observe it contrastingly compared to other phases [23–25].

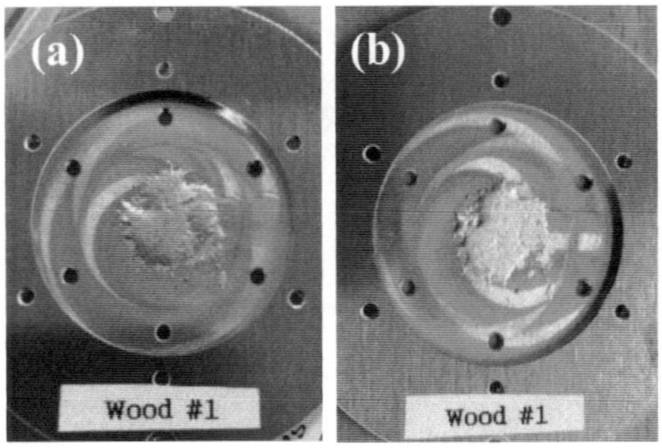

Figure 2. (a) Wood's metal alloy sample (Wood#1) was used in its initial state. **(b)** Sample (Wood#1) was melted again in a draft chamber and was solidified again by slow cooing. Outer line of the wetted islands was kept as almost the same by this process.

A shot of 0.1457 g was melted on the Al-alloy (A2017) substrate and was cooled slowly in air. We found that the wetting of the Wood's metal is good enough to the Al-alloy substrate without using any flux after trials of several times (see figure 2(a)). We denote this sample as Wood#1. Since an inner circle of the substrate has a diameter of 30 mm and a thickness of 1 mm, one can evaluate an approximate thickness by assuming a density of ~9.7 g/cm^3 if necessary.

The initial state of Wood#1 was examined by neutron-transmission imaging with a high collimator ratio (*L/D* = 1863) for 28 h. With the use of the new cryostat, we can remove (mount, replace) the sample during the same beam time. The Wood#1 sample was removed from the cryostat. The sample was heated to make it in a melting state in a draft chamber for safety. After heating to 105°C, the Wood #1 sample was cooled slowly down to 50°C. Figure 2(b) shows a photo of Wood's metal alloy sample (Wood#1) in the re-solidified state. We found that an outer contour is almost the same with that of the Wood #1 in the initial state (see figure 2).

We conducted the neutron transmission imaging of the re-solidified Wood#1 for 24 h with *L/D* = 1863. As discussed in the next section, we can see a drastic change between the initial state and the re-solidified state while the sample contour looks similar to each other.

3. Results and discussions

3.1. Neutron transmission images

Figure 3(a) shows the neutron transmission image of Wood's metal alloy sample (Wood#1) in the initial state taken for 28 h. At the bottom-left corner of figure 3(a), the fine needle-like patterns are visibly interspersed. Oppositely, we notice that there are not only dense particles at the top-right area but also tiny fine needles at the surrounding areas. The size of needles is the minimum width of 25 um (see the inset of figure 3(a) for zoomed image) and the length up to 4.5 mm. In our previous work [20], we identified very fine needle patterns of 10-μm width using the Wood's metal sample while the image was obtained by placing a sample at the cryogenic temperature. This is to avoid the blurring of the image to

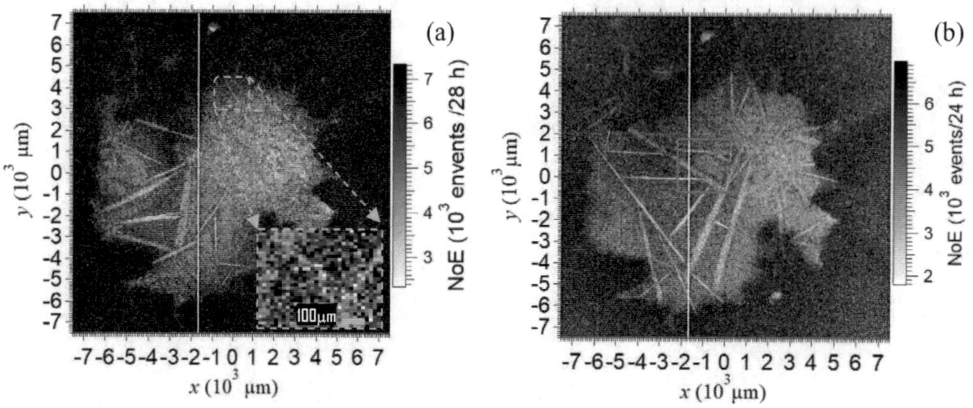

Figure 3. Neutron-transmission images of Wood's metal alloy sample. **(a)** Wood#1 in the initial state and **(b)** Wood#1 in the re-solidified state. Neutron images were constructed by neutron energies from 0.5 meV to 100 keV.

placing the sample at the very vicinity of the neutron detector. We found that it was possible to confirm the fine needle-like structures by using a room-temperature sample with our new cryostat. This indicates the usefulness of the present cryostat system. We note that a vertical bright line at x =1638.4 μm is not intrinsic, but comes from a defect of the detector. While the image of figure 3(a) is not quite homogeneous over the sample, we consider that this comes from the trial processes of heating and cooling in sample preparation.

It is an advantage of the present cryostat system that we can replace the sample very quickly during the same beam time. After removing the Wood#1 sample from the cryostat, we heated to melt the Wood's metal sample on the substrate and cooled it again to re-solidify for stimulating a formation of a different morphology of the sample. We conducted the neutron transmission imaging for 24 h with proton beam power of 812 kW at the operating temperature of 5 K. The image consists of a pixel size of 13×16 μm^2 (in average), which was processed by combining four pixels of 3×3.75 μm^2. The neutron transmission imaging of the Wood's metal was done by another group [26]. Present work revealed the images of the Wood's-metal phases more clearly as reported by Kamioka and Kataoka [23].

3.2. Comparison with the SEM and EDX images

Figure 4(a) is the magnified image of neutron transmission image of red-rectangle box in figure 3(b) of the Wood's metal in re-solidified state. This was compared with the image taken by a scanning electron microscope (SEM) and the image taken by the energy-dispersive X-ray spectroscopy (EDX). We found that the neutron transmission image provided much more details of the microstructures of sample than others. The observation is not only from the sample surface but also from buried internal structural components under the surface. For instance, all the needle-like patterns seen in SEM and EDX images well appeared in neutron transmission image, too. The width of the needles in SEM and EDX tends to be much smaller than that in the neutron transmission image. Some needles in figure 4(a) cannot be seen in figure 4(b) (compare green dashed rectangles to each other), because they were covered with Pb and Bi overlayers. A small dark island appearing in SEM image in an orange-dashed line figure 4(b) cannot be found in figure 4(a). This area is supposed as a eutectic phase consisting of Cd and Sn as was revealed by the EDX analyses in figure 4(c), (d), (e), and (f)). However, a small dark area in an orange-dashed line in SEM image of figure 4(b), of which the corresponding area in the neutron transmission image is not visible compared to surrounding areas (in a yellow dashed line) in figure 4(a). The yellow area is considered as an aggregate of very fine patterns under the surface. The results implicate that the neutron

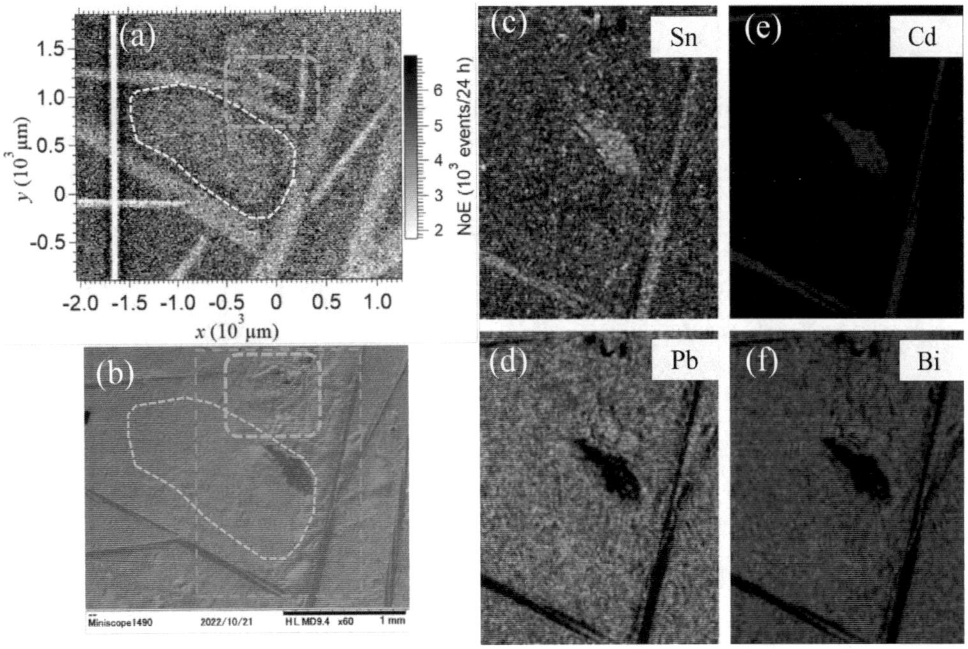

Figure 4. The comparison of scanning electron microscope (SEM) image, energy-dispersive X-ray spectroscopy (EDX) image and neutron transmission image; (a) Magnification neutron-transmission image of Wood's metal alloy corresponds to red-rectangle box of figure 3(b); (b) SEM image of Wood's metal alloy sample at the same area of neutron image in figure 3(a); (c) EDX image of Sn, (d) EDX image of Pb, (e) EDX image of Cd, and (f) EDX image of Bi for the blue-dashed line rectangle in SEM image of figure 3(b).

transmission image has an advantage compared with the SEM and EDX techniques, i.e., neutrons can investigate not only the surface property but also the detailed internal structures of materials.

We found the appearance of visible many needles of phase-I with different orientations of morphology in the Wood's metal as a consequence of solidification process in a low-melting temperature enuretic alloy. The new cryostat has much more space for sample mounting than the cryogenic sample system. This can be used for sample rotation for three-dimensional tomography in the future. We consider that the Wood's metal is a good candidate for demonstrating the three-dimensional imaging of randomly-oriented fine Cd needles with a computed tomography.

4. Summary

We built a new cryostat of the superconducting neutron imaging system to be able to mount a room-temperature sample for imaging with the superconducting CB-KID detector. Practical usefulness of the system has been demonstrated to image a microscopic morphology of the low-melting-temperature metal (the Wood's metal) placed in the outside of the cryostat at room temperature. One of the four phases is a Cd-rich fine needle, which was clearly observed with the new cryostat system. We successfully observed an impressive change in fine morphology of the phases after re-solidification of the Wood's metal sample within the same beam time. The neutron transmission image was compared with the images taken by SEM and EDX method. Thanks to the new cryostat, the CB-KID method becomes worth utilizing various purposes in material sciences.

Acknowledgments
This work is partially supported by Grant-in-Aid for Scientific Research (A) (No. JP21H04666) and Grant-in-Aid for Early-Career Scientists (No. JP21K14566) from JSPS. The neutron irradiation experiments at the Materials and Life Science Experimental Facility (MLF) of the J- PARC were conducted under the support of MLF project programs (No. 2021P0501, No. 2022A0198).

References
[1] Tengattini A, Lenoir N, Andò E and Viggiani G 2021 Neutron imaging for geomechanics: A review *Geomechanics for Energy and the Environment* **27** 100206–19
[2] Kardjilov N, Manke I, Woracek R, Hilger A and Banhart J 2018 Advances in neutron imaging *Materials Today* **21** 652–72
[3] Tengattini A, Lenoir N, Andò E, Giroud B, Atkins D, Beaucour J and Viggiani G 2020 NeXT-Grenoble, the Neutron and X-ray tomograph in Grenoble *Nucl Instrum Methods Phys Res A* **968** 163939–11
[4] Taylor A, Dunne M, Bennington S, Ansell S, Gardner I, Norreys P, Broome T, Findlay D and Nelmes R 2007 A Route to the Brightest Possible Neutron Source? *Science* **315** 1092-1095
[5] Bateman J E, Rhodes N J and Stephenson R 2002 A high-resolution, imaging neutron beam monitor using gas microstrip technology *Nuclear Instruments and Methods in Physics Research A* **477** 365–371
[6] Matsubayashi M, Faenov A, Pikuz T, Fukuda Y and Kato Y 2010 Neutron imaging of micron-size structures by color center formation in LiF crystals *Nucl Instrum Methods Phys Res A* **622** 637–641
[7] Kojima T, Katagiri M, Tsutsui N, Imai K, Matsubayashi M and Sakasai K 2004 Neutron scintillators with high detection efficiency *Nucl Instrum Methods Phys Res A* **529** 325–8
[8] Pérez M, Martínez E D, Lipovetzky J, Marín J, Sofo Haro M, Alcalde Bessia F, Gómez Berisso M and Blostein J J 2021 High spatial resolution neutron detection technique based on Commercial Off-The-Shelf CMOS image sensors covered with NaGdF_4 nanoparticles *Journal of Instrumentation* **16** P08008
[9] Barbagallo M, *et al.,* 2013 Thermal neutron detection using a silicon pad detector and ^{6}LiF removable converters *Review of Scientific Instruments* **84** 033503–7
[10] Tremsin A S, McPhate J B, Vallerga J v., Siegmund O H W, Hull J S, Feller W B and Lehmann E 2009 Detection efficiency, spatial and timing resolution of thermal and cold neutron counting MCP detectors *Nucl Instrum Methods Phys Res A* **604** 140–3
[11] Ishida T, *et al.,* 2008 Superconducting MgB_2 thin film detector for neutrons *J Low Temp Phys* **151** 1074–9
[12] Koyama T and Ishida T 2018 Electrodynamic theory for the operation principle of a superconducting kinetic inductance stripline detector *J Phys Conf Ser* **1054** 012055
[13] Swihart J C 1961 Field solution for a thin-film superconducting strip transmission line *J Appl Phys* **32** 461–9
[14] Vu T D, *et al.,* 2019 Temperature dependent characteristics of neutron signals from a current-biased Nb nanowire detector with ^{10}B converter *J Phys Conf Ser* **1293** 012051
[15] Vu T D, *et al.*, 2020 Kinetic inductance neutron detector operated at near critical temperature *J Phys Conf Ser* **1590** 012036
[16] Shishido H, Miki Y, Yamaguchi H, Iizawa Y, The Dang V, Kojima K M, Koyama T, Oikawa K, Harada M, Miyajima S, Hidaka M, Oku T, Soyama K, Suzuki S Y and Ishida T 2018 High-Speed Neutron Imaging Using a Current-Biased Delay-Line Detector of Kinetic Inductance *Phys Rev Appl* **10** 044044
[17] Iizawa Y, Shishido H, Nishimura K, Vu T D, Kojima K M, Koyama T, Oikawa K, Harada M, Miyajima S, Hidaka M, Oku T, Soyama K, Aizawa K, Suzuki S Y and Ishida T 2019 Energy-resolved neutron imaging with high spatial resolution using a superconducting delay-line kinetic inductance detector *Supercond Sci Technol* **32** 125009

[18] Vu T D, *et al.,* 2022 Narrow-area Bragg-edge transmission of iron samples using superconducting neutron sensor *J Phys Conf Ser* **2323** 012028

[19] Shishido H, *et al.,* 2022 Neutron Imaging toward Epithermal Regime using a Delay Line Current-Biased Kinetic-Inductance Detector *J Phys Conf Ser* **2323** 12029

[20] Vu T D, *et al.,* 2021 Practical tests of neutron transmission imaging with a superconducting kinetic-inductance sensor *Nucl Instrum Methods Phys Res A* **1006** 165411

[21] Shishido H, *et al.,* 2017 Neutron detection using the superconducting Nb-based current-biased kinetic inductance detector *Supercond Sci Technol* **30** 094003

[22] Kojima K M, Murakami T, Takahashi Y, Lee H, Suzuki S Y, Koda A, Yamauchi I, Miyazaki M, Hiraishi M, Okabe H, Takeshita S, Kadono R, Ito T, Higemoto W, Kanda S, Fukao Y, Saito N, Saito M, Ikeno M, Uchida T and Tanaka M M 2014 New μsR spectrometer at J-PARC MUSE based on Kalliope detectors *J Phys Conf Ser* **551** 012063

[23] Kamioka H and Kataoka A 1997 Properties of Elastic Waves in a Polycrystalline Solid during Melting and Solidification Processes (I)-Dilatational Wave in Wood's Alloy 地震 **2** 463–75

[24] Saiki Y, Kubo T, Kuwabara M and Yang J 2006 Acoustic-cavitation-based production of foamed metallic material *Japanese Journal of Applied Physics, Part 1: Regular Papers and Short Notes and Review Papers* **45** 4793–9

[25] Belchukl M, Waw D and Dryden J 1990 Determining a Constitutive Equation for Creep of a Wood's Metal Model Material *19th Canadian Fracture Conference, Ottawa, Ontario* pp 189–95

[26] Trtik P and Lehmann E H 2016 Progress in High-resolution Neutron Imaging at the Paul Scherrer Institut-The Neutron Microscope Project *J Phys Conf Ser* **746** 012004.

Crystalline Orientation of CaF_2 window determined by Neutron Transmission Imaging using a Delay Line Current-Biased Kinetic-Inductance Detector

Hiroaki Shishido[1,2], The Dang Vu[3,4], Kazuya Aizawa[4], Kenji M. Kojima[3,5], Tomio Koyama[3], Kenichi Oikawa[4], Masahide Harada[4], Takayuki Oku[4], Kazuhiko Soyama[4], Shigeyuki Miyajima[6], Mutsuo Hidaka[7], Soh Y. Suzuki[8], Manobu M. Tanaka[9], Shuichi Kawamata[3] and Takekazu Ishida[3]

[1]Department of Physics and Electronics, Graduate School of Engineering, Osaka Metropolitan University, Sakai, Osaka 599-8531, Japan
[2]The Center for Research and Innovation in Electronic Functional Materials, Osaka Metropolitan University, Sakai, Osaka 599-8570, Japan
[3]Division of Quantum and Radiation Engineering, Osaka Metropolitan University, Sakai, Osaka 599-8570, Japan
[4]Materials and Life Science Division, J-PARC Center, Japan Atomic Energy Agency, Tokai, Ibaraki 319-1195, Japan
[5]Center for Molecular and Materials Science, TRIUMF and Stewart Blusson Quantum Matter Institute, University of British Columbia, Vancouver, BC, V6T 2A3 and V6T 1Z4, Canada
[6]Advanced ICT Research Institute, National Institute of Information and Communications Technology, 588-2 Iwaoka, Nishi-ku, Kobe, Hyogo 651-2492, Japan
[7]Advanced Industrial Science and Technology, Tsukuba, Ibaraki 305-8568, Japan
[8]Computing Research Center, Applied Research Laboratory, High Energy Accelerator Research Organization (KEK), Tsukuba, Ibaraki 305-0801, Japan
[9]Institute of Particle and Nuclear Studies, High Energy Accelerator Research Organization (KEK), Tsukuba, Ibaraki 305-0801, Japan

E-mail: shishido@omu.ac.jp

Abstract. The combination of a high speed two-dimensional neutron detector and an intensive pulsed neutron source provides not only neutron transmission imaging but also information on crystal structures, orientations and constituent elements by analyzing neutron transmission spectra. The delay-line current-biased kinetic-inductance detector (CB-KID) is a two-dimensional superconducting neutron detector with high spatial and temporal resolutions, and multi-hit tolerance. We demonstrated that the delay-line CB-KID with a ^{10}B neutron conversion layer can be applied for neutron transmission measurements up to 100 keV. We observed two-different dip structures in the transmission spectra in a CaF_2 single crystal, i.e., one is from the Bragg diffraction and the other is from nuclear resonance absorption. We consider that some duplicated structures in resonance absorption dips are originating from a double-bunch structure of proton pulses for producing pulsed neutrons in the accelerator.

1. Introduction

Several superconducting detectors have well been established to exhibit high performance [1, 2, 4, 5]. High sensitivities of superconducting nanowire single photon detectors (SNSPD) [1] and transition edge sensors (TES) [2] are ensured by a steep resistance change at the superconducting transition. Based on the SNSPD, a two-dimensional single-photon imager is developed using the delay-line method [3]. The Cooper pair breaking and an associated change in kinetic inductance is applied to microwave kinetic inductance detectors (MKID) [4, 5], which can operate in the superconducting state with no bias current.

Superconducting detectors are applicable not only for the detection of electromagnetic waves but also for the detection of other quantum beam such as neutron beams [6, 7, 8]. Because of high permeability of neutrons, a neutron-conversion layer is employed to detect them by superconducting detectors. B layer and $Li_2B_4O_7$ single crystal were employed as neutron-capture layers in TES [6] and superconducting-tunnel-junction (STJ) [7, 8] neutron detectors, respectively.

We developed a superconducting neutron detector, named as the current-biased kinetic-inductance-detector (CB-KID), with a ^{10}B neutron-conversion layer [9, 10]. The CB-KID senses a time derivative of the kinetic inductance induced by a transient change in the Cooper pair density. Therefore, the CB-KID operates over a wide temperature range below a critical temperature T_c with high operating speed even at low bias currents.

In the pulsed-neutron source, a high temporal resolution of a detector can provide neutron energy resolution by using the time-of-flight (ToF) method. Recent developments of high intensity pulsed neutron sources accelerate intensive development for a detector with spatial and temporal resolutions, such as micro-channel-plate detector [11], Li-glass scintillator pixel-type detector [12], gas electron multiplier [13], and micro-pixel chamber-based neutron imaging detector [14].

We reported the CB-KID with a delay-line technique works as an energy-resolving two-dimensional neutron detector [15]. We achieved a high spatial resolution of 16 μm in neutron transmission imaging [16] and transmission measurements up to 10 eV [17], while neutron detection efficiency was estimated as 1% at 25 meV [18].

Neutron transmission exhibits dip structures induced by the nuclear resonance absorption [19] or the Bragg diffraction from single crystalline samples [20]. The former, being specific to nuclear species, behaves like a fingerprint. The later is induced by the Bragg reflection for each Miller index and can be related to information on the lattice constant, the crystal orientation, and the mosaicity of the crystals.

In the present work, we measured neutron transmission spectra and images of a CaF_2 single crystal with an unknown crystalline orientation by using the delay-line CB-KID. We observed Bragg-dips and assigned Miller indices assuming azimuthal and longitudinal angles of a CaF_2 crystal. Dip structures in transmission spectrum induced by nuclear resonance absorption of F nuclei in its natural abundance were observed.

2. Experimental methods

2.1. Delay-line current-biased kinetic inductance detector

Figure 1(a) schematically shows the measurement system of the delay-line CB-KID. The main part of CB-KID is fabricated on a Si substrate sequentially: Superconducting Nb ground plane, X- and Y-meander lines with 0.9 μm in width and 0.6 μm in spacing, and ^{10}B neutron conversion layer are stacked alternatively separated by SiO_2 insulating layers. X- and Y-meander lines with thickness of 50 nm are orthogonal to each other, and are folded 10,000 times for every segment length h = 15.1 mm, reaching total length l=151 m. Long total length allow us to consider meander lines not only as detection areas but also as delay lines for signal propagation. It was fabricated in a clean room for analog-digital superconductivity (CRAVITY) at the National

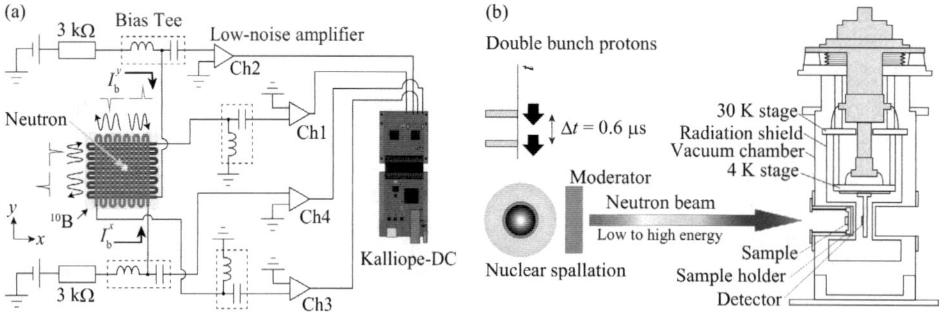

Figure 1. (a) Schematic illustration of the measurement system of the delay-line current-biased kinetic-inductance detector(CB-KID) and (b) the cryogenic system. Incident neutrons are captured at the ^{10}B layer, and pulse signals propagate toward both ends along X- and Y-meander lines. Signals from both ends of meander lines were processed by a time-to-digital converter in a Kalliope-DC readout circuit. Nuclear spallation of Hg nuclei induced by the collision of double-bunch protons emits intense neutron beams. Pulsed neutrons traveled from the source moderator through the beamline to irradiate a CaF_2 sample. The CB-KID was cooled down to 5.0 K by using a Gifford–McMahon refrigerator to keep it in a superconducting state.

Institute of Advanced Industrial Science and Technology (AIST). A ^{10}B layer with a thickness of 220 nm was deposited by an electron-beam deposition technique under an ultra-high vacuum.

Incident neutrons react with ^{10}B nuclei at the topmost layer of the detector and are converted into ^{7}Li particles and α particles. A part of kinetic energy of a charged particle is transferred to Cooper pairs in superconducting Nb meander lines at each local hot spot. Voltage pulses are generated by the combination of DC bias currents on meander lines and transient reductions of Cooper pair density at hot spots, and is propagate toward both ends through meander lines as electromagnetic-waves [21]. The DC bias currents for X and Y meander lines were fed at I_b^x=30 μA and I_b^y=40 μA.

Signals in channels 1 to 4 denoted by Ch1, Ch2, Ch3, and Ch4 were amplified by four differential ultra-low-noise amplifiers (SA-430 F5 by NF Corporation), and then were received by a time-to-digital converter (TDC) in the Kalliope-DC readout circuit, which has 1 ns-sampling multichannel TDC board [15, 22], and a 2.5 GHz sampling digital oscilloscope (Teledyne LeCroy HDO8108A).

The hot-spot position (x, y) is determined using a delay-line method as follows:

$$x = \text{ceil}\left[\frac{(t_{\text{Ch4}} - t_{\text{Ch3}})v_x}{2h}\right]p, \; y = \text{ceil}\left[\frac{(t_{\text{Ch2}} - t_{\text{Ch1}})v_y}{2h}\right]p. \quad (1)$$

Here, t_{Ch1}, t_{Ch2}, t_{Ch3}, and t_{Ch4} are signal timestamps received in Ch1, Ch2, Ch3, and Ch4 (see Fig. 1), respectively; $p = 1.5\,\mu$m is a repetition pitch; $v_x = 8.148 \times 10^7$ m/s and $v_y = 6.477 \times 10^7$ m/s are signal propagation velocities for X and Y meander lines at $T = 5.0$ K, respectively.

When multiple events occur simultaneously in the CB-KID device, one should identify a true signal quartet to determine the hot spot position. Coincidence check is possible by comparing event occurrence times $t_0^x(t_0^y)$ in the $X(Y)$ meandler line, where $t_0^x(t_0^y)$ can be estimated from a set of t_{Ch3} and t_{Ch4}(t_{Ch1} and t_{Ch2}) independently as follows:

$$t_0^x = \frac{t_{\text{Ch3}} + t_{\text{Ch4}}}{2} - \frac{l}{2v_x}, \; t_0^y = \frac{t_{\text{Ch1}} + t_{\text{Ch2}}}{2} - \frac{l}{2v_y}. \quad (2)$$

One can conclude that a selected signal quartet originates from the same event when t_0^x and t_0^y are the same within a temporal resolution. Thus, the delay-line CB-KID has a multi hit tolerance. More details are described in previous studies [15].

2.2. Cryogenic system for CB-KID

The cryogenic system is schematically illustrated in Fig. 1(b). The delay-line CB-KID was cooled down below superconducting transition temperature of Nb by using a Gifford–McMahon (GM) refrigerator and was held in vacuum in an aluminum-alloy vacuum can. The detector temperature was controlled at 5.0 K using a temperature controller (Model 54 by Cryogenic Control Systems Inc.).

There is no rigid contact between the GM refrigerator and detector to reduce mechanical vibration of the detector. Both the vacuum chamber and radiation shield concave in front of and behind the detector. The CaF_2 test sample was inserted inside of the pocket at room temperature in air and was kept at a distance of 21 mm from the detector. It was covered by 60 μm thick aluminum foil to fix on a 1 mm thick sample holder made of an aluminum alloy. In this cryogenic refrigerator system, the sample can be exchanged by keeping the detector in a superconducting state because a sample is placed outside of the vacuum chamber. Neutron beams were irradiated to the Si substrate side of the detector after passing through the sample.

2.3. Neutron imaging with pulsed neutrons

Neutron irradiation experiments were carried out at BL10 of the Material and Life science experimental Facility (MLF), J-PARC [23]. Pulsed neutrons are created by the nuclear spallation of Hg nuclei by collision with protons. Two proton bunches collide Hg nuclei with a time difference of 0.6 μs every 40 ms. It is so-called double-bunch structure in proton beams. Emerging neutrons are moderated at the moderator and travel from the moderator to the detector through the beamline with the length L=14 m at BL10. The facility supplies trigger signals in synchronous to nuclear spallation, while those contain a possible delay time t_{delay}. Time of flight (ToF) for the beamline is proportional to the neutron wavelength λ as

$$\lambda = \frac{2\pi\hbar}{m_{\mathrm{n}}L}(t - t_{\mathrm{delay}}), \tag{3}$$

where m_{n} is a neutron mass, $\hbar$ is the reduced Plank constant, and t is an experimental ToF based on the synchronized trigger. The neutron energy E is also represented as

$$E = \frac{m_{\mathrm{n}}}{2}\left(\frac{L}{t - t_{\mathrm{delay}}}\right)^2. \tag{4}$$

Experiments were performed with the neutron source size D = 100 mm, collimation ratio L/D=140 and neutron beam power of 711 kW.

3. Results and discussion

3.1. Neutron transmission image

Figure 2(a) shows an optical photograph of the test sample of an ultraviolet grade CaF_2 optical window with 15 mm$\phi\times$2 mm^{t}(Edmund Optics). Neutrons, so called open beam, were measured without a sample to characterize the detector. The obtained image with major pixel sizes of 7.5 μm$\times$6 μm or 9 μm$\times$6 μm is shown in Fig. 2(b). Here detection events corresponding to the incident neutron energy range of $0.6 \leq E \leq 300$ meV were summed up to construct the image. The image contains spatial distribution of the detection efficiencies, neutron flux and other characteristics of the detector. The number of events at edge region is slightly lower than that

Journal of Physics: Conference Series **2545** (2023) 012020 doi:10.1088/1742-6596/2545/1/012020

around the center region. A fine grid pattern results from non-uniform pixel sizes of the image. In addition, two pairs of diagonal lines and a line missing at around x = -1.65 mm are appeared. These may be due to defects or crosstalks between X and Y meander lines.

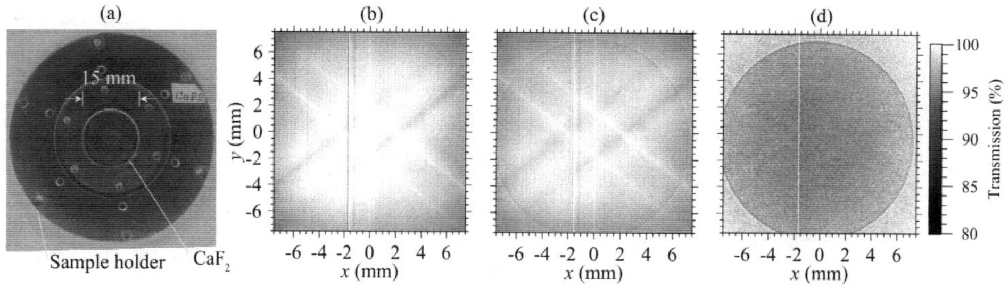

Figure 2. (a) The optical photograph of the CaF_2 sample on the holder. (b)Neutron transmission image without sample, (c) transmission image with the CaF_2 sample, and (d) normalized transmission image of the CaF_2 sample. Neutron transmission images are constructed by neutrons in the energy range of $0.6 \leq E \leq 300\,\text{meV}$.

Figure 2(c) represents a neutron transmission image of the CaF_2 sample in the neutron energy range of $0.6 \leq E \leq 300\,\text{meV}$. In addition to the same pattern observed in Fig. 2(b), a circular edge of the CaF_2 sample is clearly seen. We divided the image shown in Fig. 2(c) by that shown in Fig. 2(b) to normalize patterns from the detector characteristics and neutron flux distribution. The resulting image is shown in Fig. 2(d). Not only a fine grid pattern but also diagonal lines are removed, while a line missing still remains. The circular edge of CaF_2 sample is clearly seen and the number of events inside of the CaF_2 sample is reduced compared with outside of the sample.

3.2. Assignment of Bragg-dip indices

We divided the sum of NoEs over the entire CaF_2 area in Fig. 2(c) by NoEs those over the same area in Fig. 2(b) and calculated the neutron transmission considering several normalization factors. Many Bragg dips appear in the neutron transmission spectrum of the CaF_2 sample, and are assigned by Miller indices as shown in Fig. 3(a). Here we assume that all observed Bragg

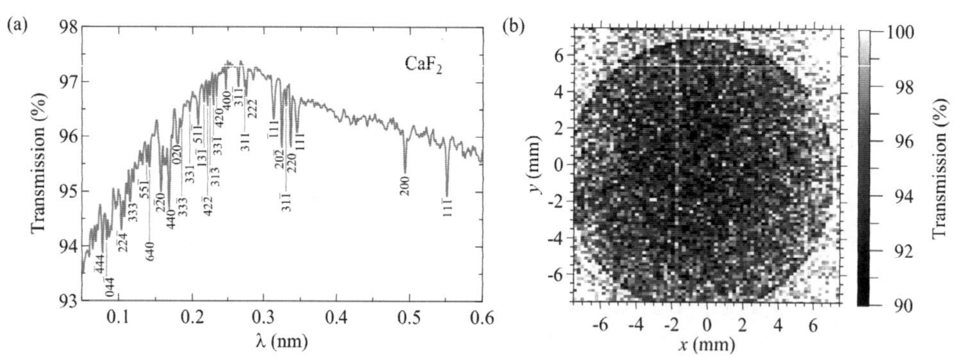

Figure 3. (a) The neutron transmission spectrum for CaF_2 as a function of wavelength λ and (b) neutron transmission image with the wavelength range of $0.549 \leq \lambda \leq 0.555\,\text{nm}$.

dips come from one single crystal of CaF_2 and the orientation of the reciprocal lattice unit vector $\boldsymbol{a}^*$ is rotated in the azimuthal angle from the opposite direction of incident neutron beams by ϕ=19.2 degrees and in the longitudinal angle from the vertical line by ψ=162.7 degrees. CaF_2 has a CaF_2-type face-centered cubic structure, and observed Bragg dips follow the extinction rule. All expected Bragg dips are observed above $\lambda > 0.3$ nm. Below $\lambda < 0.3$ nm, some Bragg dips are hard to discriminate each other because they are very close to each other.

To assign Bragg dips, first we assumed two trial Bragg dips at longer λ for Miller indices 200 and 220. Second we calculate ϕ and ψ from the 200 and 220 Bragg dip positions. Third we calculate Bragg-dip positions based on ϕ and ψ by using following equations. Finally, we compare experimental and simulated Bragg dips to confirm agreements. If agreements are not satisfactory, the assumption of 200-dip and/or 220-dip assignments was changed to other two dips to repeat the calculation procedure.

The reciprocal vector $\boldsymbol{G}(h, k, l)$ in the cubic crystal structure is expressed as follows:

$$\boldsymbol{G}(h,k,l) = \frac{1}{a}(h\cos\psi\cos\phi - k\sin\phi + l\sin\psi\cos\phi,\, h\cos\psi\sin\phi + k\cos\phi + l\sin\psi\sin\phi,\, -h\sin\psi + l\cos\psi),$$

where h, k, and l are Miller indices, and a is the lattice constant. The inner product between the inverse of the incident vector $-\boldsymbol{k}$=(-1/λ, 0, 0) and $\boldsymbol{G}$ is represented as follows:

$$-\boldsymbol{G}\cdot\boldsymbol{k} = \frac{-h\cos\psi\cos\phi + k\sin\phi - l\sin\psi\sin\phi}{a\lambda} = \frac{\sqrt{h^2+k^2+l^2}}{a\lambda}\cos\left(\frac{\pi}{2}-\theta\right).$$

Thus,

$$\sin\theta = \frac{-h\cos\psi\cos\phi + k\sin\phi - l\sin\psi\cos\phi}{\sqrt{h^2+k^2+l^2}}. \tag{5}$$

According to the Bragg's law, a Bragg dip position at wavelength $\lambda(h,k,l)$ in the cubic structure is represented as

$$\lambda(h,k,l) = \frac{2a\sin\theta}{\sqrt{h^2+k^2+l^2}}. \tag{6}$$

By substituting $\sin\theta$ of Eqn. (5) into Eqn. (6), we obtain:

$$\lambda(h,k,l) = \frac{2a(-h\cos\psi\cos\phi + k\sin\phi - l\sin\psi\cos\phi)}{h^2+k^2+l^2}. \tag{7}$$

Then, we obtain the formula of $\lambda(2,0,0)$ and $\lambda(2,2,0)$ from Eqn. (7) as follows:

$$\begin{cases} \lambda(2,0,0) = -2a\cos\psi\cos\phi \\ \lambda(2,2,0) = a(-\cos\psi\cos\phi + \sin\phi). \end{cases} \tag{8}$$

One can determine ψ and ϕ by solving simultaneous Eqns. (8) as follows:

$$\phi = \arcsin\left[\frac{2\lambda(2,2,0) - \lambda(2,0,0)}{a}\right], \tag{9}$$

$$\psi = \arccos\left[-\frac{\lambda(2,0,0)}{a\cos\phi}\right]. \tag{10}$$

The wavelengths $\lambda(h,k,l)$ for other indices can be calculated by substituting the ψ and ϕ values thus determined from Eqn. (9) and Eqn. (10) into Eqn. (7).

Figure 3(b) shows a neutron transmission image with the wavelength range of $0.549 \leq \lambda \leq 0.555$ nm, which comprehend the $11\bar{1}$ dip, and the major pixel size as $202.5 \times 160.5\,\mu\text{m}^2$. The uniform CaF_2 sample image is clearly seen. It indicates that the entire sample is a single crystal and the crystal orientation is the same throughout the sample.

3.3. Resonance absorptions of F nuclei

The neutron transmission spectrum of CaF_2 up to 100 keV summing over the whole sample area is shown in Fig. 4(a). The solid line represents a simulated one of CaF_2 containing all natural Ca and F nuclei [24]. Two pairs of dips appears as indicated by red and blue arrows in the experimental spectrum (see both Fig. 4(a) and (b)). Double bunch in nuclear spallation is responsible for the appearance of the duplicated resonance dips. A pair of dips represented by red and blue arrows appears at 600 ns interval, and hence corresponds to the primary and the secondly proton-beam pulses causing nuclear-spallation reactions, respectively. We consider that a pair of nuclear-resonance-absorption dips at higher neutron energies is produced by F nuclei and double-bunch neutrons.

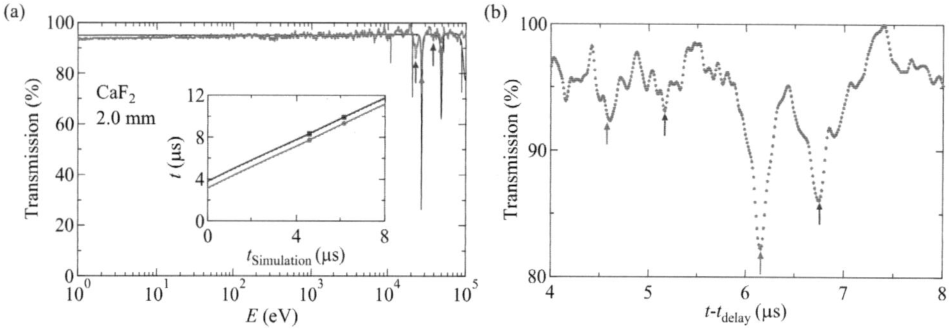

Figure 4. Neutron transmissions as a function of (a) neutron energy E and (b) calibrated time of flight. The solid line in (a) represents a simulated transmission spectrum. The inset in (a) shows comparison between ToF of detected dips and those evaluated the resonance energies from the database [24].

To confirm this consideration, the inset of Fig. 4(a) shows the comparison between ToF of detected dips and those cited from the database [24]. Linear fitting coefficients for each pair of dips are very close to one as expected from Eqn. (3): 1.00 and 0.99 for red circles and blue squares, respectively. Differences in ToFs between dips indicated by red and blue arrows are 0.6 μs, as same as the expected time delay between the primary and secondly nuclear spallation reaction (see Fig. 4(b)). E was calibrated according to Eqn. (4) with $t_{\text{delay}} = 3.16\,\mu$s obtained from the linear fitting for red circles, and the resulting energies at dips are in good agreements with simulated ones shown by the solid line in Fig. 4(a).

4. Summary

We demonstrated that the delay-line CB-KID can be used for Bragg-dip analysis and neutron transmission measurements with slow neutrons. All observed Bragg-dips are well assigned by Bragg reflections of a CaF_2 single crystal. Neutron transmission spectra exhibit dip structures induced by nuclear resonance absorption between F nuclei and neutrons. Two pairs of dips may correspond to double bunch in nuclear spallation, in which the primary and the secondly nuclear spallation reactions occur with a 600 ns interval at J-PARC. The delay-line CB-KID has a potential for applying the neutron transmission analysis and imaging down to slow neutron region.

Acknowledgments

This work is partially supported by Grant-in-Aid for Scientific Research (Nos. JP16H02450, JP21H04666, JP21K14566) from JSPS. The neutron irradiation experiments at the Materials and Life Science Experimental Facility (MLF) of the J-PARC were conducted under the support of MLF project program (No. 2021P0501).

References

[1] Natarajan C M, Tanner M G and Hadfield R H 2012 *Supercond. Sci. Technol.* **25** 063001
[2] Irwin K D 1995 *Appl. Phys. Lett.* **66** 1998
[3] Zhao Q-Y, Zhu D, Calandri N, Dane A E, McCaughan A N, Bellei F, Wang H-Z, Santavicca D F and Berggren K K 2018 *Nat. Photonics* **11** 247
[4] Day P K, LeDuc H G, Mazin B A, Vayonakis A and Zmuidzinas J 2003 *Nature* **425** 817
[5] Zmuidzinas J 2012 *Annu. Rev. Condens. Matter Phys.* **3** 169
[6] Merlo V, Salvato M, Cirillo M, Lucci M, Ottaviani I, Scherillo A, Celentano G and Pietropaolo A 2015 *Appl. Phys. Lett.* **106** 113502
[7] Nakamura T, Katagiri M, Ukibe M, Ikeuchi T and Ohkubo M 2004 *Nucl. Instrum. Methods Phys. Res. A* **520** 67
[8] Nakamura T, Katagiri M, Chen Y E, Ukibe M and Ohkubo M 2006 *Nucl. Instrum. Methods Phys. Res. A* **559** 766
[9] Ishida T, Yoshioka N, Narukami Y, Shishido H, Miyajima S, Fujimaki A, Miki S, Wang Z and Hidaka M 2014 *J Low Temp Phys* **176** 216
[10] Shishido H, Miyajima S, Narukami Y, Oikawa K, Harada M, Oku T, Arai M, Hidaka M, Fujimaki A and Ishida T 2015 *Appl. Phys. Lett.* **107** 232601
[11] Tremsin A S, McPhate J B, Kockelmann W, Vallerga J V, Siegmund O H W, Feller W B, 2011 *Nucl. Instrum. Methods Phys. Res. A* **633** S235
[12] Sato H, Takada O, Satoh S, Kamiyama T, Kiyanagi Y, 2010 *Nucl. Instrum. Methods A* **623** 597
[13] Uno S, Uchida T, Sekimoto M, Murakami T, Miyama K, Shoji M, Nakano E, Koike T, Morita K, Satoh H, Kamiyama T, Kiyanagi Y, 2012 *Phys. Procedia* **26** 142
[14] Parker J D, Harada M, Hayashida H, Hiroi K, Kai T, Matsumoto Y, Oikawa K, Segawa M, Shinohara T, Su T, Takada A, Zhang S, Tanimori T, Kiyanagi Y, 2016 In Proceedings of the IEEE Nuclear Science Symposium, Medical Imaging Conference and Room-Temperature Semiconductor Detector Workshop (NSS/MIC/RTSD), Strasbourg, France, 29 October-6 November
[15] Shishido H, Miki Y, Yamaguchi H, Iizawa Y, Vu T D, Kojima K M, T. Koyama, Oikawa K, Harada M, Miyajima S, Hidaka M, Oku T, Soyama K, Aizawa K, Suzuki S Y and Ishida T 2018 *Phys. Rev. Appl.* **10** 044044
[16] Iizawa Y, Shishido H, Nishimura K, Vu T D, Kojima K M, Koyama T, Oikawa K, Harada M, Miyajima S, Hidaka M, Oku T, Soyama K, Aizawa K, Suzuki S Y and Ishida T 2019 *Supercond. Sci. Technol.* **32** 125009
[17] Shishido H, Nishimura K, Vu T D, Aizawa K, Kojima K M, Koyama T, Oikawa K, Harada M, Oku T, Soyama K, Miyajima S, Hidaka M, Suzuki S Y, Tanaka M M, Kawamata S and Ishida T 2021 *IEEE Trans. Appl. Supercond.* **31** 9
[18] Vu T D, Shishido H, Kojima K M, Koyama K, Oikawa K, Harada M, Miyajima S, Oku T, Soyama K, Aizawa K, Hidaka M, Suzuki S Y, Tanaka M M, Malins A, Machida M and Ishida T 2020 *Supercond. Sci. Technol.* **34** 015010
[19] Festa G, Cippo E P, Martino D D, Cattaneo R, Senesi R, Andreani C, Schooneveld E, Kockelmann W, Rhodes N, Scherillo A, Kudejova P, Biro K, Duzs K, Hajnalf Z and Gorini G 2015 *J. Anat. At. Spectrom.* **30** 745
[20] Malamud F, Santisteban J R, 2016*J. Appl. Cryst.* **49** 348
[21] Swihart J C 1961 *J. Appl. Phys.* **32** 461
[22] Kojima K M, Murakami T, Takahashi Y, Lee H, Suzuki S Y, Koda A, Yamauchi I, Miyazaki M, Hiraishi M, Okabe H, Takeshita S, Kadono R, Ito T, Higemoto W, Kanda S, Fukao Y, Saito N, Saito M, Ikeno M, Uchida T and Tanaka M M 2014 *J. Phys.:Conf. Ser.* **551** 012063
[23] Oikawa K, Maekawa F, Harada M, Kai T, Meigo S, Kasugai Y, Ooi M, Sakai K, Teshigawara M, Hasegawa S, Futakawa M, Ikeda Y and Watanabe N 2008 *Nucl. Instrum. Methods Phys. Res. A* **589** 310
[24] Shibata K, Iwamoto O, Nakagawa T, Iwamoto N, Ichihara A, Kunieda S, Chiba S, Furutaka K, Otuka N, Ohsawa T, Murata T, Matsunobu H, Zukeran A, Kamada S and Katakura J 2011 *J. Nucl. Sci. Technol.* **48** 1

Propagation in Superconducting Niobium Rectangular Waveguide in the 100 GHz band

Taku Nakajima[1], Kazuji Suzuki[1], Takafumi Kojima[2], Yoshinori Uzawa[2], Masayuki Ishino[3], and Issei Watanabe[4]

[1] Institute for Space-Earth Environmental Research, Nagoya University, Furo-cho, Chikusa-ku, Nagoya, Aichi 464-8601, Japan

[2] National Astronomical Observatory of Japan, 2-21-1, Osawa, Mitaka, Tokyo 181-8588, Japan

[3] Kawashima Manufacturing Co., Ltd., 1-3-5, Higashi-Ikuta, Tama-ku, Kawasaki, Kanagawa 214-0031, Japan

[4] National Institute of Information and Communications Technology, 4-2-1 Nukui-Kitamachi, Koganei, Tokyo 184-8795, Japan

E-mail: `nakajima@isee.nagoya-u.ac.jp`

Abstract. Rectangular waveguides that use superconducting materials for walls are expected to reduce transmission loss compared to normal metal waveguides. However, research on superconducting waveguides has been limited, particularly on the fabrication and measurement of physical models in the mm/sub-mm band. We attempted to fabricate a niobium (Nb) waveguide and measure its propagation. Because Nb is known to be a difficult-to-machine material, we searched for the best cutting tools and machining conditions. Consequently, we fabricated a waveguide with a surface accuracy of 0.3 μm rms using an end mill with a diameter of 1 mm. For the measurement of waveguide propagation, the resonator method was adopted, and the resonance characteristics were measured at room (298 K) and cryogenic (4.8 K) temperatures in the 100 GHz band. The resonance property at 298 K is consistent with the shapes predicted from literature-based conductivity. However, at 4.8 K, that of superconducting Nb is significantly changed. We used electromagnetic analysis to reproduce these properties, and the electrical conductivity and propagation loss were calculated. The conductivity and loss were computed as 1.8×10^{11} S/m and 0.05 dB/m, respectively. We observed a slight transmission loss and verified the effectiveness of the superconducting waveguides.

1. Introduction

Superconducting waveguides, whose walls are made of superconductors, are effective in reducing the propagation loss, depending on the electrical conductivity of the wall material, compared with normal metal waveguides. To date, researchers have studied theories of wave propagation and signal attenuation in superconducting waveguides based on the approximation perturbation method (e.g., [1–4]). Wang et al. (1994), Yalamanchili et al. (1995), and Ma (1995) suggested the importance of alternative boundary conditions based on the penetration depth of the Meissner effect. They found that the dominant propagation mode was switched from TE_{11} to TE_{10} and from TE_{10} to TE_{11} for circular and rectangular waveguides, respectively. However, Yasshin et al. (2001) presented an experimental investigation of the wave propagation mode of a superconducting circular waveguide. They concluded that the effect of superconductivity on propagation is significantly small, and that the dominant mode remains in TE_{11}, which is the same as that of normal metal waveguides [5].

Yeap et al. (2015) applied a new complex conductivity, which expands the basic complex conductivity theory derived by Mattis and Bardeen (1958) [6], based on recent works by Noguchi et al. (2012; 2013) [7–8], for the calculation of wave propagation in the WR-10 Niobium (Nb) waveguide [9]. They suggested that the superconducting waveguide exhibits a low but finite loss from the cut-off to the gap frequency. According to their calculations, the lowest attenuation of TE_{10} mode in WR-10 Nb waveguide is approximately 0.25 Np/m (=2.17 dB/m) around 100 GHz. This value is as large as that of normal metal waveguides (e.g., Copper and Aluminum alloy), indicating that superconducting waveguides are ineffective in reducing propagation loss.

However, research on superconducting waveguides has been limited, especially on the fabrication and measurement of physical models in the mm/sub-mm band. In this study, to investigate the propagation loss of a superconducting rectangular waveguide in the 100 GHz band, we attempted to fabricate a Nb waveguide and measure its propagation.

2. Fabrication Test

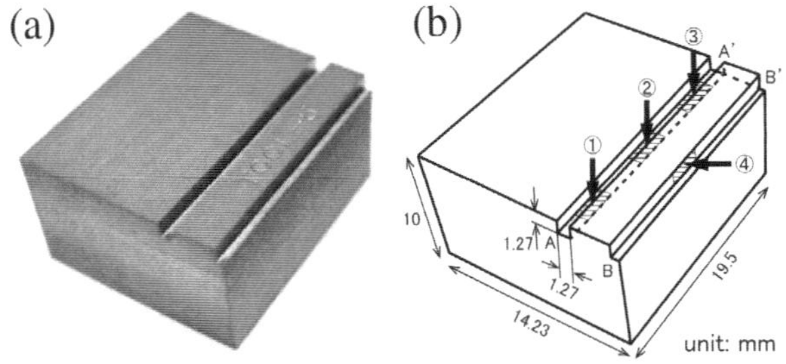

Figure 1. (a) Photograph of a niobium test block to fabricate straight waveguides. (b) Dimensions of the test block and waveguides. Encircled numbers and arrows are areas of the surface roughness measurement and its direction, respectively.

Because bulk Nb is considered as a difficult-to-machine material, we searched for the best cutting tools and machining conditions for end milling. We fabricated a half-height WR-10 (1.27 mm×1.27 mm) straight waveguide in the 19.5 mm square split-block using four types of end mills (for aluminum, heat-resistant aluminum/titanium alloy, resin, and diamond-like carbon (DLC) coating) for comparison of the state of cutting and finished surface accuracy (Fig.1(a)). The surface roughness was measured at four regions in two directions using a non-contact 3D surface measurement device. As shown in Fig.1(b), three areas for the bottom face of the end mill in waveguide A–A′ and one for the side surface in B–B′ were measured. The waveguide of B–B′ was cut in half at the center line of the bottom face after processing to measure the side wall surface roughness.

Consequently, we found that the surface roughness was not the smallest but the most stable cutting in the case with DLC coating end mill, and the others were easily broken in the middle of the process. The finished surface accuracy was approximately 0.537 μm and 0.146 μm (rms) for the bottom and side walls, respectively, using the DLC coating end mill with a 1 mm diameter. The surface roughness of the side wall was approximately 1/3 smaller than that of the bottom surface.

3. Results

3.1. Design and fabrication of resonator

To estimate the propagation loss of the superconducting Nb waveguide, we adopted the resonator method to measure its transmission characteristics, as described in Section 3.2. The waveguide resonator was designed based on a band-stop filter [10]. In this circuit, the 19.5 mm length waveguide and small box structure (1.27 mm×2.54 mm×1.63 mm) are connected through the

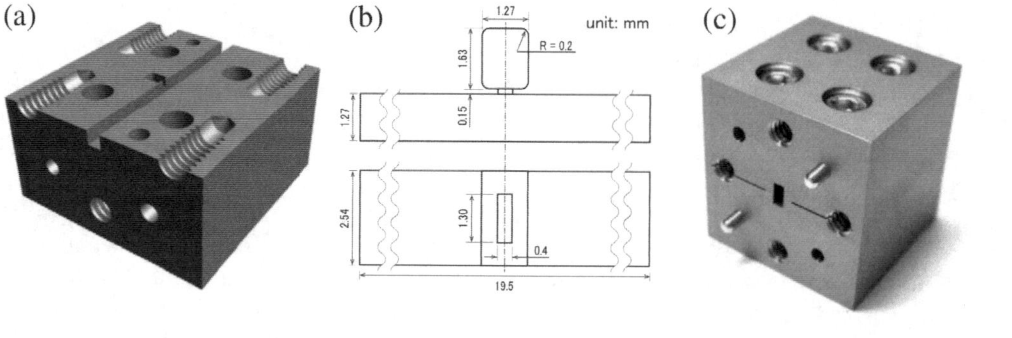

Figure 2. (a) 3D image of a split block. (b) Dimensions of waveguide resonator. (c) Photograph of assembled Nb waveguide resonator block.

iris window (0.4 mm×1.30 mm×0.15 mm), and the half-height waveguide circuit is dug in each of the two split-block (Fig. 2(a) and (b)). The surface roughness of this circuit was measured at 0.346 μm on the bottom surface; therefore, we predicted a roughness of 0.094 μm on the side wall. In addition, UG-387 flanges were fabricated at either end of the waveguide after assembly (Fig.2(c)).

The fundamental mode (TE_{10}) transmission characteristics, that is, the insertion loss of this resonator circuit in the frequency range, from 94 to 100 GHz, was simulated using HFSS software[1] (Fig. 3). The red and blue lines show the electrical conductivities of 6.54×10^6 S/m at room temperature (300 K) [11] and the assumption of a perfect conductor (PEC) as a superconductor at cryogenic temperature (4 K), respectively. According to this calculation, a change in the resonance amplitude and frequency depending on the electrical conductivity is expected. Therefore, we can estimate the electrical conductivity based on the simulated

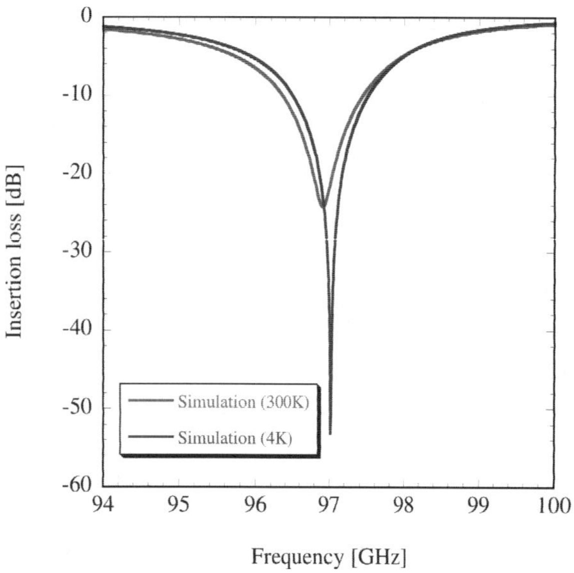

Figure 3. Simulated frequency characteristics of the designed resonator under the assumption at room temperature (300 K; red line) and cryogenic temperature (4 K; blue line).

[1] ANSYS HFSS ⟨https://www.ansys.com/products/electronics/ansys-hfss⟩

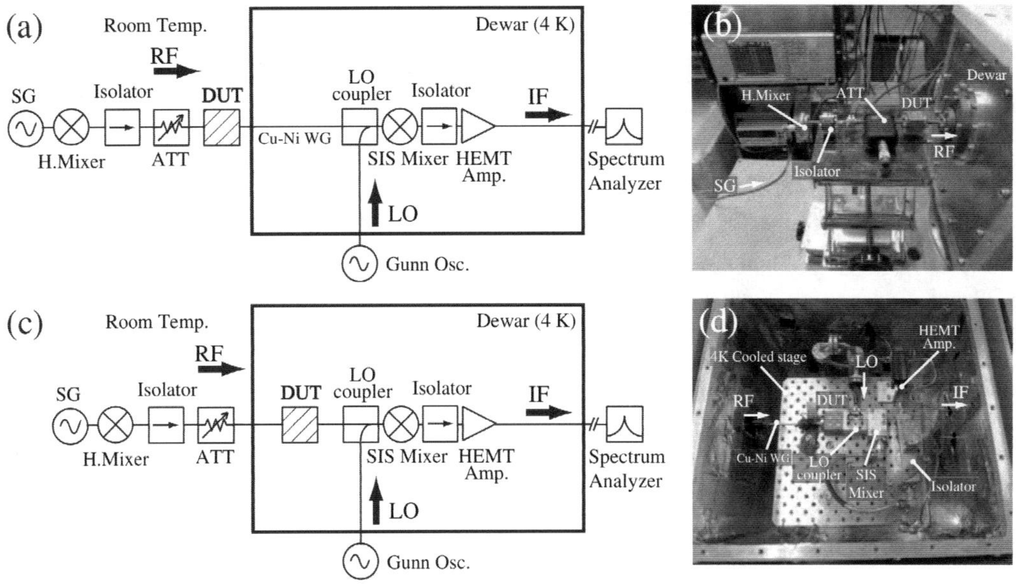

Figure 4. (a) Block diagram of measurement system for room temperature. (b) Photograph of the DUT located outside of the dewar. (c) Block diagram of measurement system for cryogenic temperature. (b) Photograph of the DUT located on the 4-K cooled stage in the dewar.

transmission characteristics in the opposite direction.

3.2. Measurement system and method
We constructed a measurement system for the transmission characteristics of the resonator at room temperature and cryogenic temperatures (Fig.4). The radio wave (RF) signal was generated by a microwave signal generator and multiplied by a harmonic mixer to millimeter wave ($\sim$100 GHz). The output of the device under test (DUT, i.e., the resonator) is transmitted to a superconductor-insulator-superconductor (SIS) mixer, which can be down-converted to an intermediate frequency (IF) signal ($\sim$8 GHz) in a 4-K dewar. The local oscillator (LO) source for down-conversion in the SIS mixer was a Gunn oscillator. When the RF signal is swept, the insertion loss of the signal peaks in the IF band is measured by a spectrum analyzer, and the transmission characteristics of the resonator are obtained from the envelope of the signal peaks.

3.3. Measurement
The resonator is located outside the dewar for the measurement at room temperature (approximately 25°C; Fig.4(a) and (b)), and it is on the 4-K cooled stage in the dewar for the measurement at cryogenic temperature (Fig.4(c) and (d)). The cooling temperature of the resonator circuit was 4.8 K, as measured by a cryogenic temperature sensor attached to the Nb resonator block. This temperature was sufficiently lower than the superconducting transition temperature of Nb (9.25 K).

The signal generator generates an RF signal from $15.66\cdots$ to $16.66\cdots$ GHz, and then multiplies by 6 using the harmonic mixer. Therefore, the frequency range of the sweeping signal to DUT is from 94 to 100 GHz. The LO and IF frequency ranges were 89 GHz and 5–11 GHz, respectively. The shape of the signals, converted to the IF frequency, was measured by the spectrum analyzer, and each signal was fitted with a Gaussian function to calculate the

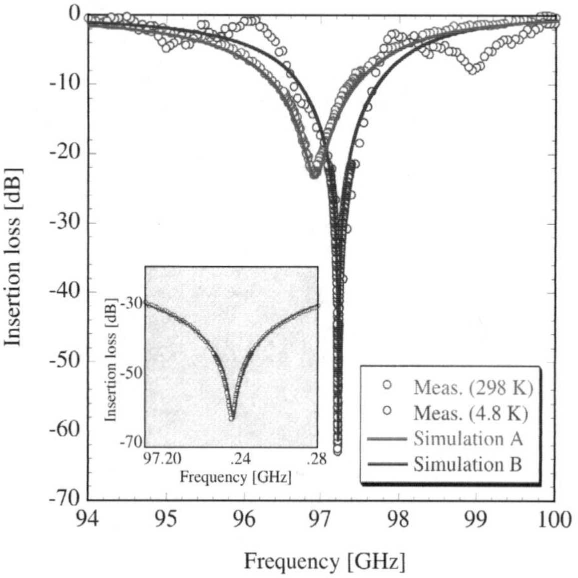

Figure 5. Measured frequency characteristics of the resonator at room temperature (298 K; red circles) and cryogenic temperature (4.8 K; blue circles). The bold red and blue lines represent simulated characteristics adopting electrical conductivity at 298 K and the best-fit curve with changing the electrical conductivity as a parameter, respectively. The inset shows the enlarged view of the resonance peak.

peak amplitude. The peak amplitudes corresponding to these frequencies are shown in Fig. 5. The red and blue circles represent the measured datasets at room temperature (298 K) and cryogenic temperature (4.8 K), respectively. These transmission characteristics represent the insertion losses of the resonator.

For the measurement at cryogenic temperatures, we considered three stages with different settings. All characteristics were obtained from 94 to 100 GHz with a frequency resolution of 50 MHz. However, the signal amplitude in the IF band around the resonance frequency was relatively weak because the insertion loss was larger than 20 dB. The measurement of this range (97.10–97.50 GHz) is changed with a 1 MHz resolution and more substantial input power. In addition, the peak of the resonance characteristics is exceptionally sharp. Therefore, the measurement from 97.225 to 97.245 GHz was set at a 100 kHz resolution to trace the shape of the peak with sufficient measurement points. The transmission characteristics in Fig.5 overlayed all the datasets.

In the transmission characteristics at 4.8 K, an amplitude periodical fluctuation of the baseline corresponding to the frequency was found. This is possibly a standing wave between the resonator and LO coupler before the SIS mixer (see Fig.4(c)). In particular, when an isolator is placed between the resonator and LO coupler in the measurement at room temperature, this fluctuation is resolved. We plan to use a cryogenic isolator to suppress the standing wave during the measurement at 4.8 K. In this analysis, the effect of this fluctuation on the shape of the resonance peak (i.e., peak amplitude and center frequency) was negligible because the bandwidth of the peak was significantly smaller than that of the standing wave.

4. Discussion

The results of the measurements were almost consistent with the expected transmission characteristics obtained by the HFSS simulation in Fig.3. The measurement at 298 K (red circles in Fig.5) is in good agreement with the simulation (bold red line in Fig.5) adopting an electrical conductivity of 6.54×10^6 S/m, which is presented in a previous work [11]. However, in the measurement at 4.8 K, an exceptionally sharp and deep resonance compared to that at room temperature was found. The peak amplitude of the output signal was less than -60 dB.

We calculated the best-fit curve at 4.8 K with changing the electrical conductivity using

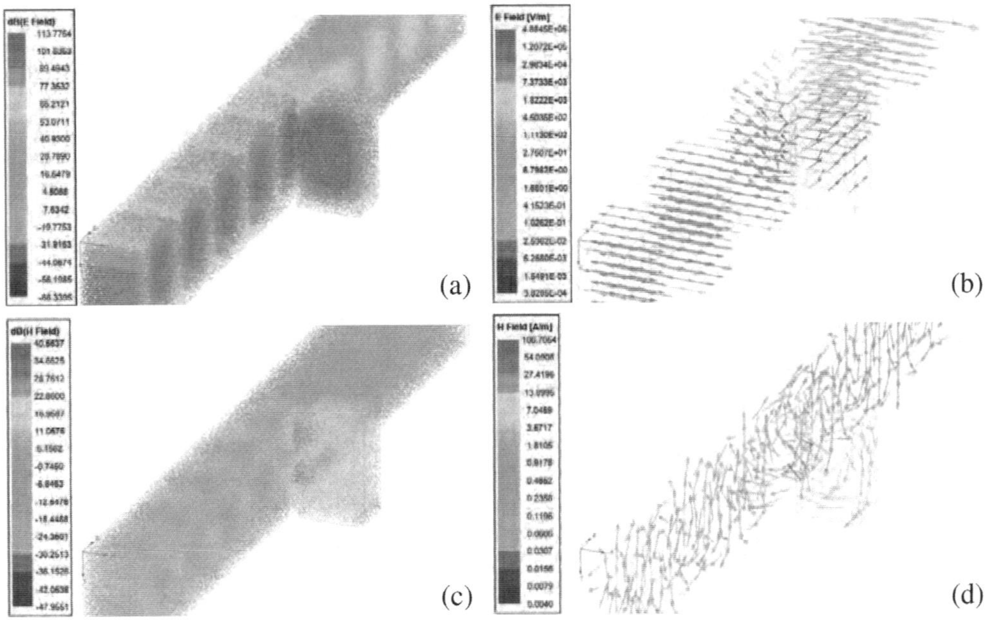

Figure 6. Simulated electric and magnetic fields of the resonator at 97.255 GHz with HFSS. (a) Electric field amplitude. (b) Electric field vector. (c) Magnetic field amplitude. (b) Magnetic field vector.

HFSS. Consequently, a value of 1.8×10^{11} S/m was obtained. The inset shows the enlarged view of the resonance peak from 97.20 to 97.28 GHz, and the simulation curve has good consistency with the measurement plots. Fig.6 shows the simulated electric and magnetic fields of the resonator at the resonance frequency at cryogenic temperature.

4.1. Measurement

The estimated transmission loss of the 1-m length waveguide at 100 GHz using this conductivity is 0.05 dB/m in the case of perfect surface accuracy of waveguide walls. This value is significantly smaller than the theoretical prediction of approximately 0.25 Np/m (= 2.17 dB/m) based on the extended Mattis-Bardeen equation [9]. Even in the case of the simulation adopting the surface roughness (0.346 μm and 0.094 μm on the bottom and side walls, respectively), the transmission loss was 0.09 dB/m. Therefore, the extended Mattis-Bardeen theory is possibly not inapplicable to the superconducting Nb rectangular waveguide propagation. A possible reason is that Noguchi-Naruse-Sekimoto's (NNS') complex conductivity, adopted in the extended Mattis-Bardeen Theory, is obtained in the measurement of a planar circuit composed of NbN thin film. Therefore, this calculation may be inapplicable to waveguides made of bulk Nb.

Although a more quantitative measurement is required for confirmation, the dominant propagation mode in the superconducting Nb rectangular waveguide may not change TE_{11} but remains in the TE_{10} mode. The cut-off wavelength of the superconducting waveguide is approximately 10% greater than that of a conventional waveguide [3]. Despite that being the case, the cut-off frequency of the TE_{11} mode in the WR-10 Nb waveguide is approximately 130 GHz, and this higher mode wave in the range of 94–100 GHz cannot propagate.

5. Summary
To investigate the propagation loss of a superconducting rectangular waveguide, we attempted to fabricate a waveguide resonator using niobium, and measure its propagation in the 100 GHz band. We obtained exceptionally sharp and deep resonance characteristics at 4.8 K compared to those at room temperature. The conductivity and loss are estimated as 1.8×10^{11} S/m and 0.05 dB/m, respectively, based on these transmission characteristics compared with the simulation by electromagnetic software. Therefore, the effectiveness of superconducting waveguides is verified to reduce propagation loss in millimeter waves.

Acknowledgments
This work was supported by a Grant-in-Aid for Challenging Research (Exploratory) from the Japan Society for the Promotion of Science (JSPS) KAKENHI Grant Number JP20K20930. The authors thank the members of the Equipment Development Support Section of the Technical Center of Nagoya University, for constructing the measurement system. Part of this study was conducted using the computational resources of the Center for Integrated Data Science, Institute for Space-Earth Environmental Research, Nagoya University.

References
[1] Winters J H and Rose C 1991 *IEEE Trans. Microwave Theory Tech.* **39** 617
[2] Wang Y -C, Qiu Z A and Yalamanchili R 1994 *Int. J. Electronics* **76** 1151
[3] Yalamanchili R, Qiu Z A and Wang Y -C 1995 *Int. J. Electron.* **78** 715
[4] Ma J 1995 *Int. J. Inf. Millim. Waves* **16** 147
[5] Yassin G, Jung G, Dikovsky V, Barboy I, Kambara M, Cardwell D A and Withington S 2001 *IEEE Microwave and Wireless Components Letters* **11** 413
[6] Mattis D C and Bardeen J 1958 *Phys. Rev.* **111** 412
[7] Noguchi T, Naruse M, and Sekimoto Y 2012 *Phys. Procedia* **36** 318
[8] Noguchi T, Naruse M, and Sekimoto Y 2013 *IEEE Trans. Appl. Superconduct.* **23** 1501404
[9] Yeap K H, Teh J S M, Nisar H, Yeong K C and Hirasawa K 2015 *Frequenz Journal of RF-Engineering and Telecommunications* **69** 111
[10] Asayama S 2003 *Int. J. Inf. Millim. Waves* **24** 1833
[11] Maglić K D 1994 *Int. J. Thermophys.* **15** 963

Evaluation of True Random Number Sequences Generated by Utilizing Timing Jitters in Superconducting Integrated Circuits

Kenta Sato[1], Naonori Sega[1], Hiroshi Shimada[1] and Yoshinao Mizugaki[1]

[1]Department of Engineering Science, The University of Electro-Communications (UEC Tokyo), 1-5-1 Chofugaoka, Chofu, Tokyo, 182-8585, Japan

E-mail: k-sato@w8-7f.ee.uec.ac.jp, y.mizugaki@uec.ac.jp

Abstract. A Hardware Random Number Generator (HRNG), which generates true random number sequences, is studied. In superconducting Single Flux Quantum (SFQ) circuitry, HRNGs that utilized timing jitters in trigger signals have been demonstrated. In this paper, we propose an HRNG with two oscillators to enhance stability against variation of oscillator frequencies. We investigate the characteristics of random numbers generated with the proposed HRNG not only by numerical simulation but also by experiments using Nb test circuits. In the numerical simulation, the probability of "1" in random number sequences has a local minimum of "1" output probability at conditions where f_{clk} (the clock oscillator frequency) and f_{gate} (the gate oscillator frequency) are approximately equal ($f_{clk} \approx f_{gate}$) for $f_{clk} \leq 60$ GHz. It is also found that pass rates for the FIPS140-2 random number tests are little affected by variations of f_{clk} at $f_{gate} = 69.2$ GHz. The random number generation rate for $f_{clk} = 60$ GHz and $f_{gate} = 43$ GHz reaches 3.98 GHz. In experimental measurements, random numbers of 1 Mbits generated by a test circuit pass the SP800-22 random number tests at $f_{clk} = 48.79$ GHz and $f_{gate} = 19.81$ GHz.

1. Introduction

Single-Flux-Quantum (SFQ) circuitry is a type of superconducting integrated circuit for signal processing [1]–[4]. Under the assumption of $Nb/AlO_x/Nb$ integration, SFQ digital circuits operate at high-speed up to sub-terahertz with power consumption about four orders of magnitude smaller than CMOS [5],[6]. Therefore, several SFQ-based digital circuits have been demonstrated to use these features [7]–[9].

Proposed applications of SFQ circuitry are not only for digital signal processing but also random number generators [10],[11]. Random number sequences are indispensable because they are used in various fields, such as cryptographic communication and the simulation of natural phenomena. There are two types of random numbers: pseudo-random numbers generated by numerical calculations such as recurrence formulas, and true-random numbers generated using random physical phenomena such as thermal noise, nuclear collapse, and chaotic dynamics. Although pseudo-random numbers are of periodicity and predictability, they have become widespread because there are established methods to implement them. In fact, many SFQ circuits generating pseudo-random numbers have been demonstrated [10],[12]–[15]. Nevertheless, it is important for some application fields such as secured information communication to ensure randomness in random number sequences. That has led to the requirement of true-random number sequences of non-periodicity and unpredictability. Therefore,

Hardware Random Number Generators (HRNGs) using an entropy source have been proposed and demonstrated.

SFQ-HRNGs, which generate true-random number sequences, utilize thermal noises in Josephson junctions and their shunt resistors. For example, SFQ-HRNGs that utilized thermal noise current to switch the junctions in a comparator were demonstrated [11],[16]. In addition, to decrease the cost and the technical difficulty, logic-based SFQ-HRNGs that utilized timing jitters (the timing fluctuation of SFQ pulses arriving at a gate) were demonstrated [17],[18].

For the latter SFQ-HRNGs, Onomi *et al.* operated the HRNGs composed of a high-frequency Josephson oscillator and a trigger signal where timing jitters in the trigger signal line were larger than the period of the high-frequency Josephson oscillator. In their HRNGs, the oscillator frequency variation directly changed the length ratio of the timing jitters and the period, which affected the quality of randomness in random number sequences.

In this paper, we propose an HRNG utilizing the timing jitters of two oscillators. We expect to enhance stability against variation of oscillator frequencies by increasing the number of referring timing jitters. We refer to this HRNG as a "Twin-Oscillator HRNG," of which the abbreviation is "TO-HRNG." We investigate the characteristics of random numbers generated with TO-HRNGs not only by numerical simulation but also by experiments using test circuits fabricated using a 10 kA/cm^2 Nb/AlO$_x$/Nb integration process.

2. Circuit Configuration and Operation Principle

Figure 1 shows a simplified block diagram of the TO-HRNG. The circuit consists of an XOR gate, a non-destructive readout memory (NDRO), a clock oscillator, and a gate oscillator. The oscillators generate SFQ pulse trains by use of the over-biasing method, in which Josephson junctions working as oscillators are set to be in the voltage state of the average voltage of V by DC current exceeding their critical currents. In this method, the repetition frequency f is proportional to the voltage V of the Josephson junctions and is calculated in a formula $f = V / \Phi_0$ where the amount of magnetic flux of an SFQ is represented as Φ_0 [19]. The high repetition-frequency SFQ pulses from the clock oscillator are transferred to the reset terminal of the NDRO as well as the clock and signal input terminals of the XOR gate. On the other hand, the other high repetition-frequency SFQ pulses from the gate oscillator are transferred to the other signal input terminal of the XOR gate. SFQ pulses from each oscillator arrive at the terminals described above with timing jitters due to thermal noise. As a result, SFQ pulse arrival timing to the XOR gate deviates from the periodicity, resulting in a random "0" or "1" sequence at the output terminal of the XOR gate. This binary sequence turns out to be true-random numbers. Then, the NDRO memorizes the random number generated at the XOR gate. Meanwhile, trigger SFQ pulses of lower frequency are fed to the clock terminal of the NDRO, and true-random numbers are obtained from the output terminal of the NDRO.

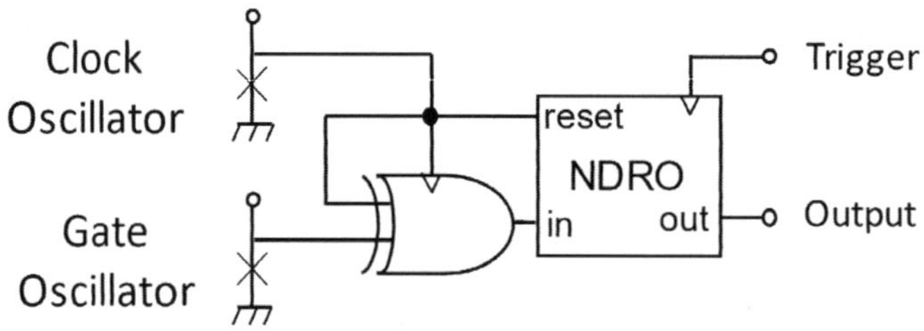

Figure 1 Simplified diagram of TO-HRNG.

The quality of random numbers of finite length is often statistically evaluated using a suite of random number tests. There are two important characteristics in random number evaluation: equiprobability and unpredictability. Equiprobability means the balance of appearance frequencies of the number of "0"s and "1"s. Unpredictability means the impossibility of supposing the next generated random number. Generally, random number tests are applied to plural random number sequences, and the rates of passing tests assess the quality.

3. Method

3.1. Simulation

The designed circuit was simulated by JoSIM [20]. The temperature condition in the simulation was 4.2 K. The input signal frequency to the trigger input, which was the generation rate of random numbers, was set at 6.31 GHz. We simulated the circuit and obtained 500 kbits random numbers with various oscillation frequencies for the clock and gate oscillators.

The random number sequences were evaluated using a test suite of FIPS PUBS 140-2 (FIPS 140-2), which was provided by the National Institute of Standards and Technology (NIST) [21]. The FIPS 140-2 test suite evaluates 20 kbits random numbers with 9 tests. In this paper, 25 random number sequences were firstly evaluated by the probability of "1", which corresponded to the Monobit test in the FIPS 140-2 suite, and then, by the pass rate for all 9 tests. Although there are more rigorous test suites such as the NIST SP 800-22[22] and TestU01[23], we did not use them in our simulations because of their long simulation time for a required large number of random numbers.

3.2. Experiment

Fabrication process of test circuits was the high-speed standard process (HSTP) [24] provided by National Institute of Advanced Industrial Science and Technology (AIST), Japan. The TO-HRNG was designed utilizing the "CONNECT" cell library [25] which was modified for the HSTP.

Fabricated test circuits were measured in a liquid He bath. We inputted a square voltage wave to the trigger input via a DC-to-SFQ converter through a 50 Ω resistor. The output SFQ pulses generated as random numbers were converted to voltage signals using an SFQ-to-DC converter. Then, the output waveforms were monitored and acquired with a digital oscilloscope via a 40-dB pre-amplifier. The trigger frequency (i.e., generation speed) was set at 1 MHz which was half of the cut-off frequency of the pre-amplifier (2 MHz). The length of the random number sequence obtained in experiments was 1 Mbits for each circuit operation.

The random number sequences were evaluated using not only the FIPS 140-2 test suite but also the SP 800-22 test suite. The SP 800-22 suite evaluates random number sequences with 15 tests. Each test calculates a p-value, which is the probability that a perfect random number generator would have produced a sequence less random than the tested sequence. A random number sequence is usually judged to pass a single test if the p-value is larger than 0.01. The length and number of random number sequences can be determined arbitrarily, and therefore in this study, we generated 50 sequences of 20 kbits and evaluated them.

4. Results and Discussion

4.1. Simulation

Figure 2 shows the gate oscillator frequency f_{gate} dependence of the probability of "1" for four clock oscillator frequencies f_{clk} at the trigger frequency f_{trig} = 6.31 GHz. The straight dashed lines indicate 48.63 % and 51.37 %, which are edges for passing the Monobit test in the FIPS 140-2 suite. The graph shows that the probability of "1" has the local minimum at $f_{clk} \approx f_{gate}$. That is because the closer each oscillator frequency was, the less probability of "1" the XOR outputs by increasing the chance of the input "00" or "11" for the XOR gate. Therefore, the probability of "1" in the output of the XOR gate became the least at $f_{clk} \approx f_{gate}$. Although that is a deterministic explanation, the TO-HRNG operated with randomness and we can see that there are influences of timing jitters. That is, the oscillator frequencies (f_{clk} and f_{gate}) for the local minimum are not always the same, such as the condition of f_{clk} = 50 GHz and f_{gate} = 60 GHz. Also, the probability of "1" does not reach 0% at the local minimum. These are attributed to timing fluctuations in the input signals to the XOR gate.

In comparison with the generation rate of 1 GHz numerically obtained for the preceding SFQ-HRNGs assuming 2.5 kA/cm^2 Nb process [17], the generation rate of 6.31 GHz was improved by 6.3 times, more than the square root of the ratio of current densities ($\sqrt{10/2.5} = 2.0$). It is because the generation rate of the TO-HRNG is free from the constraints of the timing jitters in the trigger signal. The highest generation rate has not been evaluated and should be done in the future.

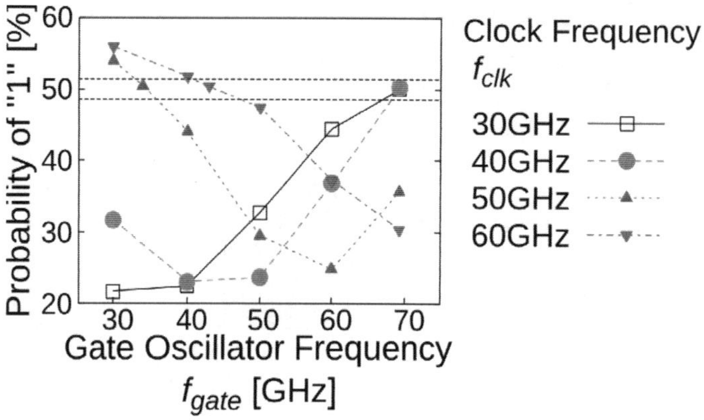

Figure 2 Gate oscillator frequency f_{gate} dependence of the average probability of getting "1" for four clock oscillator frequencies f_{clk}. The straight dashed lines at 48.63 % and 51.37 % indicate edges for acceptance region of the Monobit test in the FIPS 140-2 suite. f_{trig} is 6.31 GHz.

Figure 3 shows the f_{clk} and f_{gate} dependence of the pass rate, *PR,* for the FIPS 140-2 suite. The conditions for PR = 100% are found in the upper left and lower right regions. For the upper left region, where $f_{clk} > f_{gate}$, the conditions for PR = 100% were dependent on both f_{gate} and f_{clk}. For the lower right region, where $f_{clk} < f_{gate}$, the conditions for PR = 100% were f_{gate} = 69.2 GHz and $f_{clk} \leq 40$ GHz. The generation rate of f_{trig} = 6.31 GHz is an order of magnitude faster than the previous SFQ-HRNGs based on timing jitters [17]. We have confirmed that the main cause for PR less than 100% is the failure of the Monobit test. One exception is the condition of f_{clk} = 60 GHz and f_{gate} = 43 GHz, where the failure of the Runs 1 test is the main cause. Runs 1 is a test that examines the number of a single "0" and a single "1" appearing alternatively (i.e., "10101010…").

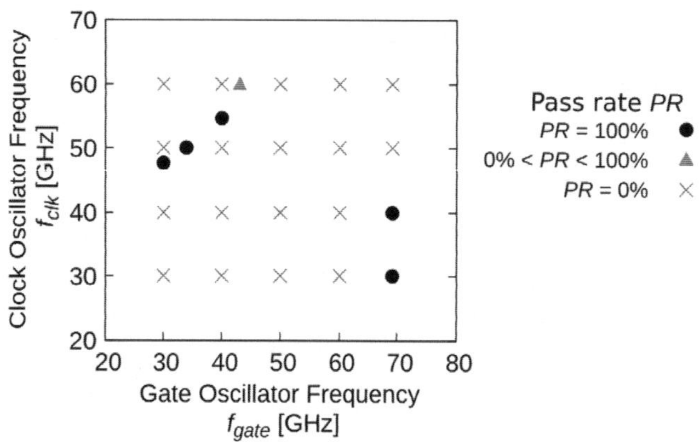

Figure 3 f_{clk} and f_{gate} dependence of three categories of the pass rate PR for the FIPS 140-2 suite. f_{trig} is 6.31 GHz.

Figure 4 shows the f_{trig} dependence of the probability of "1" and the pass rate for the FIPS 140-2 suite at f_{clk} = 60 GHz and f_{gate}= 43 GHz. The straight dashed lines indicate 48.63 % and 51.37 %. It shows that the TO-HRNG passes the FIPS 140-2 suite with *PR* of 100% up to f_{trig} = 3.98 GHz. For f_{trig} of 6.31 GHz and above, the *PR* value rapidly decreases, although the probability of “1” stays in the range between 48.63% and 51.37% and passes the Monobit test. It is due to the failure of the Runs 1 test. It suggests the maximum generation rate of the TO-HRNG where the generated random number sequences begin to deteriorate.

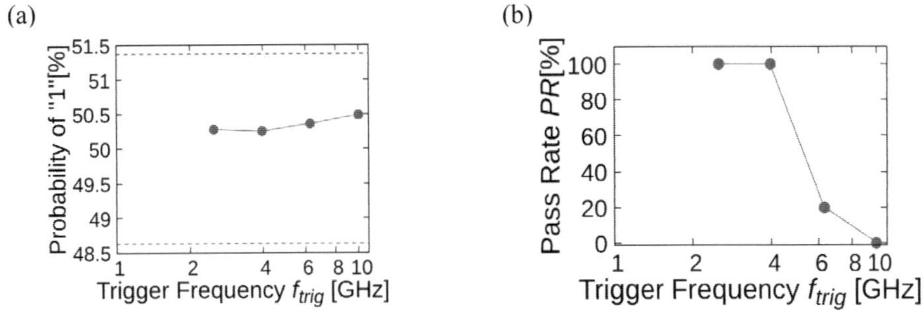

Figure 4 f_{trig} dependence of (a)the probability of “1” and (b)the pass rate for the FIPS 140-2 suite at f_{clk} = 60 GHz and f_{gate}= 43 GHz.

4.2. Experiment

Figure 5 shows a micrograph of a test circuit fabricated using the AIST HSTP. The total number of junctions was 152.

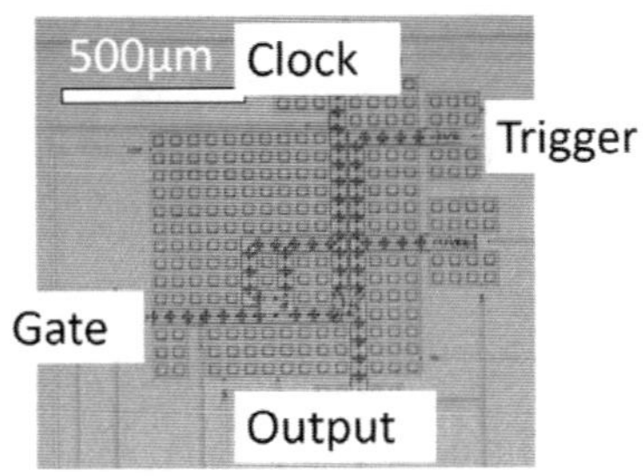

Figure 5 Photomicrograph of a fabricated TO-HRNG.

Figure 6 shows an example of circuit operation. The input trigger signal was a 1 MHz square wave. The output voltage waveform represents a random number sequence of "100010110010".

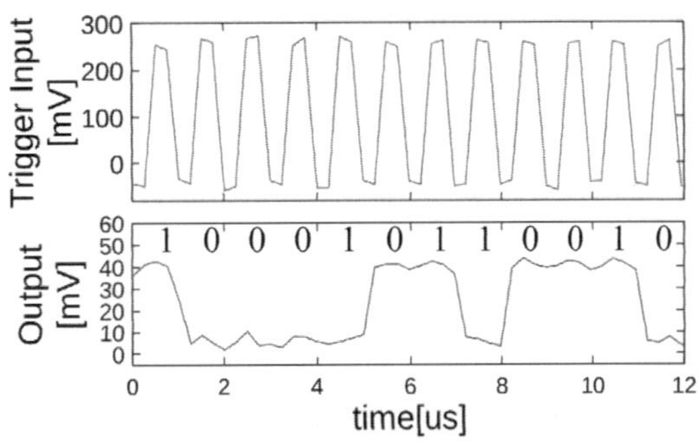

Figure 6 Measured voltage waveforms of TO-HRNG operation at f_{clk} = 48.79 GHz and f_{gate} = 19.81 GHz. The trigger signal frequency was 1 MHz, which signal was fed to a DC-to-SFQ converter via a 50 Ω resistor on chip. The output signal from an SFQ-to-DC converter was amplified by a 40-dB pre-amplifier.

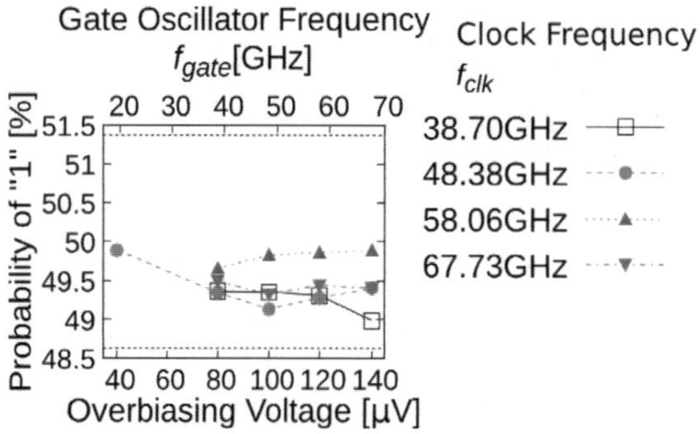

Figure 7 Gate oscillator frequency f_{gate} dependence of the average probability of getting “1” for four clock oscillator frequencies f_{clk}. The straight dashed lines at 48.63 % and 51.37 % indicate edges for acceptance region of the Monobit test in the FIPS 140-2 suite.

Figure 7 shows the gate oscillator frequency f_{gate} dependence of the probability of "1" for four clock oscillator frequencies f_{clk} at the trigger frequency f_{trig} = 1 MHz. We firstly obtained the random number sequences at the right region, where f_{gate} > 30 GHz. Then we inferred the frequency condition of the probability of "1" being almost 50% and obtained the random number sequences under the condition of f_{clk} = 48.79 GHz and f_{gate} = 19.81 GHz. The probabilities of "1" shown in figure 7 are depended on f_{gate} much less than the simulation results shown in figure 2, which suggests that the timing jitters in experiments could be induced by external noises and be larger than the situation assumed in the simulation. Timing jitters induced by external noises could vanish some synchronization effects, which makes the probability of "1" be close to 50% even for the condition of $f_{clk} \approx f_{gate}$.

Figure 8 shows the f_{clk} and f_{gate} dependence of the pass rate, *PR*, for the FIPS 140-2 suite. It shows that there were no conditions for *PR* = 0% and suggests that the random number generation of the TO-HRNG was not deteriorated but was eventually improved by external noises.

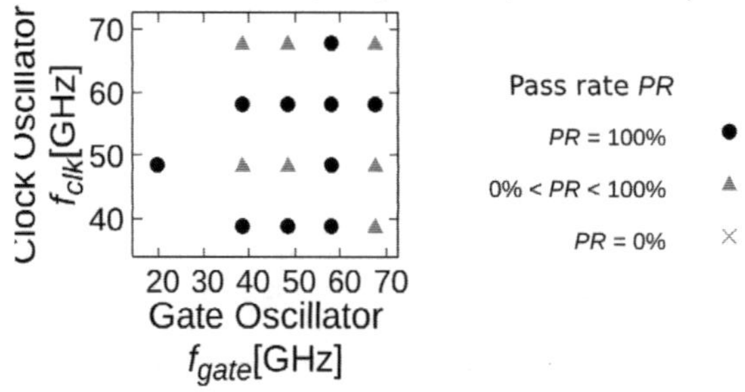

Figure 8 f_{clk} and f_{gate} dependence of three categories of the pass rate *PR* for the FIPS 140-2 suite. f_{trig} is 1 MHz.

Table 1 shows the results of the SP 800-22 suite evaluating 50 random number sequences of 20 kbits obtained under the frequency conditions of f_{clk} = 48.79 GHz and f_{gate} = 19.81 GHz. It should be noted that tests with an asterisk in the test name (DFT, OverlapppingTemplate, and Universal) are known to include some problems [26]. "DFT" and "OverlappingTemplate" have some deviations from the theoretical distribution, whereas "Universal" has some mistakes in the theoretical model. In Table 1, p-values greater than 0.0001 indicate sufficient randomness, and those tests are judged to be passed. Table 1 shows that the generated random number sequences passed 9 out of 12 tests excluding the problematic tests. We add a few words that three failed tests (ApproximateEntropy, RandomExcursions, and Random-ExcursionsVariant) require longer lengths and larger sample sizes of the random number sequences than we generated. The SP 800-22 suite recommends testing 1000 random number sequences of 1 Mbits length, while this study used 50 random number sequences of

Table 1 Results of the SP 800-22 suite for 50 random number sequences of 20 kbits length generated by TO-HRNG.

Test name	P-value	Fail Rate [%]	Pass
Frequency	0.61631	0	Success
BlockFrequency	0.07572	2	Success
CumulativeSums	0.53415	0	Success
Runs	0.41902	2	Success
LongestRun	0.21331	2	Success
Rank	0.04011	4	Success
*DFT	0.28967	0	Success
NonOverlappingTemplate	0.14734	1.5	Success
*OverlappingTemplate	0.00176	0	Success
*Universal	0.00000	100	Fail
ApproximateEntropy	0.00020	16	Fail
RandomExcursions	N/A	N/A	Fail
RandomExcursionsVariant	N/A	N/A	Fail
Serial	0.53415	2	Success
LinearComplexity	0.93572	4	Success

20 kbits.

It is reported that a CMOS oscillator-based HRNG requires at least 20 oscillators for passing the SP 800-22 suit [27]. Therefore, we presume that the TO-HRNG can generate true random numbers with less oscillators than the CMOS oscillator-based HRNG.

5. Conclusion

We proposed a twin-oscillator hardware random number generator circuit referred to as a TO-HRNG. It utilized timing jitters in each signal line from two oscillators. By increasing the number of referring timing jitters from one source to two sources, stability against variation of oscillator frequencies was expected to be improved. Random numbers the circuit generated were evaluated by random number test suite of the FIPS 140-2 for JoSIM simulation and by the SP 800-22 suite for measurements of Nb/AlO$_x$/Nb integrated circuits.

In the simulations, the probability "1" and the pass rate of FIPS 140-2 were evaluated for the various frequencies of the gate oscillator f_{gate} and the clock oscillator f_{clk}. The probability of "1" had a local minimum at $f_{clk} \approx f_{gate}$. The pass rate of the FIPS 140-2 suite being 100 % was obtained for both conditions of $f_{clk} > f_{gate}$ and $f_{clk} < f_{gate}$. The main cause of the FIPS 140-2 failures was the Monobit test corresponding to the probability of "1", but there were also failures at Runs 1 due to effects possibly caused by the frequency of the trigger input.

In the measurements of Nb test circuits, the probability of “1” and the pass rate of FIPS 140-2 were influenced by increasing timing jitters. In addition, the test circuits were performed to obtain 50 random number sequences of 20 kbits. Evaluation using the SP 800-22 suite showed the random number sequences passed 9 out of 12 tests excluding 3 problematic tests. Obtained lengths of the random number sequences were too short for the failed tests.

Acknowledgement

The authors thank Prof. T. Onomi, Fukuoka Institute of Technology, for the initiation of the present work, and the lab members in The University of Electro-Communications (UEC Tokyo) for their fruitful discussion and technical support. We are also grateful for the stable supply of liquid helium from the Coordinated Center for UEC Research Facilities, UEC Tokyo. The circuits used in this work were fabricated using the Nb HSTP in the CRAVITY, AIST, Japan. This work was partly supported by JSPS KAKENHI Grant Number JP20H02201, and by VLSI Design and Education Center (VDEC), the University of Tokyo in collaboration with Cadence Design Systems. Nation-wide Cooperative Research Projects of the RIEC, Tohoku University, are also acknowledged.

References

[1] Nakajima K and Onodera Y 1976 *J. Appl. Phys.* **47** 1620–7
[2] Nakajima K and Onodera Y 1978 *J. Appl. Phys.* **49** 2958–63
[3] Mukhanov O, Semenov V and Likharev K 1987 *IEEE Trans. Magn.* **23** 759–62
[4] Likharev K K and Semenov V K 1991 *Trans. Appl. Supercond.* **1** 3–28
[5] Hayakawa H, Yoshikawa N, Yorozu S and Fujimaki A 2004 *Proc. IEEE* **92** 1549–63
[6] Katam N K, Kawa J and Pedram M 2019 *Proc. Design, Autom. Test Eur. Conf. Exhib.* 1781–7
[7] Tanaka M, Matsuzaki F, Kondo T, Nakajima N, Yamanashi Y, Fujimaki A, Hayakawa H, Yoshikawa N, Terai H and Yorozu S 2004 *IEEE Int. Conf. Solid-State Circuits* **1** 298–529
[8] Yamanashi Y, Kainuma T, Yoshikawa N, Kataeva I, Akaike H, Fujimaki A, Tanaka M, Takagi N, Nagasawa S and Hidaka M 2010 *IEICE Trans. Electron.* **E93-C** 440–4
[9] Nagaoka I, Tanaka M, Inoue K and Fujimaki A 2019 *IEEE Int. Conf. Solid-State Circuits* 460–2
[10] Kang J H, Przybysz J X, Martinet S S, Worsham A H, Miller D L and McCambridge J D 1997 *IEEE Trans. Appl. Supercond.* **7** 2673–6
[11] Yamanashi Y and Yoshikawa N 2009 *IEEE Trans. Appl. Supercond.* **19** 630–3
[12] Akahori A, Takeuchi N, Mori N, Suzuki Y, Furuta F, Fujimaki A and Hayakawa H 2001 *IEEE Trans. Appl. Supercond.* **11** 521–4
[13] Zhou X, Xu S, Rott P, Mancini C A and Feldman M J 2001 *IEEE Trans. Appl. Supercond.* **11** 617–20
[14] Yamada T, Maezawa M and Urano C 2015 *Physica* C **518** 85–8
[15] Mizugaki Y, Mutoh Y, Urai Y, Sawada K and Watanabe T 2016 *IEEE Trans. Appl. Supercond.* **26** 1300504
[16] Sugiura T, Yamanashi Y and Yoshikawa N 2011 *IEEE Trans. Appl. Superconduct.* **21** 843–6
[17] Onomi T and Mizugaki Y 2020 *IEEE Trans. Appl. Supercond.* **30** 130305
[18] Sato K, Sega N, Somei Y, Shimada H, Onomi T and Mizugaki Y 2022 *IEICE Trans. Electron.* **E105-C** 296–9
[19] Bunyk P I, Oliva A, Semenov V K, Bhushan M, Likharev K K and Lukens J E 1995 *Appl. Phys. Lett.* **66** 646–8
[20] Delport J A, Jackman K, Roux P l and Fourie C J 2019 *IEEE Trans. Appl. Supercond.* **29** 1300905
[21] National Institute of Standards and Technology (NIST) 2002 *Federal Information Processing Standards Publication*
[22] Rukhin A, Soto J, Nechvatal J, Smid M, Barker E, Leigh S, Levenson S, Vangel M, Banks D, Hechert N, Dray J, Vo S and Bassham L 2010 *Natl. Inst. Stand. Technol. Spec. Publ. 800-22rev1a* (NIST)
[23] L'Ecuyer P, Simard R 2007 *ACM Transactions on mathematical Software* **33** 1–40
[24] Takeuchi N, Nagaswa S, China F, Ando T, Hidaka M, Yamanashi Y and Yoshikawa N 2017 *Supercond. Sci. Technol.* **30** 035002
[25] Yorozu S, Kameda Y, Terai H, Fujimaki A, Yamada T and Tahara S 2002 *Physica* C **378-381** 1471–4
[26] Yoshida H, Murakami T and Kawamura S 2012 *IEICE Tech. Rep.* **112** 13–8
[27] Şarkışla M A and Ergün S 2018 *IEEE APCCAS* 274–8

Accuracy Improvement of Polynomial Computing RSFQ Circuits Based on Stochastic Computing by Partial Duplication

Moeka Tsuji and Nobutaka Kito
Graduate School of Engineering, Chukyo University, Aichi, Japan
E-mail: t32205m@m.chukyo-u.ac.jp

Abstract. This study proposes a design method for RSFQ stochastic computing (SC) circuits. The method generates circuits with improved accuracy compared to a previously proposed method. This improvement in accuracy is realized using partial duplication. SC requires a large number of clock cycles to obtain a calculation result owing to lengthy observation of the output bitstream. The proposed method can either enhance the accuracy while maintaining the latency or mitigate the latency in clock cycles while maintaining the same accuracy. The design method duplicates a small part of the original circuit considering effectiveness for accuracy improvement. Therefore, the number of additional gates in circuits designed with the proposed method is negligible. The evaluation results indicate that, compared to circuits without duplication, the mean absolute error of circuits including one duplicated part designed with the method is reduced to approximately 60%.

1. Introduction

Rapid-single-flux-quantum (RSFQ) circuits [1] are attracting interest as ultra-fast and low-power computing devices. These circuits represent logic values using voltage pulses. In RSFQ circuits, each logic gate is a clocked gate. Specifically, each logic gate receives clock pulses at its clock input terminal and outputs a calculation result as the presence or absence of a pulse for each clock period.

Stochastic computing (SC) [2] has recently drawn academic interest in both CMOS and superconducting circuit designs. SC is suitable for applications in which minor errors can be tolerated, such as neural network processing. A value in SC is represented by a ratio of 1s in a bitstream on a signal line. In RSFQ circuits, a value in SC is represented by the number of pulses on a signal line during observed clock periods. Moreover, in SC, multiplication can be realized with an AND gate. Accordingly, a small arithmetic circuit can be realized with SC. There have been studies on RSFQ-based random number generators [3][4][5] necessary for realizing SC circuits. Previously, several schemes based on this technology have been proposed: superconducting SC-based circuits [6] using Adiabatic Quantum-Flux-Parametron (AQFP) technology, SC-like RSFQ multiplication and matrix multiplication circuits [7] [8], and a design method of RSFQ SC circuits for calculating polynomials [9]. However, the number of clock cycles required for a calculation required for a calculation increases significantly with the calculation precision.

RSFQ logic circuits have been designed as serial processing circuits [10][11][12]. However, circuits using serial processing tend to experience considerable latency in clock cycles. To address this drawback, circuits using parallel processing have been actively developed [13] [14]. In this study, we utilize

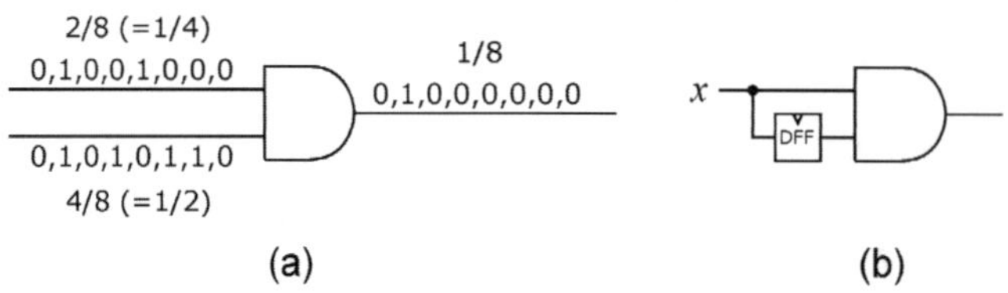

Figure 1. Example of multiplication with an AND gate. (a) demonstrates $\frac{2}{8} \times \frac{4}{8}$ in SC. (b) is a squarer circuit. A D flip-flop is inserted to remove a correlation.

duplication for SC circuits to increase parallelism, in other words, preparing copies of an original circuit and observing their output bitstreams either to improve the accuracy while maintaining the number of clock cycles for a calculation or to reduce the number of clock cycles while maintaining the accuracy.

This study proposes a design method of SC circuits utilizing partial duplication. The method extends the design method of calculation circuits of polynomials in [9] to improve the accuracy through duplication. A variety of useful arithmetic functions can be approximated by polynomials using the Taylor expansion. Although duplication can improve accuracy, naive duplication greatly increases the circuit area. This method reduces the area overhead of duplication by sharing an internal part between the original and copies.

The method in [9] designs a circuit for calculating a polynomial based on a modified polynomial. The method transforms a polynomial considering available operations in SC, such as multiplication with an AND gate and subtraction-from-one with a NOT gate. It produces a modified formula for creating circuits. The modified formula can be calculated by a circuit in which AND gates and NOT gates are alternately connected. An AND gate is located at the output of a circuit designed with the modified formula. The proposed method duplicates only the AND gate at the output of the circuit. In the designed circuit, duplicated units share the majority of the original unit.

Several circuits computing polynomials have been designed using the proposed method and evaluated through simulation. The mean square error of circuits, including one duplicated unit, was reduced to approximately 60% compared to circuits without duplication when the same number of clock cycles was used for a calculation.

The remainder of this paper is organized as follows. The following section introduces SC and describes the realization of SC with RSFQ circuits. Section 3 proposes the accuracy improvement of polynomial computing RSFQ circuits based on SC by partial duplication. Section 4 shows the evaluation results. Finally, Section 5 concludes this paper.

2. Preliminaries

2.1. Stochastic Computing

Stochastic computing (SC) is a calculation method [2] that utilizes probabilities to perform calculations. A bitstream, which consists of 0s and 1s and represents numerical value using probability, is called a stochastic number (SN). A numerical value of an SN is expressed by the ratio of 1s in the SN. Although a large number of gates are employed to form a multiplier for binary numbers, an AND gate performs multiplication for SNs. Figure 1(a) illustrates an example of multiplication for SNs. The figure illustrates a bitstream consisting of 8 time-frames for each line. A numerical value corresponding to a bitstream is the ratio of 1s in 8 time-frames. As depicted in the figure, the ratio at the output of the AND gate is $\frac{1}{8}$ as the result of $\frac{2}{8} \times \frac{4}{8}$. Other gates can perform arithmetic operation in SC. A NOT gate can implement subtraction from 1 because it inverts each bit in a bitstream, and the ratio of 1s in the input bitstream is equal to the ratio of 0s in the output bitstream.

SC inherently causes computational error. In addition, correlations at the gate inputs increase the error. In other words, if bits that depend on the same bit are fed to inputs of a gate, the error of an SC

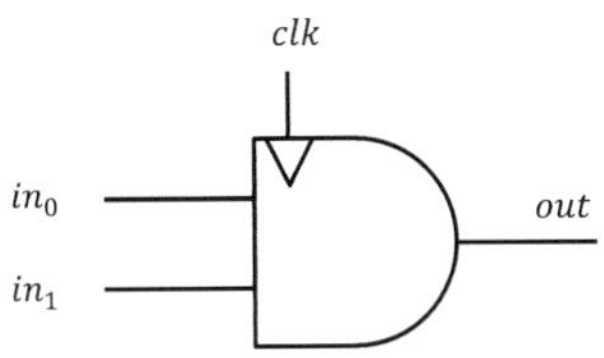

Figure 2. AND gate in RSFQ circuits.

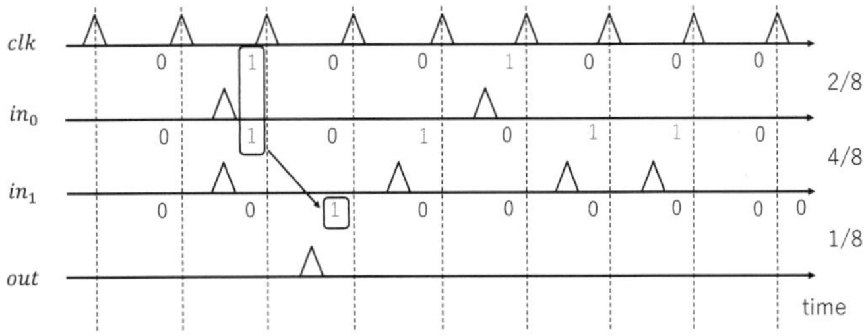

Figure 3. Example of an operation of an SC multiplication circuit implemented as RSFQ circuits.

circuit increases. For circuits without feedback loops, correlation can be eliminated through insertion of memory elements such as D flip-flops (DFFs) in the signal line. Figure 1(b) is a squarer circuit and is an example of inserting a DFF to remove a correlation. If the DFF was not inserted, the circuit would calculate x in SC instead of x^2.

2.2. *RSFQ circuits*

RSFQ circuits represent logic values using voltage pulses. In RSFQ circuits, each logic gate has a clock input terminal, as illustrated in Figure 2, and clock pulses are fed to each logic gate in a circuit. A gate input's logic value is evaluated by referring to the clock pulses fed to the gate. The presence of a pulse during a period between adjacent clock pulses is treated as logic 1, whereas the absence of a pulse during the period is interpreted as logic 0. Each logic gate outputs a calculation result triggered by a clock pulse. Therefore, each input and output logic value is separated by clock pulses. Thus, a sequence of input logic values and a sequence of output logic values can be seen as a bitstream.

Figure 3 illustrates the behavior of an SC circuit implemented as an RSFQ circuit. The operation in this figure corresponds to the operation in Figure 1.

2.3. *Design method of RSFQ SC circuits [9]*

A design method of the RSFQ SC circuits was proposed in [9]. This method produces computing circuits for polynomials of the following form, in which $b_i (0 \le i \le n)$ are integers and $1 \le b_0 < \cdots < b_n$, and $0 < a_n \le \cdots \le a_0 \le 1$.

$$\sum_{i=0}^{n}(-1)^i a_i x^{b_i} = a_0 x^{b_0} - a_1 x^{b_1} + \cdots + (-1)^n a_n x^{b_n} \tag{1}$$

The method's input is an objective polynomial and constant probability values of available SNs, which are named $c_0, \ldots, c_{l-1}$. The method comprises three main steps: first, the objective polynomial is transformed into the following parenthesized expression. Through the transformation, each arithmetic part in parentheses can be implemented as an SC circuit with NOT and AND gates.

$$\sum_{i=0}^{n}(-1)^i a_i x^{b_i} = a_0 x^{b_0}\left(1 - \frac{a_1}{a_0}x^{b_1-b_0}\left(1 - \frac{a_2}{a_1}x^{b_2-b_1}\cdots\left(1 - \frac{a_{n-1}}{a_{n-2}}x^{b_{n-1}-b_{n-2}}\left(1 - \frac{a_n}{a_{n-1}}x^{b_n-b_{n-1}}\right)\right)\cdots\right)\right)$$

Figure 4 depicts the structure of the designed circuit. The coefficient-generation circuit in the figure is designed in the second step. It generates bitstreams corresponding to the coefficients in the transformed formula from the given bitstreams. In the figure, we illustrate the structure when the number of given bitstreams is 1, i.e., $l = 1$. The circuit obtained through the two steps is an initial circuit. This initial

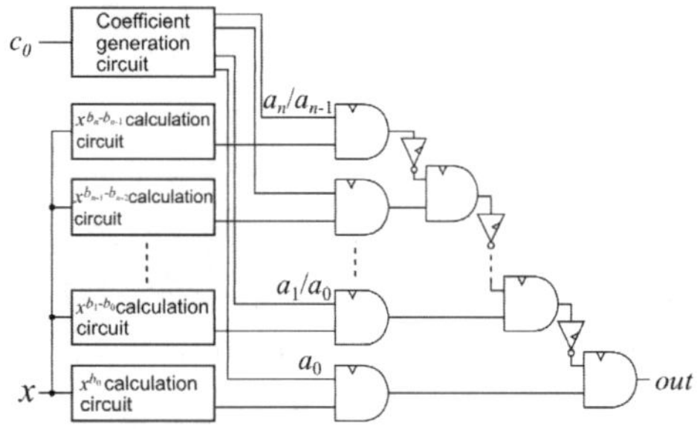

Figure 4. Structure of RSFQ SC circuits in [9].

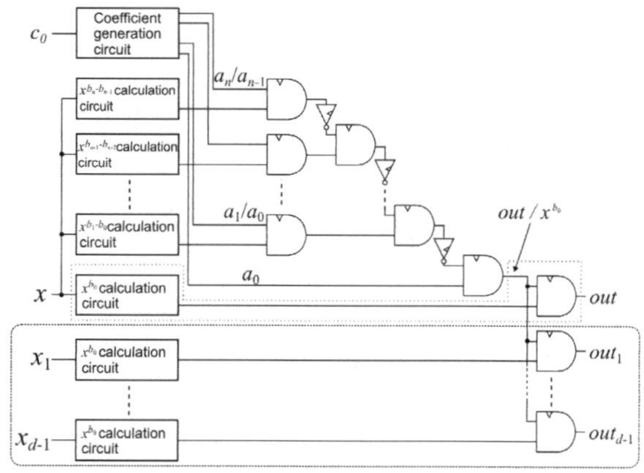

Figure 5. Structure of accuracy improved SC circuits by partial duplication. d represents the degree of duplication.

circuit is designed with clock-follow-data clocking scheme; in other words, the latching function of gate inputs is disabled.

The inputs of several gates are correlated in the initial circuit. Consequently, calculation error is significant. Therefore, to eliminate correlations and lower the calculation error of the circuit, the method inserts memory elements for signal lines in the final step. In particular, this method utilizes the latching function of gate inputs to remove the correlations and minimizes the number of inserted DFFs.

A designed circuit has one output terminal, and a large number of observation cycles for the output is necessary to obtain a calculation result. Therefore, the desirable strategy involves either reducing the number of observation cycles while maintaining the same accuracy or improving the accuracy while maintaining the original number of observation cycles.

3. Accuracy Improvement of Polynomial Computing SC Circuits by Partial Duplication

We propose a design method of polynomial calculating circuits with improved accuracy. This method is based on [9] and improves accuracy by partial duplication, i.e., duplication of a small part of the original circuit. The input of the method is the identical to that of the method in [9]. Namely, the input is the objective polynomial and constants of given bitstreams. Polynomial (1) is treated, but the circuit structure illustrated in Figure 5 is utilized instead of the structure depicted in Figure 4. In the figure, d represents the degree of duplication. The circuit uses d independent bitstreams of x as the inputs, and the output of the circuit is obtained as d bitstreams. The result is the ratio of 1s on the d bitstreams. After removing correlations in a manner similar to the method in [9], we obtain the circuit with partial duplication.

We consider the part for duplication to improve the accuracy of the calculation. Naively duplicating the original circuit can improve accuracy as the number of output bits increases in each clock cycle. However, the naive method increases the circuit size drastically. Therefore, the proposed method does not duplicate the entire circuit. In Figure 5, the numerical value of the output SN of the AND gate depicted with the arrow is out/x^{b_0}. This value of the SN can be expressed as follows:

$$a_0 - a_1 x^{b_1-b_0} + a_2 x^{b_2-b_0} - \cdots + (-1)^n a_n x^{b_n-b_0}.$$

The power of x is scaled with $a_i (0 < i \leq n)$, which is less than 1 and is generally small in practical functions. The power of x for $x \leq 1$ and the scaling implies that the bitstream fed from x exerts little influence at the AND gate output. As revealed by this observation, duplicating the calculation circuit of the SN is not an effective strategy for improving the accuracy of the result. Hence, we only duplicate

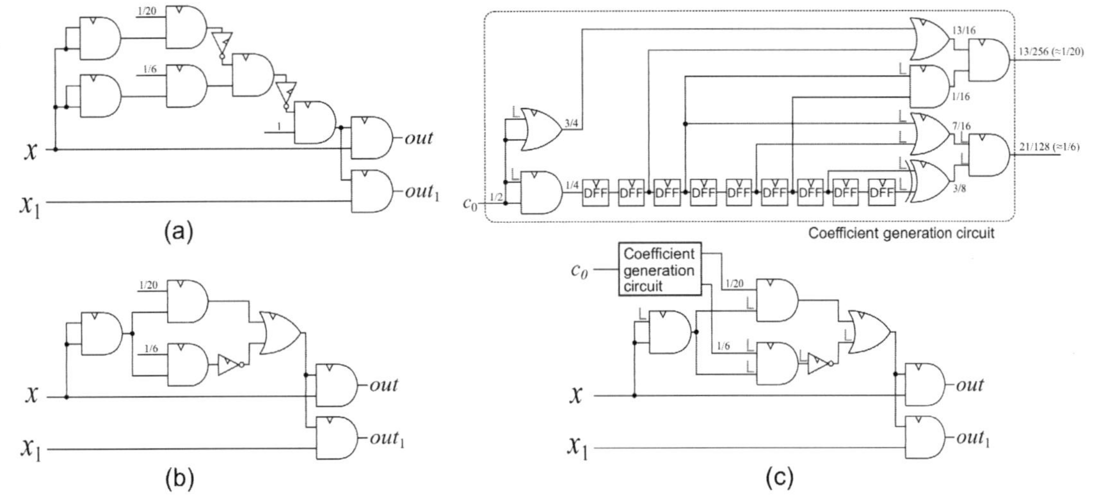

Figure 6. Example of a circuit design for polynomial $x - \frac{1}{6}x^3 + \frac{1}{120}x^5 (\approx \sin(x))$. (a) is obtained from Figure 5 with $d = 2$. (b) is the simplified circuit of (a). The coefficient-generation circuit is omitted in (a) and (b). (c) is the resultant circuit after removing correlations when $c_0 = 1/2$. "L" represents the use of the latching function of gate inputs.

the small part encircled using broken red lines in the figure and share the SN for the duplicated parts. Consequently, the area overhead of the proposed design method is negligible.

As an example, we consider $x - \frac{1}{6}x^3 + \frac{1}{120}x^5$, which is the Maclaurin expansion of the sine function. Figure 6 presents the circuits for the polynomial. Figure 6(a) depicts the circuit obtained from Figure 5. This circuit is designed for $d = 2$, and it is simplified in (b). Gates calculating the identical function are unified and useless gates are removed. DeMorgan's law is also used to reduce the number of NOT gates. As elucidated in this circuit, a circuit is obtained with partial duplication by adding AND gates at the output positions for polynomials of the following form. In other words, the first term of the polynomial is a_0x.

$$a_0x - a_1x^{b_1} + a_2x^{b_2} - \cdots + (-1)^n a_n x^{b_n}$$

The OR gate in Figure 6(b) calculates $1 - \frac{1}{6}x^2 + \frac{1}{120}x^4$ in SC. The scaling factors, $1/6$ and $1/120$, are small. The bitstream fed from x exerts little influence at the output of the OR gate as mentioned previously. Therefore, the circuit designed with the proposed method, which does not duplicate the part for calculating the value, functions effectively.

In Figure 6(b), certain correlations exist at the gate inputs. For example, the AND gate, which is expected to calculate x^2 in SC, calculates logic AND of the same logic value, implying that the calculation result of the AND gate is x in SC. These correlations result in a significant calculation error. We remove these correlations by leveraging the method in [9]. The resultant circuit is illustrated in Figure 6(c). In the figure, the coefficient-generation circuit generates bitstreams of $1/6$ and $1/20$ from the bitstream of $1/2$. The entire circuit is basically designed using the clock-follow-data clocking scheme. The latching function is enabled at the gate inputs with "L" markings. The outputs can be used as inputs of a circuit designed with the proposed method. Note that the output bitstreams of a designed circuit correlate because the outputs of the duplicated part are calculated using the signal from the original circuit. If the correlation in bitstreams increases calculation error, we can insert components, such as DFFs, to mitigate or remove the correlation.

Table 1. Evaluation Results of Mean Square Error

Functions	cycle	Without duplication	Proposed $d = 2$	$d = 3$	$d = 4$	$d = 5$
	32	0.0053 (1.00)	0.0032 (0.61)	0.0027 (0.51)	0.0022 (0.42)	0.0020 (0.38)
	64	0.0029 (0.56)	0.0018 (0.34)	0.0015 (0.28)	0.0013 (0.25)	0.0012 (0.23)
$x - \frac{1}{6}x^3 + \frac{1}{120}x^5$	128	0.0016 (0.31)	0.0010 (0.19)	0.0008 (0.16)	0.0007 (0.14)	0.0007 (0.13)
($\approx \sin(x)$)	256	0.0007 (0.14)	0.0005 (0.09)	0.0004 (0.07)	0.0003 (0.06)	0.0003 (0.06)
	512	0.0004 (0.07)	0.0003 (0.05)	0.0002 (0.04)	0.0002 (0.03)	0.0002 (0.03)
	32	0.0053 (1.00)	0.0033 (0.62)	0.0027 (0.50)	0.0022 (0.41)	0.0020 (0.38)
	64	0.0025 (0.46)	0.0016 (0.31)	0.0013 (0.24)	0.0011 (0.21)	0.0010 (0.19)
$x - \frac{1}{3}x^3 + \frac{2}{15}x^5$	128	0.0012 (0.23)	0.0008 (0.15)	0.0006 (0.12)	0.0005 (0.10)	0.0005 (0.09)
($\approx \tanh(x)$)	256	0.0006 (0.11)	0.0004 (0.07)	0.0003 (0.06)	0.0003 (0.05)	0.0003 (0.05)
	512	0.0003 (0.05)	0.0002 (0.04)	0.0002 (0.03)	0.0001 (0.02)	0.0001 (0.02)
	32	0.0055 (1.00)	0.0032 (0.57)	0.0022 (0.39)	0.0018 (0.33)	0.0016 (0.29)
	64	0.0028 (0.51)	0.0015 (0.28)	0.0012 (0.21)	0.0010 (0.18)	0.0009 (0.16)
$x - \frac{1}{12}x^3 + \frac{1}{120}x^5$	128	0.0014 (0.25)	0.0008 (0.14)	0.0006 (0.11)	0.0005 (0.09)	0.0005 (0.08)
($\approx$ 4sigmoid(x) − 2)	256	0.0007 (0.13)	0.0004 (0.07)	0.0003 (0.06)	0.0003 (0.05)	0.0002 (0.04)
	512	0.0004 (0.07)	0.0002 (0.04)	0.0002 (0.03)	0.0001 (0.02)	0.0001 (0.02)

4. Evaluation results

The improvement in accuracy of circuits designed using the proposed method was evaluated. The following three polynomials were employed:

$$x - \frac{x^3}{6} + \frac{x^5}{120}, \quad x - \frac{x^3}{3} + \frac{2x^5}{15}, \quad \text{and} \quad x - \frac{1}{12}x^3 + \frac{1}{120}x^5.$$

The first and second polynomials are Maclaurin expansions of sine and hyperbolic tangent functions, respectively. The third polynomial is the Maclaurin expansion of a scaled sigmoid function without the constant term and corresponds to 4sigmoid(x) − 2.

The circuits are designed for $c_0 = \frac{1}{2}$. Software simulation was performed by utilizing Python code to evaluate the designed circuits. The outputs of circuits were calculated for x ranging from 0 to 1 for every 1/1000. The mean square error, the average of squares of the difference between the actual polynomial result and the obtained result, and the mean absolute error, the average of absolute differences between the actual polynomial result and the obtained result, were evaluated. Tables 1 and 2 summarize the result. Circuits designed with [9] and the proposed circuits up to $d = 4$ were investigated. The cycle is the number of clock cycles for a calculation. The circuits designed with the proposed method output $32 \times d$ bits when observing the output for 32 cycles. The figures in parenthesis represent the ratio with respect to the error of 32-cycle operations of the circuits without duplication.

Tables 1 and 2 indicate that partial duplication improves accuracy. In 32-cycle operations, compared to circuits without duplication, the mean square error of partially duplicated circuits with $d = 2$ was reduced to approximately 60%. From the central limit theorem, mean square error, which takes the same value as the variance when the average of calculation results and the actual result is the same, is expected to decrease inversely proportional to the number of observed bits. For example, a calculation with 64 cycles is expected to be about half the error of a calculation with 32 cycles. Thus, as d increases, the mean square error of circuits designed with the proposed method is expected to be reduced to $1/d$ with the same number of cycles. Though the error of $d = 2$ was slightly larger than the expected error, i.e., 50%, the proposed partial duplication method improved the accuracy by adding a small number of gates. The mean absolute error in Table 2 shows a similar trend. In other words, though the error of the partially

Table 2. Evaluation Results of Mean Absolute Error

Functions	cycle	Without duplication	Proposed $d=2$	$d=3$	$d=4$	$d=5$
$x-\frac{1}{6}x^3+\frac{1}{120}x^5$ ($\approx \sin(x)$)	32	0.0574 (1.00)	0.0443 (0.77)	0.0393 (0.69)	0.0359 (0.63)	0.0338 (0.59)
	64	0.0431 (0.75)	0.0328 (0.57)	0.0295 (0.51)	0.0273 (0.48)	0.0259 (0.45)
	128	0.0314 (0.55)	0.0248 (0.43)	0.0220 (0.38)	0.0201 (0.35)	0.0191 (0.33)
	256	0.0210 (0.37)	0.0165 (0.29)	0.0144 (0.25)	0.0134 (0.23)	0.0127 (0.22)
	512	0.0155 (0.27)	0.0122 (0.21)	0.0109 (0.19)	0.0100 (0.17)	0.0096 (0.17)
$x-\frac{1}{3}x^3+\frac{2}{15}x^5$ ($\approx \tanh(x)$)	32	0.0565 (1.00)	0.0450 (0.80)	0.0396 (0.70)	0.0356 (0.63)	0.0338 (0.60)
	64	0.0384 (0.68)	0.0309 (0.55)	0.0276 (0.49)	0.0254 (0.45)	0.0238 (0.42)
	128	0.0281 (0.50)	0.0220 (0.39)	0.0195 (0.35)	0.0179 (0.32)	0.0168 (0.30)
	256	0.0189 (0.34)	0.0155 (0.27)	0.0138 (0.24)	0.0128 (0.23)	0.0121 (0.21)
	512	0.0135 (0.24)	0.0110 (0.20)	0.0097 (0.17)	0.0087 (0.15)	0.0084 (0.15)
$x-\frac{1}{12}x^3+\frac{1}{120}x^5$ ($\approx 4\mathrm{sigmoid}(x)-2$)	32	0.0586 (1.00)	0.0442 (0.75)	0.0366 (0.62)	0.0337 (0.58)	0.0314 (0.54)
	64	0.0417 (0.71)	0.0307 (0.52)	0.0268 (0.46)	0.0244 (0.42)	0.0230 (0.39)
	128	0.0290 (0.49)	0.0219 (0.37)	0.0189 (0.32)	0.0173 (0.30)	0.0161 (0.27)
	256	0.0206 (0.35)	0.0157 (0.27)	0.0137 (0.23)	0.0125 (0.21)	0.0116 (0.20)
	512	0.0154 (0.26)	0.0109 (0.19)	0.0096 (0.16)	0.0087 (0.15)	0.0081 (0.14)

duplicated circuits with $d = 2$ was slightly larger than those without duplication with double observation cycles, the partial duplication reduced the error.

5. Conclusion

We have proposed a design method of SC circuits that improves the accuracy by partial duplication. This method reduced the number of gates to be added for duplication through sharing of most of the internal parts between the original part and duplicated parts. By considering the effectiveness to improve accuracy, we selected the part to be duplicated.

The evaluation results demonstrated that the error of the circuits calculating Maclaurin expansions of practical functions was smaller than that of the circuit without duplication with the same number of clock cycles.

References

[1] Likharev K and Semenov K 1991 *IEEE Transactions on Applied Superconductivity* **1** pp 3-28
[2] Alaghi A and Hayes J P 2013 *ACM Trans. Embed. Comput. Syst.* **12** ISSN 1539-9087
[3] Kang J, Przybysz J, Martinet S, Worsham A, Miller D and McCambridge J 1997 *IEEE Transactions on Applied Superconductivity* **7** pp 2673-2676
[4] Yamanashi Y and Yoshikawa N 2009 *IEEE Transactions on Applied Superconductivity* **19** pp 630-633
[5] Sato K, Sega N, Somei Y, Shimada H, Onomi T, Mizugaki Y 2022 *IEICE Transactions on Electronics* **E105.C** pp 296-299
[6] Cai R, Ren A, Chen O, Liu N, Ding C, Qian X, Han J, Luo W, Yoshikawa N and Wang Y 2019 *46th International Symposium on Computer Architecture (ISCA 2019)* pp 567-578
[7] Kito N, Odaka R and Takagi K 2019 *IEICE Transactions on Electronics* **E102.C** pp 607-611
[8] Kito N and Takagi K 2021 *Journal of Physics: Conference Series* **1975** 012025
[9] Wada K and Kito N 2022 *Journal of Physics: Conference Series* **2323** 012032
[10] Yamanashi Y et al. 2007 *IEEE Transactions on Applied Superconductivity* **17** pp 474-477
[11] Tanaka M, Akaike H, Fujimaki A, Yamanashi Y, Yoshikawa N, Nagasawa S, Takagi K, Takagi N 2011 *IEEE Transactions on Applied Superconductivity* **21** pp 792-796
[12] Peng X, Xu Q, Kato T, Yamanashi Y, Yoshikawa N, Fujimaki A, Takagi N, Takagi K, Hidaka M 2015 *IEEE Transactions on Applied Superconductivity* **25** Art. ID 1301106
[13] Nagaoka I, Ishida K, Tanaka M, Sano K, Yamashita T, Ono T, Inoue K and Fujimaki A 2021 *IEEE Transactions on Applied Superconductivity* **31** Art. ID 1302505

[14] Nagaoka I, Tanaka M, Inoue K and Fujimaki A 2019 *IEEE International Solid-State Circuits Conference (ISSCC)* pp 460-462

Journal of Physics: Conference Series **2545** (2023) 012025 doi:10.1088/1742-6596/2545/1/012025

Demonstration and thermal equilibrium analysis of a 10 kJ capacity energy storage coil made of $MgB_2$with liquid hydrogen indirect cooling

R Inomata[1], T Onji[1], T Yagai[1], Y Makida[2], T Shintomi[2], T Komagome[3], N Hirano[4], T Hamajima[5]

[1]Sophia University, 7-1 Kioi, Chiyoda, Tokyo 102-8554, Japan
[2]High Energy Accelerator Research Organization (KEK), Oho 1-1, Tsukuba, Ibaraki 305-0801, Japan
[3]Mayekawa Manufacturing Co., Ltd., 3-14-15 Botan, Koto, Tokyo 135-8482, Japan
[4]National Institute for Fusion Science, Oroshicho 322-6, Gihu, Toki 509-5292, Japan
[5]Tohoku University, 6-6-05 Aramaki Aza Aoba, Aoba-ku, Sendai, Miyagi 980-8579, Japan
E-mail: r-nomata-9k6@eagle.sophia.ac.jp

Abstract. We have demonstrated an advanced superconducting power conditioning system, in which a superconducting magnetic energy storage (SMES) device, a generator based on a fuel cell (FC), and an electrolyzer are used to compensate for electricity fluctuations over a wide frequency range, combined with a liquid hydrogen storage system to both cool the SMES and provide pure hydrogen gas to the FC and other gas-dependent systems. To manufacture the coils for the SMES, we used MgB_2, whose critical temperature is below the boiling temperature of hydrogen. We developed a 10 kJ SMES coil system indirectly cooled by liquid hydrogen using thermosyphon passive heat exchange to isolate the flammable hydrogen from the electrical components. We performed a successful demonstration of this system for both DC and AC currents ramped at different rates. In the present study, we use computer simulations involving heat balance equations to evaluate the stability of the system. The results obtained are expected to lead to the design of future large-capacity energy storage systems, such as the MJ class, which offer comparable performance to conventional NbTi SMES devices.

1. Introduction

Environmental problems have now become an urgent worldwide concern. To mitigate such problems, hydrogen, which does not produce carbon dioxide when burned, is one of the most attractive energy sources. Various applications of hydrogen energy, such as in power generation facilities and electric vehicles powered by fuel cells, are being considered. However, moving towards a hydrogen economy requires the storage and transport of a large amount of hydrogen. The most promising approach is the use of liquid hydrogen, which requires a far smaller volume than the gaseous form. This is especially important in Japan, where space for constructing large storage tanks is limited. Liquid hydrogen can also be used a refrigerant for superconducting power devices. However, due to its high flammability, it is not generally used for purposes that would bring it into contact with electrical components. It is therefore important to develop methods to allow liquid hydrogen to be used as a refrigerant in such

Journal of Physics: Conference Series **2545** (2023) 012025 doi:10.1088/1742-6596/2545/1/012025

situations. Development of a liquid hydrogen coolant supply chain would be expected to have synergistic effects across a wide range of industries.

For renewable energy sources such as solar power and wind power, the amount of energy produced fluctuates depending on environmental conditions, and a means must be developed to compensate for such fluctuations. To achieve this and minimize time lags, a power storage device is required that can be charged and discharged over a wide range of response times. In the present study, we investigated the use of a superconducting magnetic energy storage (SMES) system, owing to its fast response, its stability over many charge-discharge cycles, and its ability to use liquid hydrogen as a refrigerant. The superconducting coils in the proposed system use MgB_2. Since the coils do not use rare or precious metals, the SMES is expected to offer cost savings over systems that use copper oxide based superconducting coils.

To demonstrate the feasibility of the MgB_2-based SMES in our advanced superconducting power conditioning system using liquid hydrogen cooling, we developed SMES coils with an overall capacity of 10 kJ. We performed DC and Triangular Pulse current charge/discharge tests with several ramp rates using indirect liquid hydrogen cooling. Here, we report the results of these experiments, together with those for simulations based on thermal equilibrium solution equations.

2. Conductor and double-pancake coil design and specifications

2.1. MgB_2 wires

In this research, we used two types of MgB_2 wires produced by Hyper Tech Research Inc. Figure 1 shows cross sections of the wires, and Table 1 lists their specifications [1]. The 30-NM consists of thirty MgB_2 filaments with a filling factor of 17%. It also contains 12% Cu to provide sufficient thermal stability to prevent burning upon quenching. The outer sheath is composed of CuNi without any coating, and Nb is used as a diffusion barrier around each MgB_2 filament to prevent chemical reaction between Mg and the Cu core. The 24-NM consists of twenty-four MgB_2 filaments with a filling factor of 11%. Both of these wires were manufactured using a powder-in-tube method with a continuous tube forming and filling process, to reduce the time required to fabricate long wires.

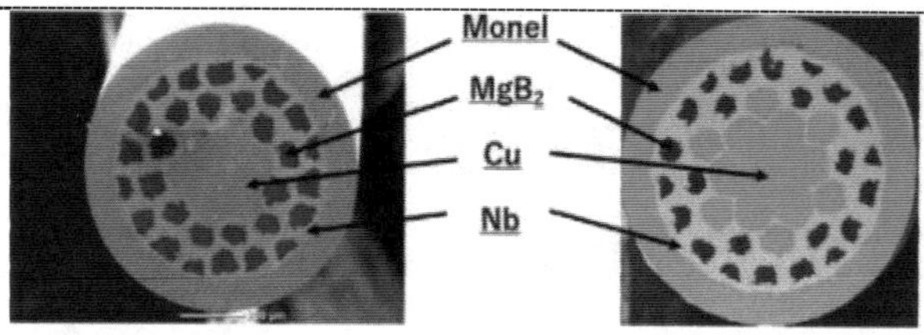

Figure 1 Cross-sectional images of (left) 30-NM and (right) 24-NM.

Table 1 Specifications of MgB_2 wires

Name of wires	**30-NM**	**24-NM**
Number of filaments	30	24
Diameter	0.83[mm]	0.84[mm]
Diameter of filament region	~0.07[mm]	~0.07[mm]
Fill factor	17[%]	11[%]
Cu ratio	12[%]	20[%]

Permissible strain (before reaction)	4[%]	4[%]
Permissible strain (after reaction)	0.23[%]	0.23[%]
Jc(20K,2.0T)	1780[A/mm^2]	

2.2. MgB_2 conductors and coils

The SMES system used in this study consists of three stacked double-pancake coils connected in series. We previously fabricated three types of MgB_2 test coils to establish the manufacturing method [2], and these coils, referred to as DP-1, DP-2 and DP-3, were used in the present study. The inner and outer diameters of the coils are identical, but the thickness and number of turns are different. Figure 2 show cross-sectional schematics of these coils, and Table 2 lists their specifications.

DP-1 was fabricated using the wind & react (W&R) method, in which the conductor and coil windings are finished before heat treatment is performed [1][3]. This conductor has eight strands and four brass wires twisted around a copper mandrel. This ensures high thermal stability because of the presence of fewer strands, and therefore a smaller amount of copper. DP-2 and DP-3 were fabricated using the react & wind (R&W) method, in which a heat treatment is performed before the winding process, and have a bending strain tolerance of 0.23%, which is lower than that for DP-1 (4%). Therefore, for DP-2 and DP-3, an internal mandrel for reducing the bending strain is not required during the winding process. DP-2 uses 24-NM, which has a higher copper content but a lower critical current. DP-3 has a higher critical current due to the increased number of 30-NM strands. To achieve this, the width of the conductor and the thickness of the coil are larger. The critical current for each type of coil was calculated in a previous study based on the number of wires and the wire deterioration rate [4]. The critical current for DP-1, DP-2 and DP-3 was calculated to be 915, 758, and 1374 A under a maximum self-generated magnetic field of 2.0 T at 20 K. In the stacked-coil system, DP-1, which has a small number of turns and undergoes less deterioration, was placed at the centre, sandwiched by DP-2 on the top and DP-3 on the bottom. Since the coil located at the centre is exposed to the largest self-generated magnetic field, this arrangement was expected to optimize the system stability. In addition, a glass-fiber-reinforced polymer plate was attached to protect the surface of the coil, together with a Cu plate to cool the coils indirectly. A large number of slits were introduced into the Cu plates to minimize the eddy current loss. The current leads were stacked NbTi and Bi2223 conductors with a width of 100 mm and a length of 15 mm.

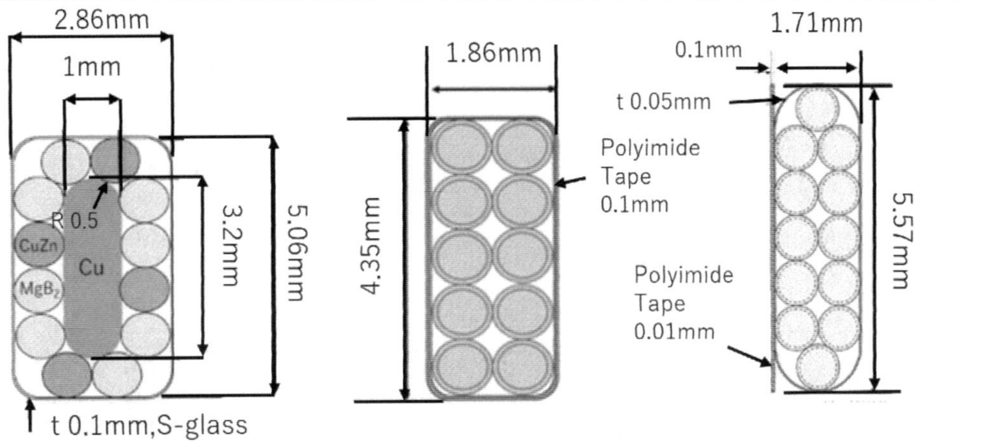

Figure 2 Schematic cross sections of (left) DP-1, (middle) DP-2 and (right) DP-3 conductors.

Table 2 Specifications of DP coils

MgB_2 coil	**DP-1**	**DP-2**	**DP-3**
Winding method	W&R	R&W	W&R
MgB_2 strand	30-NM	24-NM	30-NM
Twist pitch of strands	51[mm]	82[mm]	82[mm]
Inner diameter	400[mm]		
Outer diameter	603[mm]		
Number of turn	72	110	110
Rated current	600[A]at20K		
Inductance	58.1[mH]		
Stored Energy	10.5[kJ]		

2.3. Cooling system

When liquid hydrogen is used as a coolant, explosion-proof construction is required. In the present study, to allow safe operation, indirect cooling using thermosiphon circulation is employed. Figure 3 shows the internal structure of the cryostat. The tank that stores the refrigerant is installed in the centre of the cryostat. The piping passes through the lower part of the refrigerant storage unit and returns to the upper part. The driving force for coolant circulation is provided automatically by the gravitational force and boiling pressure of the coolant without any mechanical pumps. The MgB_2 coils are placed below the tank. Copper plates with eddy-current termination slits for heat conduction are connected to the piping. This indirectly cools the coils. Since the volume inside the cryostat is vacuum, each part of the device is subject to adiabatic conditions except for the parts with contacting Cu plates.

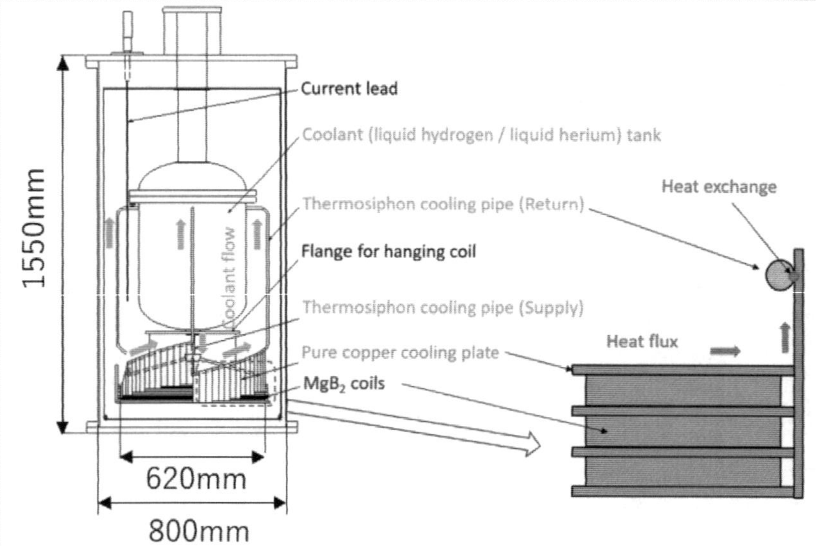

Figure 3 Schematic of internal structure of cryostat. The thick green arrows show the refrigerant flow and the red arrows show the heat flux.

3. Experimental details

Liquid hydrogen was circulated in the cooling device, cooling the coil and transforming it to a superconducting state. It took about 4 hours to store the liquid hydrogen in a tank at room temperature, and another 4 hours for the coils to reach liquid hydrogen temperature. When the coil temperatures measured by sensors reached around 21 K, we evaluated the performance of the coils under DC and AC currents. The location of the sensors is shown in Figure 5.

3.1. DC excitation tests

The DC current test was performed by ramping the current from 0 A to 231, 288, 300, 320, or 346 A, and then returning it to 0 A with no holding time. The ramp rate was varied up to 50 A/s, which is the maximum rate for the current source used in this experiment. The Joule loss associated with DC currents of 231, 288, and 346 A is equivalent to that for triangular plus currents of 400, 500, and 600 A, respectively.

3.2. Repetitive Triangular Pulse excitation tests

The purpose of the MgB_2 coil in this study is compensation for fluctuating renewable energy. As a basic experiment, the coil was excited by a triangular current with constant sweep speed. Repetitive Triangular Pulse excitation tests were first conducted at the maximum coil current of 600 A. Additional tests were performed with stepwise decreases of the maximum current from 500 to 400 A at a fixed sweep rate of 50 A/s, and with a maximum current of 450 A with a stepwise increase in sweep rate from 10 to 50 A/s.

4. Experimental results

4.1. DC excitation test results

No quenching occurred after 30 min of excitation at a current lower than 320 A, whereas quenching occurred after several hundred seconds at a current of 320 A or more. Figure 4 shows the change in coil temperature during the 320 A excitation test. The inset indicates that current shut-down was caused by an exponential increase in voltage. We installed a quench detection system with a detection voltage of 2 V, as determined from the copper rate in the coils so that coil is not burned. Figure 5 shows a photograph of the coils, where the temperature measurement points are indicated. In Figure 4, the temperature of terminal N, which is the DP-2 terminal connected to the negative inlet of the power supply (copper, RRR=10, width=100 mm, length=120 mm), continuously rose even after current shut-down. This is thought to be because the heat-insulated terminal generates Joule loss, initiating DP-2 quenching after a specific time delay depending on the transport current, and the heat generated inside the coil is transported to the terminal after a delay of several hundred seconds.

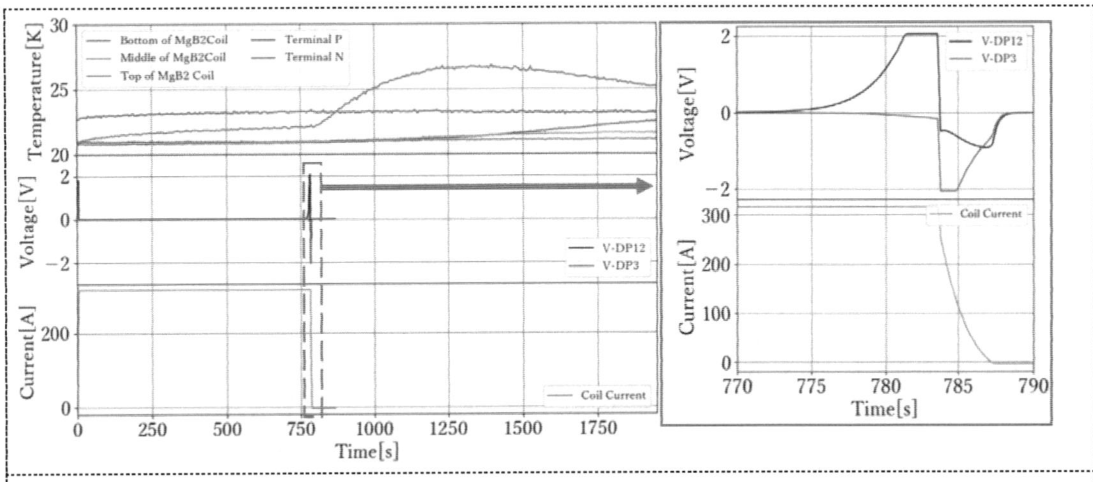

Figure 4 Change in coil temperature and voltage during 320 A DC excitation test.

Figure 5 Coil temperature measurement points. The sensor that measures the temperature of the coil is installed on the inner diameter of the coil from above the resin.

4.2. Repetitive Triangular Pulse excitation results

For the first time, we succeeded in energizing the coils at a rated current of 600 A at 20 K for four pulse periods, using design and fabrication processes for constructing energy storage devices. Quenching occurred for a current of 420 A for a sweep rate of 50 A/s, and for a sweep rate of 35 A/s or more for a maximum current of 450 A. Figures 6 and 7 show the test results for continuously changing current. Similar to the DC excitation test, the temperature of terminal N temperature rose, and the voltage increase at DP1-2 was significant enough to trigger current shut-down for coil protection. Also, detection voltage is 2V as same as DC excitation tests.

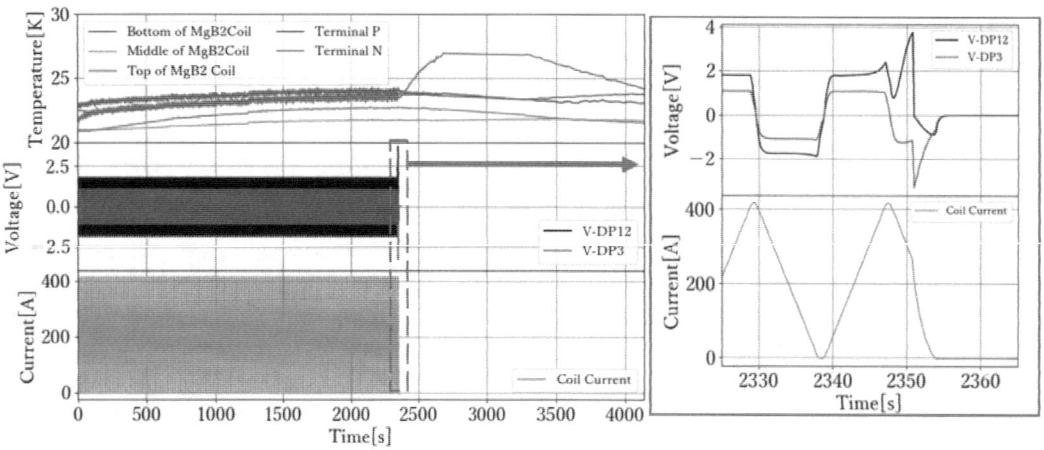

Figure 6 Change in coil temperature and voltage in 420 A for sweep rate of 50 A/s Repetitive Triangular Pulse excitation test.

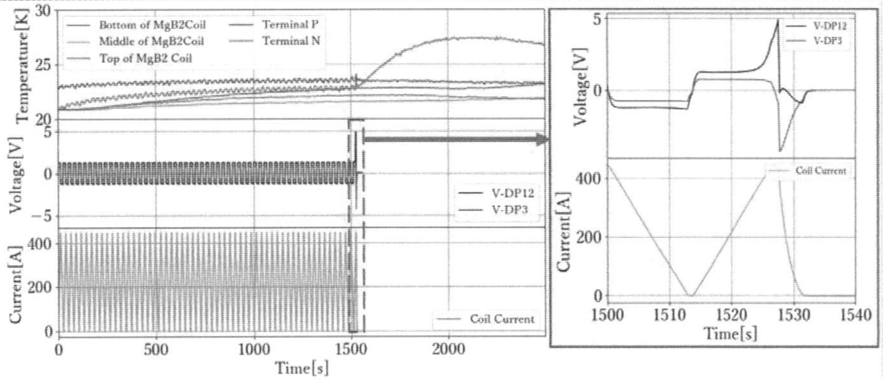

Figure 7 Change in coil temperature and voltage in 450 A for sweep rate of 35 A/s Repetitive Triangular Pulse excitation test.

4.3. Loss calculation

The loss was calculated from the experimental data, and the results are shown in Figure 8. The quenching threshold was about 3 W in the DC excitation test, and about 4 W in the Repetitive Triangular Pulse excitation test. Therefore, it can be inferred that Joule loss contributed more to quenching. To briefly summarize the test results, due to the adiabatic conditions at the terminal and the larger heat capacity than that for a conductor that is thermally connected to the terminal, most heat generation and conduction to initiate quenching in the conductor used in DP-2 could have been taken place around the terminal. To clarify the temperature rise around the connection between DP-2 and the terminal, we performed a numerical analysis of heat conduction based on the two-dimensional thermal-equilibrium equation.

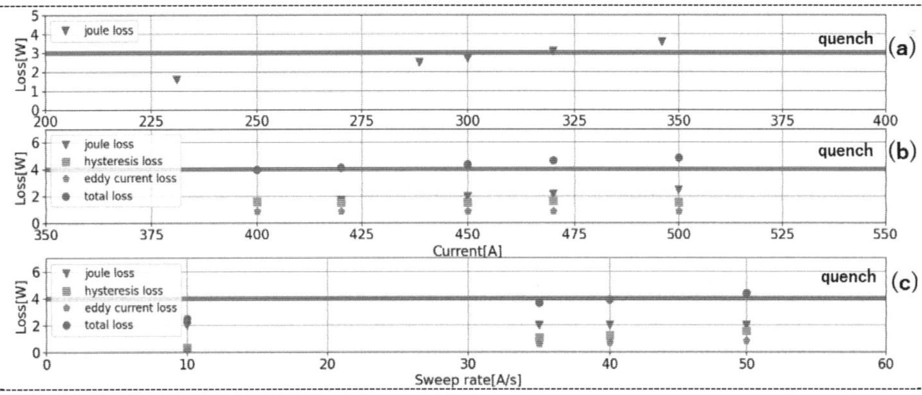

Figure 8 Losses during (a) DC excitation test, (b) 450 A,35 A/s Repetitive Triangular Pulse excitation test, (c) 420 A,50 A/s Repetitive Triangular Pulse excitation test.

5. Theoretical analysis and discussion

5.1. Quenching location

Terminal N, which experienced a noticeable temperature increase in this experiment, is attached to DP-2. It has the lowest critical current among the three DPs, as estimated from the properties of the wires manufactured by Hyper Tech Research Inc. We therefore considered that quenching occurred due to the low rated-current margin for DP-2 and the adiabatic conditions at the terminals.

5.2. Thermal equilibrium analysis

To investigate the reason that quenching took place, the heat diffusion equation was applied to the internal coil temperature. The calculations were performed using the following heat diffusion equation:

$$\gamma C\frac{\partial T}{\partial t}=\frac{\partial}{\partial x}\left[k_x(T)\frac{\partial T}{\partial x}\right]+\frac{\partial}{\partial y}\left[k_y(T)\frac{\partial T}{\partial y}\right]+G(T,I_t)-Q(T,t)+G_d(x,t)\left[\frac{W}{m^3}\right] \quad (1)$$

where γ is the mass density of Cu, C is the specific heat of Cu, and k_x and k_y are the heat transfer coefficients along the x and y axes for the coil, respectively. G_d represents heat generation due to the electrical resistance of the normal metal, Q is a cooling term representing the removable heat flux, and G is the term that approximates the heat generated per unit volume of the superconductor when the MgB_2 coil turn to the normal or when the energizing current exceeds the critical current and a voltage emerges, also we assume that the current flows through the Cu in the conductor. These terms are calculated using the following formula.

$$G=\rho_{Cu\_MgB2}\frac{I_{op}^2}{A_{Cu}A_{tot}}f \quad f=\begin{cases}0 \quad for \quad T\leq T_{cs}\\ \frac{T-T_{cs}}{T_c-T_{cs}} \quad for \quad T_{cs}<T<T_c\\ 1 \quad for \quad T\geq T_c\end{cases} \quad (2)$$

$$Q=a_f\frac{p_e}{A_{Cu}}\Delta T \quad (3)$$

$$G_d=\rho_{Cu\_terminal}\left(\frac{I_{op}}{S_{Cu}}\right)^2 \quad (4)$$

Here, I_{op} is the transport current, $\rho_{Cu\_MgB_2}$ is the resistivity of Cu in the MgB_2 conductor (RRR=150), A_{tot} is the total area of the conductor, $A_{Cu}$is the areal cross section of copper in the conductor, T_{cs} is the current sharing temperature, T_c is the critical temperature, a_f is a cooling factor, p_e is the coolingperimeter, $\rho_{Cu\_terminal}$ is the resistivity of terminal N (RRR=10) and $S_{Cu}$is the areal cross section of the terminal. Resistivity of Cu is RRR dependent, so it is different depends on analysis area. Figure 9 shows the analysis model. In this analysis, the MgB_2 coil is modelled as a rectangle with a width of 103 mm and a length is 300 mm. In this calculation, a simple model was used to roughly investigate the cause of the delayed temperature increase in terminal N temperature after quenching. In particular, the coil close to the terminal is modelled as a rectangle to simulate anisotropic heat transmission in both the axial (along the conductor) and radial (across the turn-to-turn insulation) directions. It is assumed that a direct current flows through each of the 55 turns of the conductor. The resistivity of the terminal is assumed to be uniform but to have a temperature dependence. The boundary conditions for the perimeter are adiabatic. The analytic model is consists of current lead, terminal N, and MgB_2 coil. Therefore, the material parameters in equation (1) are different for each part of the model. In the current lead, there is no heat source, so G and G_d are ignored. Terminal N is in the adiabatic condition, and heat source of it is only heat generation due to the electrical resistance of the normal metal, so G and Q are ignored. Heat source inside MgB_2 coil is flux flow voltage and then heat from Cu matrix, so G_d is ignored.

5.3. Calculation results and discussion

The results of the analysis are shown in Figure 10. In this analysis, it took about 430 s for quenching to occur. The terminal temperature increased to 40 K, which was higher than that for the DC excitation test. However, the temperature rise that was experimentally observed after quenching did not occur. In the excitation test, it was considered that the temperature of the quenched coil was higher than that of the terminal, so the heat generated in the coil could flow into the terminal, causing a temperature increase. When the calculations were performed with a uniform terminal resistance, the terminal temperature was higher than that of the coil. This suggests that the resistance of the terminal is not uniform and the large contact resistance between the coil and the terminal caused significant heat generation in the coil, which means that highly localized resistance might trigger quenching, causing a larger heat flux to return to the terminal.

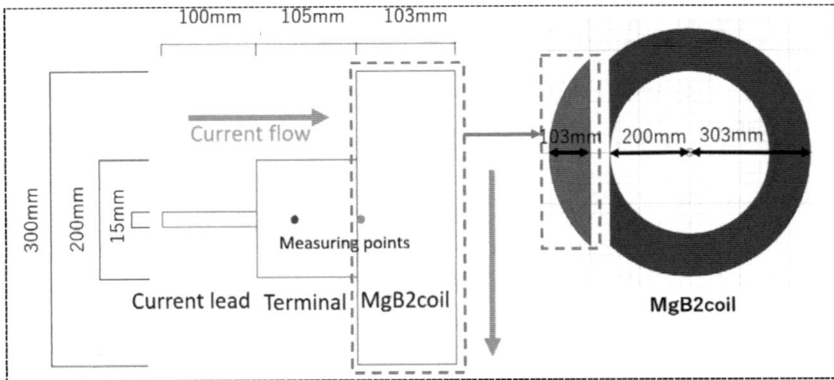

Figure 9 Analysis model. The orange arrows show the current flow, and the green and purple dots are measurement points.

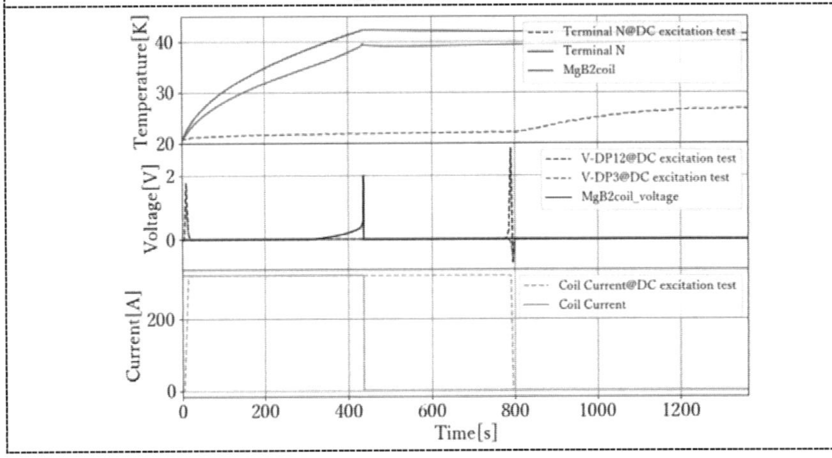

Figure 10 Analysis results for change in coil temperature and voltage.

6. Conclusion

We manufactured a MgB_2 coil with a capacity of 10.5 kJ and tested its excitation at liquid-hydrogen temperature using indirect thermosiphon cooling. We succeeded in energizing the coils at the rated current of 600 A for four pulse periods before quenching occurred, possibly due to unexpected heat generation around the DP-2 coil. We investigated the temperature change in the MgB_2 coil using a thermal equilibrium analysis, which revealed that the terminal resistance was not uniform and that there may be a large contact resistance between the terminal and the DP-2 coil. It is necessary to refine the calculation model and boundary conditions to emulate the experimental conditions more precisely, and further numerical studies are needed in order to clarify internal thermal fluctuations and realize larger-scale SMES coil design for practical uses.

References

[1] Komagome T, et al. 2020 Development of MgB_2 Rutherford Cable for Large-scale SMES Coils Cooled by Liquid Hydrogen. *TEION KOGAKU.* **55** 36–43

[2] Yagai T, et al. 2018 Development of design for large scale conductors and coils using MgB_2 for Superconducting Magnetic Energy Storage Device. *Cryogenics.* **96** 75–82

[3] Tomita M, et al. 2018 Superconding Properties of a Prototype Pancake Coil Using a MgB_2 Rutherford-Type Stranded Conductor. *2018 IEEE Trans. Appl. Supercond*, **28** 5700604

[4] Yagai T, et al.2021 Development of Large Currentt Capacity MgB_2 Conductors as a Potential Application of Large-Scale SMES Magnets. *TEION KOGAKU.* **56** 269–276

Test operation of 500 m /100 MW Ishikari high-temperature superconducting DC transmission line to demonstrate the required cost reduction and robustness

Noriyuki Inoue[1,2], Noriko Chikumoto[2] , Hirofumi Watanabe[2] , Yuri Ivanov[2], Tomoaki Hino[2], Akio Sato[1,3] , Toru Sawamura[1,4] , Hisashi Yamamoto[1,5], Katsunori Sumi[1,2] , Kiyoshi Okuno[2] and Osamu Motojima[1,2]

[1] Ishikari Superconducting DC Power Transmission System, c/o Chubu University 1200 Matsumoto-cho, Kasugai, Aichi, Japan

[2] Chubu University, 1200 Matsumoto-cho, Kasugai, Aichi , Japan

[3] JFE Steel Corporation, 2-2-3 Uchisaiwaicho, Chiyoda-ku, Tokyo, Japan

[4] SAKURA internet Inc., 1-12-1 Umeda, Kita-ku, Osaka, Osaka, Japan

[5] JGC CORPORATION, 2-3-2 Minatomirai, Nishi-ku, Yokohama, Kanagawa, Japan

inoue@isc.chubu.ac.jp

Abstract. This paper presents the results of the Ishikari Line 1 restart test conducted in 2021 to collect basic data for cost reduction, improved efficiency, and confirm the integrity of the equipment. The Ishikari Superconducting DC transmission (SCDC) demonstration test facility was constructed in 2013 in Ishikari, Hokkaido. It comprises the world's longest class of 500 m (Line 1) and 1000 m (Line 2) SCDC systems. In 2015, Line 1 successfully demonstrated the world's first power transmission from a solar power plant to an Internet data center. The restart test was conducted approximately six years after shutdown, and a strong earthquake occurred during the shutdown period. The restart test operation confirmed the integrity and robustness of the facility, which had experienced a major earthquake. Cooling tests and electrical tests were conducted, and useful data were obtained, including efficient cooling operation and the integrity and shunt current of the superconducting cables. Based on these results, we aim to promote the social implementation of SCDC by compiling a database for the quantitative evaluation of reliability, safety, and cost.

1. Introduction

Superconducting DC transmission (SCDC) can reduce the power losses during transmission to one-tenth of the losses occurring in conventional AC transmission. Therefore, it is a key technology in achieving carbon neutrality (CN) [1, 2]. The advantages of SCDC primarily include its ability to transmit power over long distances at a low voltage and the compact geometry of the superconducting cables. Owing to their compactness, underground cables that are robust against natural disasters, such as typhoons and earthquakes, can be installed at low costs and negligible environmental impact [3]. SCDC is expected to address several problems, including insufficient capacities of existing transmission lines, which may result as renewable energy consumption increases. Thus, innovative power networks can be developed.

However, the social implementation of SCDC may require further preliminary research and the creation of a new business model based on the resulting database. Accordingly, as the first step in the process, this paper describes the results of the Ishikari Line 1 restart test conducted in September 2021 to collect basic data for cost reduction and improved efficiency and to confirm integrity of the equipment after approximately six years of shutdown. Based on these results, we aim to promote the social implementation of SCDC by compiling a database for the quantitative evaluation of reliability, safety, and cost.

2. Results of the Ishikari project and restart operation

2.1 Achievements to Date and Objectives of Reactivation

The Ishikari superconducting DC power transmission system (I-SPOT) was developed in January 2014 as a collaborative innovation partnership (CIP) under the authority of the ministry of economy, trade, and industry (METI). Its primary objective is to implement research activities on SCDC and related technologies, including design, construction, and performance demonstrations, at large-scale facilities. The Ishikari SCDC demonstration test facility was constructed in 2013 in Ishikari, Hokkaido. It comprises the world's longest class of 500 m (Line 1) and 1000 m (Line 2) cable systems [4–6]. Table 1 lists the main specifications of Ishikari SCDC Lines 1 and 2. Based on demonstration tests conducted on Lines 1 and 2 of the SCDC system throughout four years, it was confirmed that SCDC can reduce the transmission loss to one-tenth of the conventional AC transmission system losses. An integral technology for this reduction was the recently developed ultra-low-loss cryogenic pipe, which achieved a heat loss of less than 1 W/m along the cable [7–8].

2.2 System Overview of Line 1

Figure 1 illustrates an overview of the Ishikari SCDC Line 1. In September 2015, we successfully demonstrated the world's first power transmission from a solar power plant to an Internet data center using Line 1, which is a 500 m SCDC system (5 kA, 20 kV, 100 MW) installed underground [9]. Following two comprehensive tests (cooling and energizing operations) in 2015, the operation of Line 1 was suspended for an extended period and restarted in September 2021, with neither cooling nor energizing conducted during the period in which it was suspended, whereas a vacuum was maintained inside the cryogenic pipe. During the shutdown period, a strong earthquake occurred (2018 Hokkaido Eastern Iburi Earthquake, September 6, 2018, magnitude-seven earthquake), and the Ishikari area experienced magnitude-five Lower earthquake. The restart operation after the suspension of operation and earthquake confirms the integrity and robustness of the facility.

3. Restart operation test of Line 1

The primary components of Line 1 include the cooling, insulation, vacuum, electrical, measurement, and monitoring systems. Figure 2 shows a cross section of the cryogenic pipe and superconductor cable. The interior of the cryogenic pipe was vacuumed, and liquid nitrogen flowed through the two inner pipes (cable and return pipes). A superconducting cable was installed inside the cable pipe. Terminal cryostats were installed at both ends of the cryogenic pipe. Subsequently, measurement signals, such as power, liquid nitrogen refrigerant, superconducting cable, and current lead voltage, were transmitted at low and room temperatures.

Table 1. Specifications of the Ishikari SCDC Line 1 and Line 2

			Line 1	**Line 2**
Length, type			Underground 500 m line	Overland 1000 m line (two 250 m lines *3)
Current, voltage			DC 5 kA,20 kV	DC 2.5 kA,20 kV
Cable	**Type**		BSCCO superconducting cable	
	Diameter		42 mm	40 mm
	Critical Current@77 K	Outer conductor	> 6.3 kA	> 3.3 kA
		Inner conductor	> 6.6 kA	> 4.3 kA
Cooling system	Refrigerator	Main(2kW@77K) *1	1	2
		Sub(1kW@77K) *2	1	2
	LN2 circulator(100 kPa - 40 l/min)		2	3

*1 Turbo-Brayton refrigerator, *2 Stirling refrigerator, *3 repaired in 2020

The following procedure was adopted for the restart test: (1) vacuum evacuation of the cryogenic pipe; (2) initial cooling of the cable and return pipes (flow of low-temperature nitrogen gas); (3) circulation of liquid nitrogen using a refrigerator and circulation pump; (4) conducted cooling and circulation tests, as well as various electrical tests; and (5) following the completion of all tests, the temperature was increased to the room temperature and the entire operation was terminated. During the restart test, cooling circulation and electrical tests were performed under various operating conditions in addition to the rated conditions.

The integrity and robustness of all system components, including the SCDC main unit, refrigerator, control equipment, vacuum system, and measurement system, have been confirmed [10, 11].

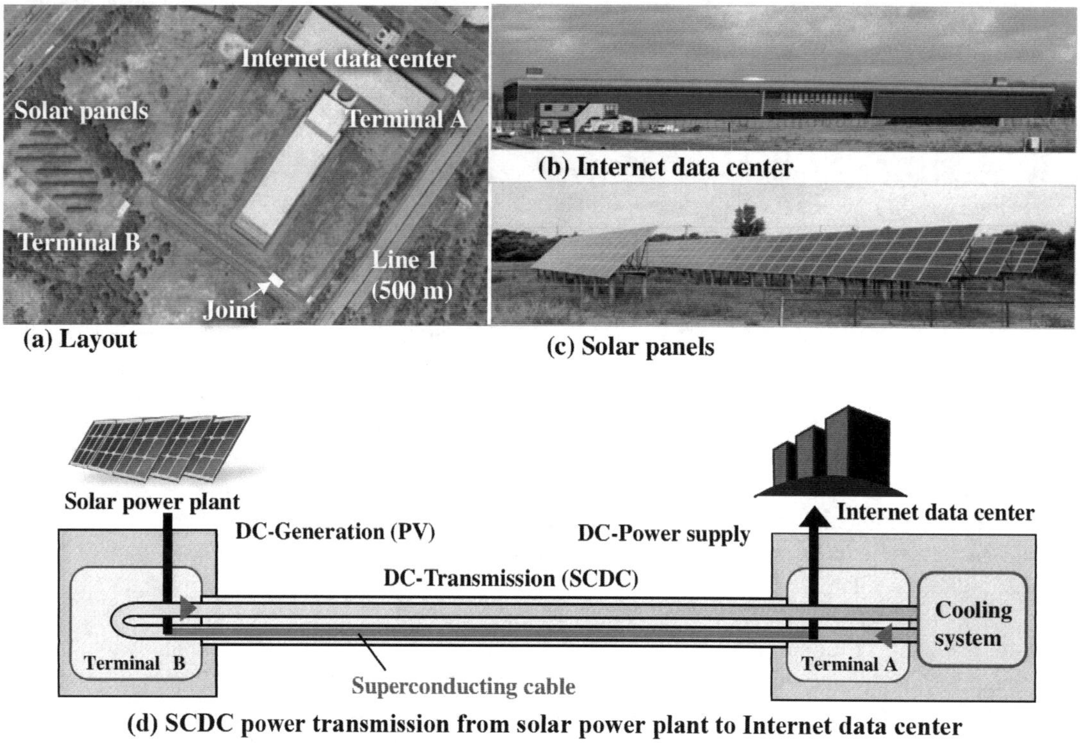

Figure 1. Overview of the Ishikari SCDC Line 1

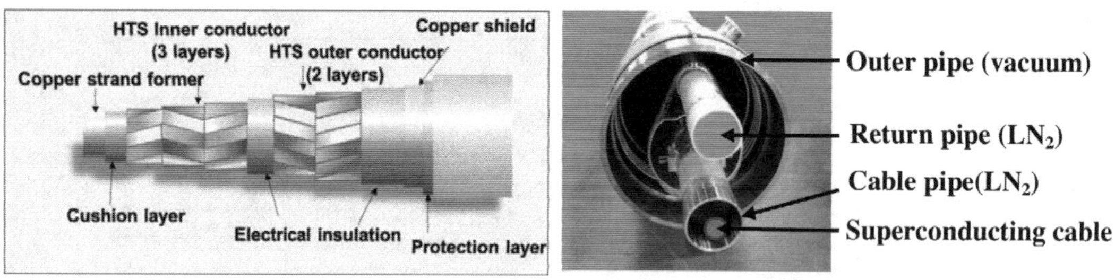

(a) Schematic of a superconducting cable (b) cryogenic pipe (type for Line 2)

Figure 2. Superconducting cable and cryogenic pipe

3.1. Refrigeration circulation test

Figure 3 illustrates the cooling system of Line 1. Based on the results of two comprehensive tests, it was thought that the entire system could be cooled using only the turbo Brayton refrigerator. Therefore, a cooling circulation test was performed using only the turbo Brayton refrigerator. This operation is a key element of the low-cost operation demonstration. The prevailing status of the system was confirmed by measuring (1) the amount of heat ingress of the system and (2) the pressure loss caused by liquid nitrogen circulation. (3) Measurements of the refrigerator capacity and efficiency were performed under partial load to obtain the basic data for an efficient operation method for the refrigerator in future systems.

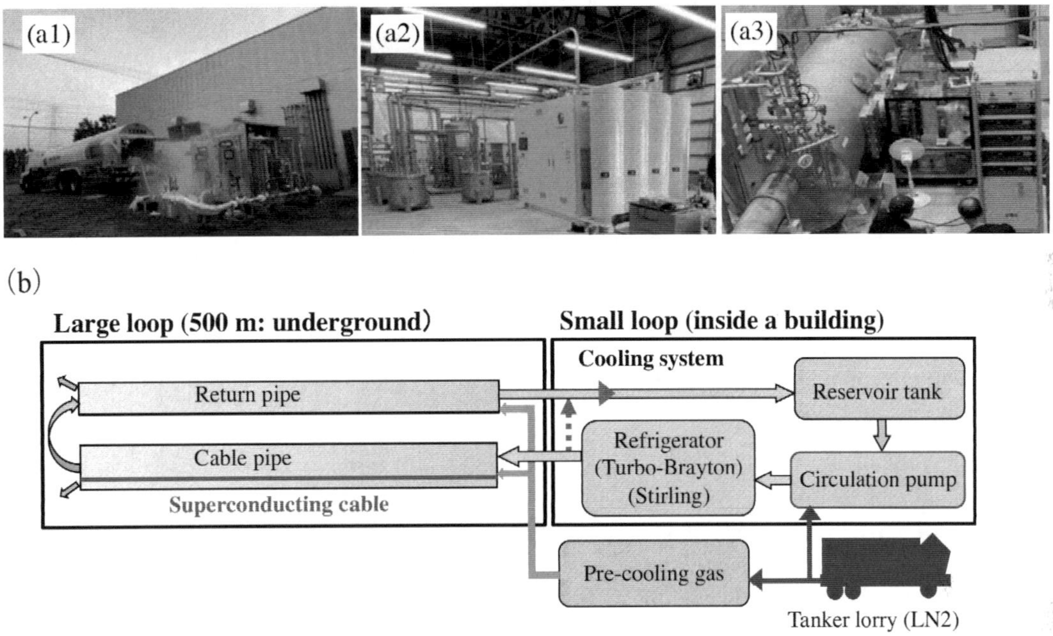

Figure 3. Operation of refrigerator for cooling superconducting cables at Ishikari SCDC line 1 (a1) liquid nitrogen supply (lorry), (a2) refrigerator (front right) and cooling system (back left), (a3) terminal and DC power supply, (b) System of pre-cooling and cooling circulation operation

3.2. Heat penetration and pressure drop

The cooling system of Line 1 is depicted in Figure 3(b). On the cable system side, the liquid temperature rise ΔT [K] from the inlet to the outlet is expressed as:

$$\Delta T = Q / (v\ A\ \rho\ Cp) \quad (1)$$

where Q [W] denotes the heat input, v [m/s] indicates the flow velocity, A [m^2] indicates the pipe cross-sectional area, ρ [kg/m^3] indicates the liquid density, and Cp [J/kg/K] refers to the specific heat of the refrigerant. W was estimated when the circulation flow rate varied between approximately 20 L/min and 40 L/min at 5 L/min intervals. For the entire cable system side, including the terminals, the overall heat ingress was approximately 1.4 kW and the pressure drop was approximately 34.4 kPa at the overall standard flow rate (40.4 L/min).

3.3. Refrigeration capacity and partial load

Refrigeration capacity measurements and partial load tests were conducted. During the tests, the liquid nitrogen circulation system was operated with the cable system side disconnected, and only the cooling system circulated. The coefficient of performance (COP) of the refrigeration system was evaluated based on the refrigeration capacity and power consumption at liquid nitrogen temperatures (72 K or 80 K).

As the liquid nitrogen temperature was changed from 72 K to 80 K, the refrigeration capacity increased from 2.3 kW to 2.7 kW, and the COP improved by approximately 20% from 0.043 to 0.053. The results provide quantitative guidance for the optimization of SCDC operating temperatures.

Generally, the efficiency of a refrigerator decreases under partial-load conditions, which is different from the full-load condition. The partial-load conditions of the system were investigated. The COP was calculated under partial load conditions by measuring the refrigeration capacity and energy consumption when the refrigerator temperature was set at 70 K, and 1 kW, 1.5 kW, and 2 kW of heat was applied to liquid nitrogen using a heater. The results show that the change in efficiency under partial-load conditions can be evaluated quantitatively.

For example, when the full load of 2.3 kW is reduced by approximately half to 1 kW, the refrigeration power consumption decreases by approximately 34%, and the COP drops by approximately 20%. Consequently, we obtained knowledge on efficiency improvement by operating the system without a partial load.

4. Electrical test

Figure 4 depicts the circuit used for measuring the shunt current of a superconducting cable. The following electrical tests were conducted with the primary objectives of confirming the integrity of the cables and stability of energization.

4.1. Superconducting cables

To evaluate the insulation performance in terms of cable integrity, capacitance and dielectric loss tangent (tan δ) measurements, insulation resistance tests and DC withstand voltage tests were conducted. The capacitance and dissipation factor (tan δ) were measured between the inner conductor and outer conductor (inner insulation) and between the outer conductor and shield (outer insulation). Moreover, it serves as a monitor for cable breakage during the initial cooling. Despite the fact that the cable had undergone a thermal history of three cooling, two temperature increases, and a six-year shutdown period, the measured values at 70 K were almost the same as those from the two previous cooling tests, which confirms that the cable was maintained in good condition. Furthermore, insulation resistance (500 VDC–1 min) and DC withstand voltage (600 VDC–10 min) tests were conducted on the inner and outer insulation layers. It was confirmed that there was no damage or aging degradation due to the thermal history and that the cables maintained decent voltage withstand performance.

4.2 Confirmation of current carrying stability

4.2.1. Rated current energization test

Under the rated temperature (70 K) and liquid nitrogen flow rate (40 L/min) conditions, an energization test was conducted at a rated current of 5 kA for 60 min. Figure 4 shows the current-carrying test circuit using Peltier current leads (PCLs). The DC power supply (maximum current of 6 kA) was placed on the terminal B side and connected to the PCLs (P, N). By shortening the PCLs between P and N on the terminal A, the current flows from the terminal B through the outer conductor, terminal A and inner conductor to the terminal B. The energizing results show that the voltage at both ends of the cable during 5 kA energization was maintained in a superconducting state at 1 μV/cm or less, which indicates that the current was energized stably.

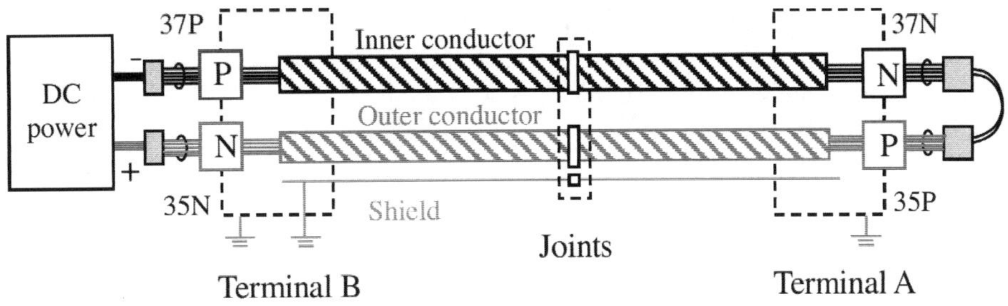

Figure 4. Test circuit for energization

4.2.2. Shunt current test

The number of current leads connected to the DC power supply was reduced to approximately 60%, 80%, and 85%, and the current was energized under the rated conditions of 70 K and 40 L/min. In the 85% reduction case, five of the 35 outer conductor current leads (N) and five of the 37 inner conductor current leads (N) were connected to the DC power supply in terminal B. These five leads were energized at 500 A, and the current distribution caused by the passage of the cable was measured using a through-current sensor attached to each PCL. The current flowing in from the five current leads at terminal B are connected via the inner conductors of the superconducting cable (37 conductors), outer conductors connected at terminal A, and back to the power supply.

Figure 5 illustrates the current distribution in 37 inner conductors of the superconducting cable in the 85% reduction case. The current connected to the five current leads at terminal B was evenly distributed at the joints. The result confirms that autonomous current-distribution equalization occurs at the joint. PCLs were originally intended; however, there was a concern that the system would become more complex. This result contributes to the consideration of replacing PCLs with normal-conducting current leads, which were designed as a countermeasure for the current imbalance between the SCDC strands. Moreover, the result will contribute to the simplification and cost reduction of SCDC systems for social implementation.

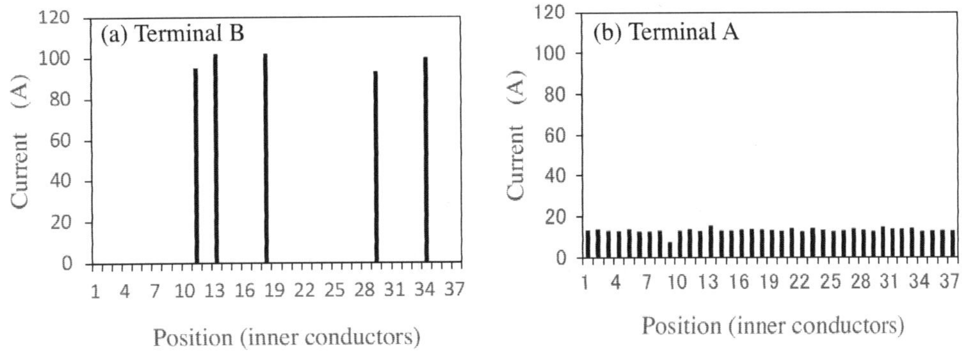

Figure 5. Results of current measurements at each PCL of the inner conductor during the current-sharing test (500 A energized). (a) Terminal B: Cathode side, inner conductor current lead (N)
(b) Terminal A: Anode side, inner conductor current lead (P)

4.2.3. Energization test at low flow rate

The stability of energization was tested at a reduced circulating flow rate of liquid nitrogen. The flow rate of liquid nitrogen was lowered to 3.6 L/min (less than 1/10 of the rated flow rate), and the current was supplied at 3 kA for 60 min. The flow rate is the lower limit for a stable circulation. The temperature of liquid nitrogen was approximately 70 K at the inlet and 77 K at the outlet, resulting in a large temperature difference of approximately 7 K between the two ends of the cable. Even under these conditions, the voltage at both ends of the superconducting cable is less than one μV/cm, maintaining the superconducting state and demonstrating that the cable can be energized stably. The data contributing to the optimization of the flow rate in relation to the energizing current were obtained. The data can also be used to respond to accidents, such as a lack of coolant (LoC).

5. Future plan for social implementation

The practical application of SCDC requires basic fundamental research to create a steady build-to-print design that is acceptable to a smart society and a new business model. Based on the success of the Line 1 restart test, there are plans to use the SCDC to conduct demonstration tests for social implementation by connecting future large-scale renewable energy generation equipment, such as wind turbines and solar power generation systems, to an Internet data center.

We plan to conduct safety tests by consolidating the results of preliminary tests. The safety tests will address the cooling and vacuum failure modes for cryogenic pipes and against high-voltage lightning strikes. The plan is to systematize accident scenarios, destructive tests, and response measures, based on the results. For lightning protection, we aim to clarify how lightning strikes on SCDCs and renewable energy facilities propagate via superconducting cables and how they affect each facility and propose protective measures. Through this fundamental research, we plan to establish international safety standards for social implementation and propose SCDC systems that can cope with natural disasters and accidents.

6. Conclusion

This paper describes the results of the Ishikari Line 1 restart test conducted in September 2021 to collect basic data for cost reduction and improved efficiency and to confirm equipment integrity following approximately six years of shutdown. Based on these results, we aim to promote the social implementation of SCDC by constructing a database for the quantitative evaluation of reliability, safety, and cost.

The integrity and robustness of all the system components, including the SCDC main unit, refrigerator, control equipment, vacuum system, and measurement system, were confirmed. For superconducting cables, the findings confirmed that autonomous uniformity of the current distribution occurred at the connections. These results are useful for replacing PCLs with normal current leads. The integrity of the superconducting cable was tested in terms of its capacitance, insulation resistance, and DC ability to withstand voltage. The superconducting state was maintained under rated conditions and the cable was energized in a stable manner. To enhance the refrigerator operation efficiency, the overall system could be cooled using only the turbo Brayton refrigerator. Accordingly, partial load conditions that reduced the refrigerator efficiency could be observed. Moreover, we demonstrated that the SCDC can be operated by circulating liquid nitrogen at high temperatures and low flow rates. Optimization of the refrigerant conditions and high-efficiency operation of the refrigeration equipment can be achieved.

The results suggest that with appropriate maintenance, the SCDC system can sufficiently maintain normal operation characteristics for several decades in practical applications.

Acknowledgments

The test operations were carried out as part of the basic infrastructure research activities of the Ishikari Superconductivity and DC Power Transmission System Technology Research Association using METI facilities managed by the association. The authors would like to thank Chancellor Atsuo Iiyoshi of Chubu University for his efforts in promoting this project, Takashi Mutoh, the First Vice-Chancellor of Chubu University, and Director of the Center for Superconductivity, for his cooperation and guidance throughout the Center for Superconductivity, and all parties involved in their cooperation in this research.

References

[1] Motojima O 2019 Current Status and Trends in Fusion Technology *Electrical Review* **104** (5) 7-15
[2] Motojima O and Iiyoshi A 2020 Research and Development of Technology for Electric Power Infrastructure Utilizing High-Temperature Superconducting Direct Current Transmission Technology to Realize a Society 5.0 Society *Electrical Review* **105** (10) 53-56
[3] Center of Applied Superconductivity and Sustainable Energy Research Chubu University https://www3.chubu.ac.jp/organization/institute/sustainable_energy/
[4] Chikumoto N and Yamaguchi S 2016 Expanding Applications of Superconductive DC Power Transmission(Japanese) *Journal of the Japan Society of Applied Physics* **85** (4) 270
[5] Chikumoto N, Watanabe H, Ivanov Y, Takano H, Yamaguchi S, Ishiyama K, Oishi Z, Watanabe M and Masuda T 2018 Second Cooling Test of 1000-m Superconducting DC Cable System in Ishikari *IEEE Transactions on Applied Superconductivity* **28** (4) 1-5
[6] Watanabe H et al. 2017 Cooling and Liquid Nitrogen Circulation of the 1000 m Class Superconducting DC Power Transmission System in Ishikari *IEEE Transactions on Applied Superconductivity* **27** (4) 1-5
[7] Motojima O, Iiyoshi A, Tendler M and Ivanov Y 2020 Proposals for the implementation of the society based on HTS DC power transmission technology *Bulletin of the Russian Academy of Sciences:* Power 5 pp. 13-20 (in Russian)
[8] Ishikari International Forum of Superconductivity Transmission Report *Organized by Chubu University and Ishikari City* June 24-26 2017 http://www.city.ishikari.hokkaido.jp/uploaded/attachment/26844.pdf
[9] Ishikari Superconductivity DC Power Transmission System Pamphlet 2022 http://i-spot.jp/wp/wp-content/uploads/2022/12/00_pamphlet_0324.pdf
[10] Ishikari Line 1 Restart test operation Report for FY2021, *ed. Ishikari superconductivity DC power transmission system* http://i-spot.jp/wp/wp-content/uploads/2022/06/saikadouhoukokusyo.pdf
[11] Inoue N, Chikumoto N, Watanabe H, Ivanov Y, Hino T, Sato A, Sawamura T, Yamamoto H, Sumi K, Okuno K and Motojima O 2020 Research and Development of Technology for Electric Power Infrastructure Utilizing High-Temperature Superconducting Direct Current Transmission Technology to Realize a Society 5.0 Society *Electrical Review* **105** (10) 53-56

Design Study of an HTS Linear Power Generator Suitable for Wave Energy Converter

Petrus Kambo[1], Yuhi Yamanouchi[1], Antomne A Caunes[1], Kota Yamaguchi[2], Mitsuru Izumi[1,2] and Tetsuya Ida[1]

[1]Tokyo University of Marine Science and Technology, Tokyo 108-8477, Japan
[2]National Institute of Technology, Toba College, Mie 517-8501, Japan

Email : pkambo3@gmail.com

Abstract. In this paper, we show a conceptual structure of a direct drive linear power generator with REBaCuO high-temperature superconducting (HTS) bulk field poles for wave energy conversion. The proposed linear power generator structure is comprised of a moving armature coil and HTS bulks concept, which we call it a dual translator power generation system. Its performance has been analyzed using a finite-element electromagnetic field analysis method. A determination of the main machine parameters for the dual translator HTS linear power generator was conducted. Simulation results were compared between our dual translator HTS linear generator structure and the use of permanent magnets (PMs) for the field poles of that structure. It was found that the electromotive forces, electromagnetic forces and output power of the dual translator HTS linear power generator structure are 1.67, 2.63 and 2.60 times superior to that of dual translator PM linear generator structure, respectively. A combination of the dual translator power generation system and HTS bulks has the potential to help us to achieve higher torque and higher output power despite low ocean wave speed, which is important to achieve practical application of ocean power generation systems.

1. Introduction

The increase in global energy demands, environmental pollution and cost of fossil fuels is forcing society to find different ways of producing energy. To solve the above-mentioned problems, renewable source of energy needs to be adopted. Ocean wave energy is a kind of renewable energy which has a high energy density than wind and solar energy [1–4]. For this reason, it has attracted attention for the past few years [5–8]. Ocean waves can be seen as the up and down near linear motion of the water surface, which can be converted into electric power using linear generators. Consequently, many linear generators have been proposed to generate electric power from wave energy [9–14]. Among many proposed linear generators, a substantial number of them uses permanent magnet (PM) for field poles [15–19].

Since the stroke lengths obtained from many wave power structures are very short, improving the output power obtained from each power generation module is a major challenge in the practical application of wave power generation. In addition, the point absorber type wave energy converter (WEC) shown in Figure 1 utilizes seawater, which is a density 840 times higher than that of air, so the linear generators for WEC used for this application must be capable of driving at low speed with high torque. In general, the frequency of waves is lower than other methods of power generation and the vertical velocity is slow, so linear power generation module for WEC generate less power. To improve

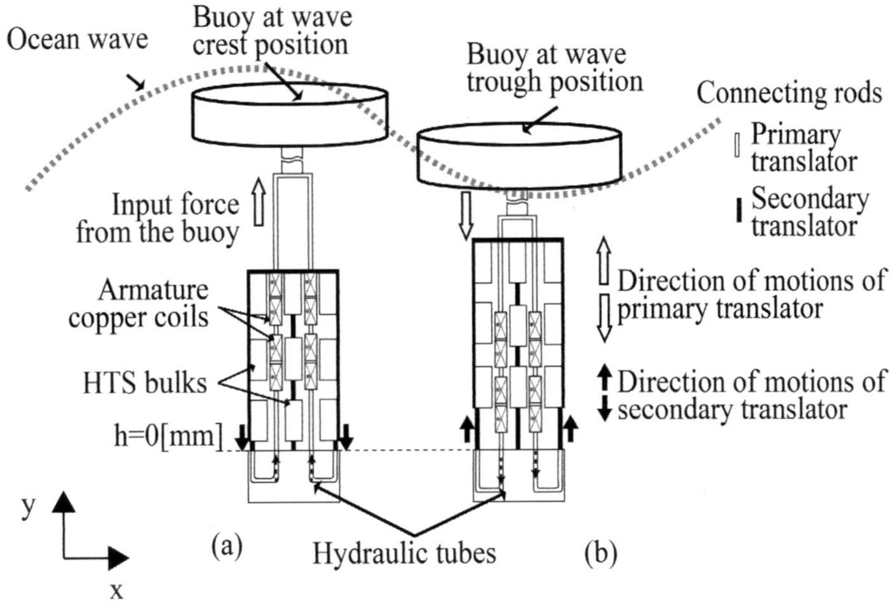

Figure 1. Single wave energy converter with a cross-sectional view of the proposed dual translator HTS linear power generator structure. (a) During crest ocean wave condition. (b) through ocean wave condition.

the output power, the size and weight of the power generating device increases, which may make it difficult to install generators and inhibit power generation operation. These are detrimental to the practical application of wave energy power generations. For this reason, there is a need to develop ocean wave power generating devices to accommodate the slow speed of ocean waves and still be able to achieve improved output power without the need to increase the weight or size of such a power generating device. In addition, because the large forces generated by ocean waves often place significant loads on the power generation mechanism of the WEC, the structural damage caused by a large loads of ocean waves must be avoided. Therefore, a combination of power generation module and the damper is required. It is wasteful to install a damper in parallel with the linear generator to receive the large force generated by the ocean waves, as it would increase the size and weight of the power generation system. Since the mechanical energy received by the damper is not used to generate electricity, another mechanism of buffering should be sought in WEC.

In this paper, we propose a new design concept of high-temperature superconducting (HTS) linear power generator structure, which we call the dual translator power generation system, to solve two key issues associated with ocean power generating devices. And we evaluate the validity of using HTS bulk magnets for the poles of that system to achieve high output power by comparing the field poles using HTS bulk material with those using PM.

2. Conceptual structure

Figure 1 shows a simple schematic of a wave energy converter with the proposed dual linear power generator structure. We plan to adopt a buoy as our primary capture of the kinetic energy from the ocean waves in our linear generator. The buoy captures kinetic energy from the horizontal and vertical motions of ocean waves, whereby the vertical motion is slow as we mentioned earlier. This slow motion is challenging for obtaining output power of practical application. In order to avoid a reduction of the output power, we came up with the idea of increasing the rate of change of the magnetic field through armature copper coils and also increasing the strength of the magnetic field, which are the key to solve two issues associated with ocean power generating devices we mentioned earlier. Unlike more common

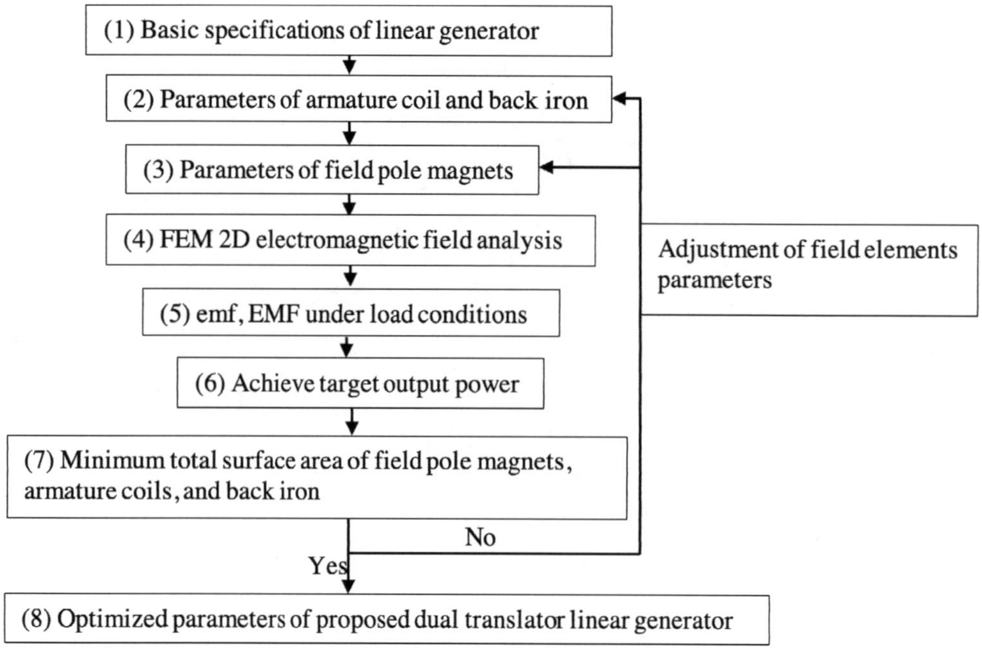

Figure 2. Design flowchart of dual translator HTS linear generator machine.

single translator power generators, where the armature coil is stationary, the new design concept called dual translator power generation system moves the armature copper coils and the HTS bulk field poles in an opposite oscillatory motion at the same time. The device is equivalent to stretching the short strokes obtained from small wave force amplitudes by a factor of two to generate electricity. For this reason, our proposed power generator has a linear type of power generation structure in which the non-stationary armature and HTS bulk field poles come and go in opposite directions to each other. This increases the electromotive and electromagnetic forces, which improves the output power and the damping ability of the WEC, through the increase in the rate of change in the magnetic field through the armature copper coils and the increase in the strength of the magnetic field. The dual translator power generation system allows for increased output power and damping ability without increasing the size or weight of the linear power generator. The high trapped magnetic flux by HTS bulks leads to a generation of a strong magnetic field in our linear generator, which increases the electromagnetic forces, which can be used as a damper to protect the linear power generation driving structure from the loads exerted upon the buoy by the ocean waves. This increase in electromagnetic forces is not suitable for land-based power generation applications, however suitable for marine power generation applications such as wave energy converters.

The operation principle of the proposed dual translator linear power generator is schematically explained in Figure 1, whereby the armature copper coils are the primary translator while the HTS bulk field poles are secondary translator. When the buoy captures the kinetic energy from the oscillatory vertical motion of the ocean waves, it directly exerts a linear force on the primary translator, and it moves in the upwards direction during wave crest ocean conditions as illustrated by the white arrows in Figure 1(a). Whereby it transfers the kinetic energy through hydraulic oils indicated by black dotted arrows into the secondary translator as represented in Figure 1(a), which then move at the same time in the downward direction until it reaches h=0[mm] as shown by the black solid arrows in Figure 1(a). During wave trough, the direction of motion of the primary translator is downwards, which forces the hydraulic oil through the hydraulic tube to move the secondary translator in the upwards direction as shown in Figure 1(b).

Table 1. Basic specifications of the dual translator linear generator

Items	HTS bulk	PM
Maximum output power [kW]	13.0	5.0
Maximum stroke length [mm]	896	
Magnetic flux density [T]	2.0	1.40
Number of field poles	27-70	
Pole pitch of field pole [mm]	66.0-82.0	
Field pole magnet type	REBaCuO	NdFeB
Armature coil type	Copper	
Diameter of copper wire [mm]	2.0	
Armature coil self-inductance [μH]	650	
Number of armature coils	6	
Airgap length [mm]	2.0	
Speed of generator [m/s]	0.30	

3. Analysis method and determination of main machine parameters

In this section, we show an analysis method of the proposed dual translator linear generator, which is depicted in the flowchart in Figure 2. First, the basic specifications of the linear generator used in step 1 are presented in Table 1. Next, the parameters of the armature coils, back iron and field pole magnets are computed from steps 2–3. In step 4, 2D electromagnetic field analysis is performed using a finite element method implemented in COMSOL Multiphysics® software. From the analysis, we obtain electromotive under loads conditions in step 5, using (1) :

$$e_{ind} = \frac{d\lambda}{dt} = N\frac{d\phi(t)}{dt} \tag{1}$$

Whereby, e_{ind}, λ, ϕ and N are induced electromotive force, total flux passing through, flux passing through a single turn coil, and number of turns in a coil. We also compute electromagnetic forces in step 5 by integrating maxwell 's stress tensor over the armature coil domain as expressed in (2) :

$$\boldsymbol{F} = \iint \overleftrightarrow{\boldsymbol{T}} \cdot \hat{\boldsymbol{n}}\, dS \tag{2}$$

Whereby, $\boldsymbol{F}, \overleftrightarrow{\boldsymbol{T}}, \hat{\boldsymbol{n}}$ and S is total electromagnetic force on the charges in a volume or domain, force per unit area acting on the surface of domain, unit vector and cross section area of the volume or domain. Target output power is computed in step 6. In step 7, we calculate the total area of back iron, armature coil and field pole magnets, i.e., the total area of the linear generator internal components, by a negative feedback control loop connected to the parameters of step 2 and 3, in order to minimize the calculated area. Finally, the optimized parameters of the linear generator are obtained.

The weight of the generator which is one of the key specifications can be determined from the 3D geometry, however the present study, which is in the pre-design phase of the specific design of the linear generator, is a 2D analysis, it is not possible to discuss the volume. Therefore, we focused on the total area of the internal components in the linear generator to use it as a proxy to infer the weight of the linear generator. The reason this indicator is valid is that the smaller the total area of the linear generator internal components indicate the lower weight of the generator. The minimum total area may correspond

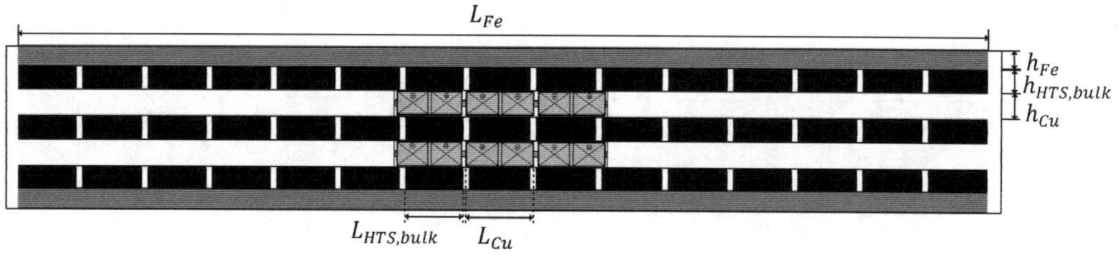

(a)

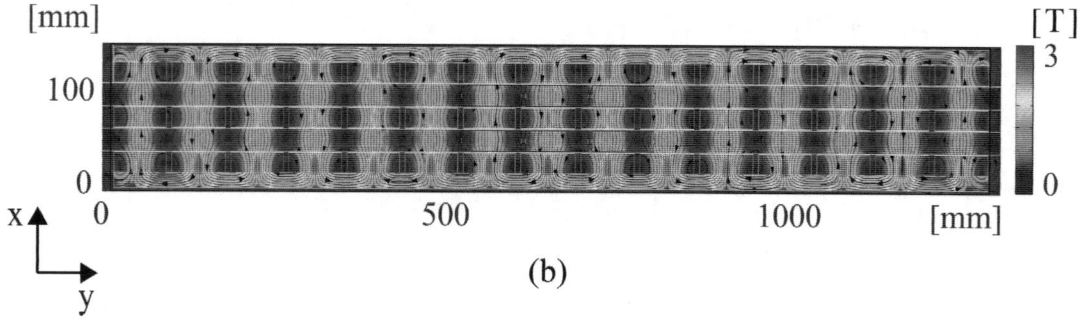

(b)

Figure 3. 2D optimized model of the dual translator HTS linear generator. (a) Schematic of the generator internal structure; orange – armature coils, black – field pole bulk magnets, gray – back iron. (b) Magnetic flux density distribution in the internal cross section of the generator, where six armature coils are located near the center of the generator.

to the smallest volume of the generator machine because the increase in the area of any shape increases its volume. Therefore, the lowest total area should indicate the lowest volume, which indicates the lowest weight. The cross section of the center of the linear generator is shown in Figure 3(a). The orange rectangles represent three sets of six armature coils corresponding to three-phase AC power, the black rectangles represent the HTS bulk, and the top and bottom gray represent the back yoke. The field pole magnets and vortex-type armature coils are 3D cylinder-shaped and are assumed to be those used in an axial-gap type synchronous rotating machine [20]. Nevertheless, this model does not consider depth dimensions as an initial step in the analysis. From this 2D perspective, the total area of the linear generator internal parts was calculated using equation (3) to obtain a weight-related index:

$$S=n_{Cu}L_{Cu}h_{Cu} + n_{Fe}L_{Fe}h_{Fe} + n_{HTS,bulk}L_{HTS,bulk}h_{HTS,bulk} \tag{3}$$

Whereby, S is the total area of the linear generator, and L_{Cu}, h_{Cu}, L_{Fe}, h_{Fe}, $L_{HTS,bulk}$, $h_{HTS,\,bulk}$ are the length and height of the armature coil, back iron, and HTS bulk for the field pole magnet, respectively. The n_{Fe}, n_{Cu}, $n_{HTS,bulk}$ factors represent the lower and upper back iron layer, number of armature coils, and number of field pole magnets. We calculated the total area by multiplying the area of each component by its respective quantity, $n_{Cu} = 6$, $n_{Fe} = 2$ and $n_{HTS,bulk} = 45$. This analysis does not take into account the depth information of each component, so it does not optimize the weight. However, we believe that such an estimate is useful as an indicator for future detailed design, so we added it to our considerations.

We used the 2D-analysis to determine the best combinations of internal parts of the linear generator that yield the lowest area. The main parameters of our linear generator, such as pole pitch and the number of field poles were determined using an iterative procedure considering, generator efficiency and the total area of the internal components in the linear generator, which was used as a proxy to infer the weight of the generator. To this end, we sought optimal values for linear generator design parameters based on Figure 4 and 5. This was done to achieve a light weight and less costly generator.

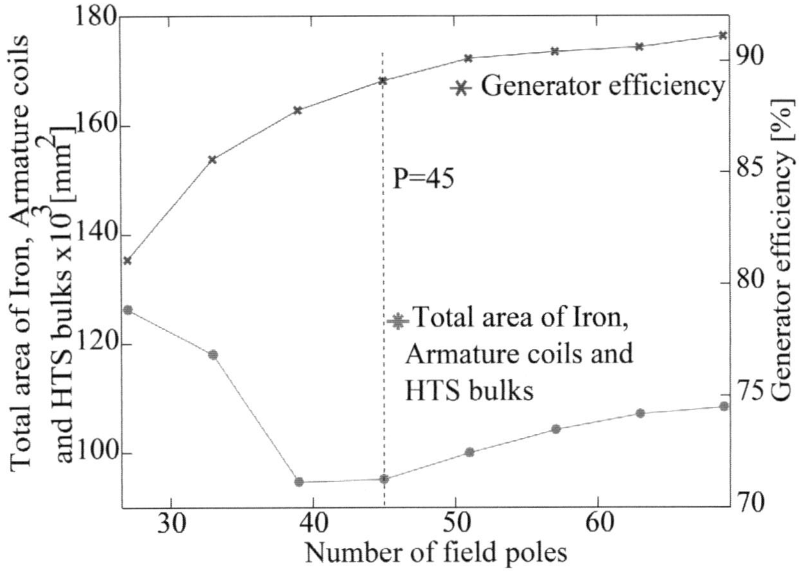

Figure 4. Total area of HTS bulks, armature coil and back iron, and linear generator efficiency with respect to the number of field poles.

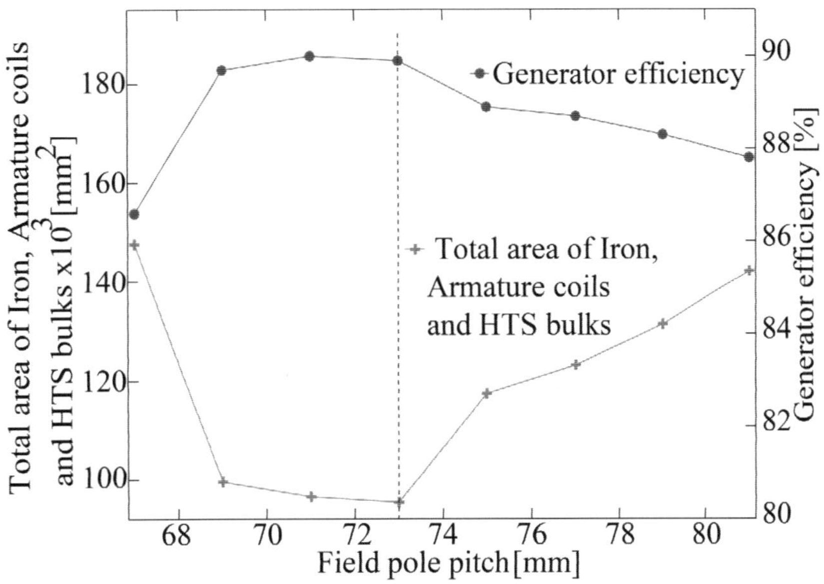

Figure 5. Total area of HTS bulks, armature coils and back iron and linear generator efficiency with respect to field pole pitch with 45 of field poles.

4. Analysis results and discussion

The result from the 2D electromagnetic field analysis yielded a final design model of our dual translator HTS linear generator shown in Figure 3 (a), with the dimensions of the armature coil at 74 mm in diameter and 20 mm in thickness and the HTS bulk field pole has a diameter of 66 mm and 19 mm of thickness. Its magnetic flux density distribution in the internal cross section is shown in Figure 3 (b). It has a magnetic flux density in the airgap of around 2 T, which was measured at the middle of a 2 mm of mechanical airgap, between surface of an armature coil and an adjacent HTS bulk with a maximum

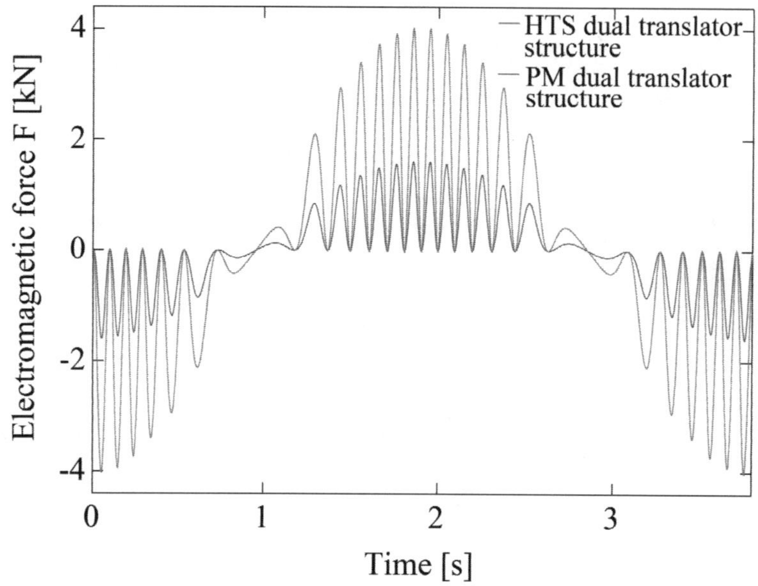

Figure 6. Electromagnetic force of the dual translator linear generator structures.

on-surface magnetic flux density of 3 T. Figure 3(b) shows six armatures, which are a bit inconspicuous in electromagnetic field analysis, in the same position as the armatures in Figure 3(a). This machine generates power by crossing an area of high magnetic flux density while they move left and right in tandem.

Figure 4 shows the effect of the field poles on the basic performance of the linear generator at the generating speed of 0.3 m/s. The difference in the number of field poles affects the volume and weight of the elements needed in the structure for power generation. The total area of the internal components in the linear generator was used as a proxy for volume showed significant differences between the number of field poles. While the number of field poles increased from 30 to 40, it was observed that the total area of the internal components in the linear generator decrease. If the number exceeded 45, the total area increased again gradually. On the other hand, the generator efficiency increases with the number of field poles increases but remains generally unchanged when it exceeds 50. This is because as the number of poles decreases, the dimensional ratio of armature coils and field poles leads to a small ratio of effective armature coil diameter, which generates induced electromotive force, and this leads to an increase in armature resistance loss. Figure 5 shows the relationship between the field pole pitch and the basic performance of our linear generator. The total area of the internal components in the linear generator was particularly smaller between 69 and 73 mm in the field pole pitch. And as the field pole pitch exceeded 73 mm, the efficiency gradually decreased. On the other hand, the power generation efficiency was relatively high at about 90% from 69 mm to 73 mm in the field pole pitches. This is because the increase and decrease in pole pitch leads to a long and thick back iron respectively, which increases iron losses and the total area of generator internal components.

Figure 6 shows the electromagnetic force waveform of an HTS dual translator linear generator structure, which is superior when compared to that of PM dual translator linear generator. Using the above electromagnetic waveform, we obtained the maximum electromagnetic force of both HTS and PM dual translator linear generators at optimum load resistance of 14 Ω. The best performance of our linear generator with HTS bulks field poles was obtained at optimum load resistance of 14 Ω. Using this model, we replaced the HTS bulks with PM for field pole. During the comparison between the HTS bulks and PM, the same load conditions is used for both PM and HTS bulk analyses in order to achieve

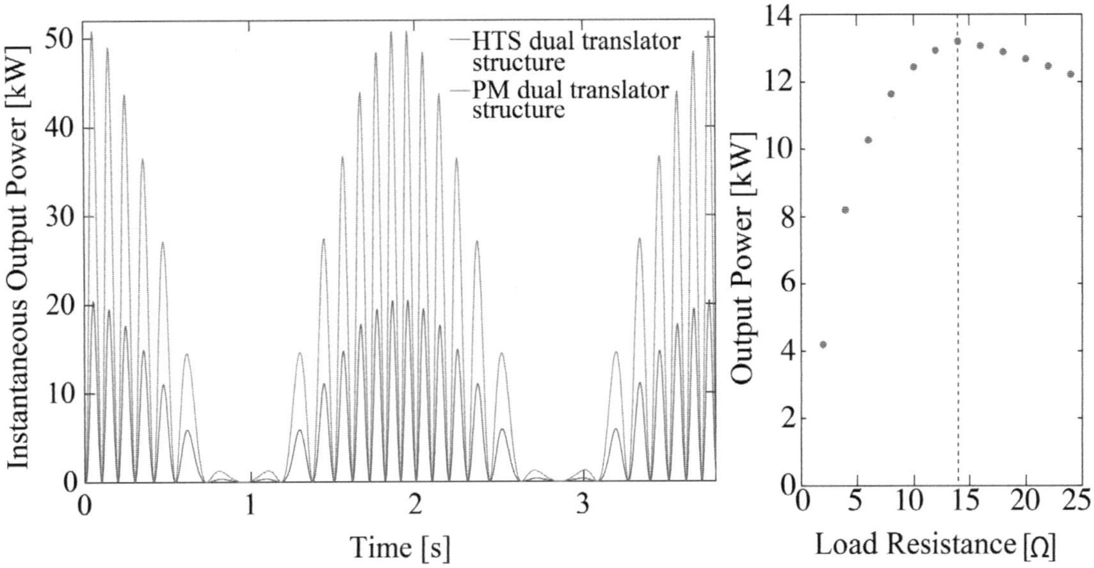

(a) **(b)**

Figure 7. Instantaneous output power of the dual translator linear power generator structures.
(a) Replacing the PM field pole with HTS increases the output by a factor of 2.60. (b) The optimum load resistance is 14Ω.

Table 2. Analysis results of a HTS dual translator linear power generator structure compared to that of PM counterpart.

Quantity	HTS bulk	PM
Total area of linear generator $W \times L$ $[m^2]$	1.66×10^{-1}	
Magnetic flux density on armature coil surface [T]	1.80	0.70
Maximum electromagnetic force [kN]	4.20	1.60
Load resistance [Ω]	14.0	
Average electromotive force [V]	348	208
Average output Power [kW]	13.0	5.0
Losses [kW]	1.60	0.60
Efficiency [%]	89.2	88.0

a fair comparison of the two types of field poles. The maximum electromagnetic force of HTS dual translator linear generator is 2.63 times that of PM counterpart as presented in Table 2, which can be used as a damper in our proposed linear generator structure for wave energy converter.

In Figure 7, we show the instantaneous output power of a dual translator HTS linear generator and the dual translator PM linear generator, which we obtain by first computing the electromotive force using equation (1). The average electromotive force of HTS dual translator linear generator is 1.67 times that of the PM model as shown in Table 2. We use the result in Figure 7 to obtain the average output power for both HTS and PM dual translator linear generator model, which we present in Table 2. The 2.60 times higher HTS dual translator linear generator can be attributed to its higher magnetic flux density on the armature coils surface as shown in Table 2, which is a result of the ability of HTS bulks to trap high magnetic flux density. The rate of change of this magnetic field on the surface of the armature coil through the armature coils increased the output power, which was caused by our proposed dual translator power generation system.

5. Conclusion

A conceptual structure of a direct drive dual translator HTS linear generator for wave energy converter was proposed. An electromagnetic field analysis was conducted to examine its validity. The main parameters of the linear generator were determined by analyzing their influence on the basic performance of our linear generator. The number of poles and pole pitch that corresponds to the lowest generator weight were obtained to be 45 and 73 mm respectively. From the results of the analysis, we compared the results of a dual translator HTS linear generator with a dual translator PM linear generator. The results show that electromotive forces, electromagnetic forces, and output power of dual translator HTS linear generator are 1.67, 2.63 and 2.60 times than that of dual translator PM linear generator. This makes the proposed linear generator structure superior in terms of improved output power and damping ability that of PM structure through the use of HTS bulk magnet.

Acknowledgement

This work was supported by JSPS KAKENHI Grant Numbers 21H01541 (2021-2024).

References

[1] Babarit A, Hals J, Muliawan M J, Kurniawan A, Moan T and Krokstad J 2012 *Renew. Energ.* **41** 44-63
[2] Polinder H, Damen, M E C, Gardner F 2005 *Trans. Electr. Power.* **15** 245–256
[3] López I, Andreu J, Ceballos S, De Alegría I M and Kortabarria I, 2013 *Renew. Sust. Energ. Rev.* **27** 413-434
[4] Sheng W 2019 *Renew. Sust. Energ. Rev.* **109** 482-498
[5] Astariz S and Iglesias G 2015 *Renew. Sust. Energ. Rev.* **45** 397-408
[6] Aderinto T and Li H 2019 *Energies* **12** 4329
[7] Bedard R 2007 *Mar. Technol. Soc. J.* **41** 44-50
[8] Foteinis S 2022 *Renew. Sust. Energ. Rev.* **162** 112448
[9] Huang L, Hu M, Liu J, Yu H, Zeng C and Chen Z 2017 *IEEE Trans. Appl. Supercond.* **27** 5201706
[10] García-Tabarés L, Hernando C, Munilla J, Torres J, Santos-Herran M, Blanco M, Sanz S, Sarmiento G, Lorenzo F G, Neri M. and Magrassi D 2022 *IEEE Trans. Appl. Supercond.* **32** 5201405
[11] Jing H, Maki N, Ida T and Izumi M 2017 *IEEE Trans. Appl. Supercond.* **27** 5202105
[12] Hong Y, Eriksson M, Castellucci V, Boström C and Waters R 2016 *IET Renew. Power Gen.* **10** 349-359
[13] Elwood D, Yim S C, Prudell J, Stillinger C, Von Jouanne A, Brekken T, Brown A and Paasch R 2010. *Renew. Energ.* **35** 348-354
[14] Faiz J. and Nematsaberi A 2017 *IET Renew. Power Gen.* **11** 1163-1176
[15] Baker N J, Raihan M A and Almoraya A A 2018 *IEEE Trans. Energy Conver.* **34** 691-700
[16] Prudell J, Stoddard M, Amon E, Brekken T K and Von Jouanne A 2010 *IEEE Trans. Ind. Appl.* **46** 2392-2400
[17] Huang L, Yu H, Hu M, Liu C and Yuan B 2013 *IEEE Trans. Magn.* **49** 1917-1920
[18] Vermaak R and Kamper M J 2011 *IEEE Trans. Ind. Electron.* **59** 2104-2115
[19] Trapanese M, Boscaino V, Cipriani G, Curto D, Di Dio V and Franzitta V 2018 *IEEE Trans. Ind. Electron.* **66** 4934-4944
[20] Watasaki M, Miki M, Felder B, Tsuzuki K, Sato R, Kase S, Izumi M, Ida T 2012 *IEEE Trans. Appl. Supercond.* **23** 8201604

Electromagnetic analysis of fully superconducting motors employing dilute gas rotor and liquid hydrogen stator cooling structure

Yutaka Terao[1], Kazuki Akasaka[1], Hiroyuki Ohsaki[1], Keiichi Okai[2] and Hideyuki Taguchi[2]

[1]Department of Advanced Energy, Graduate School of Frontier Sciences, The University of Tokyo, 5-1-5 Kashiwanoha, Kashiwa, Chiba, 277-8561, Japan

[2]Environmentally Sustainable Engine Team, Aviation Environmental Sustainability Innovation Hub, Aviation Technology Directorate, Japan Aerospace Exploration Agency, 7-44-1 Jindaiji-Higashi, Chofu, Tokyo 182-8522, Japan

e-mail: yterao8934@g.ecc.u-tokyo.ac.jp

Abstract. This paper deals with fully superconducting motors (FSCMs) applied in the electric aircraft propulsion systems; the stator employs the MgB_2-superconductor (SC) armature windings cooled at 20 K by using liquid hydrogen, while the rotor has the field windings made of REBCO SC wires cooled below 50 K via heat conduction and convective heat transfer by using dilute gas. The cooling structure of rotor can be simplified by adopting this type of design. The authors evaluated the optimal design for FSCM by using analytical equations and loss analysis via FEM. The FSCM characteristics were compared between ones using electromagnetic shield made of stainless steel (SUS) and ones using the shield made of copper (Cu); the results show that the output density using the SUS shield achieved 18.3 kW/kg, and that in terms of the loss estimation via FEM, both the motors had the efficiency over 98.5%.

1. Introduction

Aviation industry has accelerated the efforts to reduce the greenhouse-gas emission based on the statement of ATAG (Air Transport Action Group), targeting the Net Zero CO_2 Emissions by 2050. One of the solutions to fulfill the statement is no doubt to electrify the aircraft propulsion systems.

Fully superconducting rotating machines (FSCMs) are one of the most effective solutions for realizing high-power density machines applicable to electric aircraft propulsion systems, since the superconducting field/armature windings have high-current density at ultra-low temperature and, moreover, can generate high magnetic field; therefore, the weights of windings and iron can be reduced and consequently the lightweight motor/generators can be realized. Thus, the authors have been proposing the FSCMs using two kinds of superconducting windings: MgB_2 armature windings and REBCO field windings [1, 2].

However, the cooling structures for the FSCMs tend to be complicated because both the armature/ field windings demand coolant, refrigerators and so on; in such a technical challenge, the authors employed the cooling method using both dilute gas and liquid hydrogen [3]. While the MgB_2 armature windings being cooled at 20 K via liquid hydrogen, the REBCO field windings are operated below 50 K by means of the heat conduction and the convective heat transfer, via dilute gas. This structure significantly contributes to simplifying the rotor cooling.

In this study, the authors determined the electromagnetic design of the FSCMs via analytical equations and finite element method (FEM) while considering both aspects of material and structure; based on the analysis results, we evaluate and discuss motor output density, losses, and so on.

2. Superconducting aircraft propulsion system architecture

The conceptual diagrams of the proposed aircraft propulsion system are as shown in Figure 1 and Table 1; the blended wing body (BWB) structure was employed to increase the propulsion-system efficiency from a standpoint of aero dynamics, namely boundary layer ingestion (BLI) [4]. Number of passengers were assumed as 150 PAX based on A320, while the maximum output power during takeoff as 44 MW. The cruising power and speed in flight were defined as 19.2 MW and Mach 0.78 at 35,000 ft, respectively.

Focusing on the propulsion system, the system architecture consists of the following five processes: firstly, the gas turbine engine and the superconducting generator (SCG) generate electric power (①); secondly, AC Power is converted to DC via converter and transferred to whole aircraft systems (②); thirdly, DC power is converted to AC power by inverter (③); fourthly, the FSCMs connected to fans between two rear wings are activated by the inverter (④); and finally, after removing the thermal losses of superconducting machines and power converters, the gas hydrogen (H_2) is transferred to the gas turbine engine as fuel (⑤). This system contributes to reduce the cooling cost dramatically and some devices including compressors can be omitted.

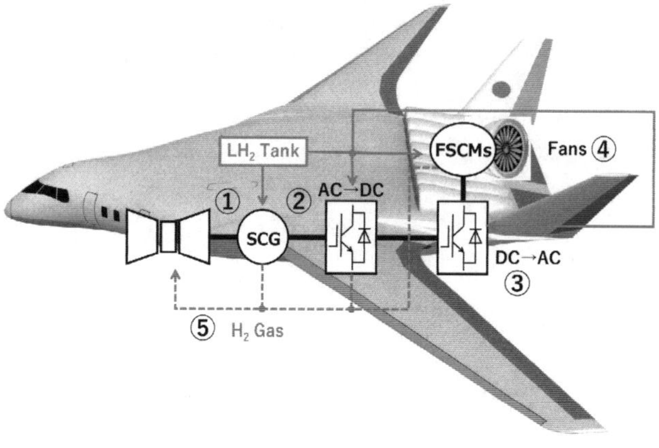

Figure 1. Conceptual diagram of aircraft propulsion system with superconducting machines

Table 1. Aircraft specifications

Passenger	150 PAX
Flight altitude	35,000 ft
Cruising speed	Mach 0.78
Max. take-off power	44 MW
Cruising power	19.2 MW
Num. of motors	8

3. Design concepts of the FSCMs using dilute gas and liquid hydrogen

Figure 2 shows conceptual illustrations of the FSCMs using two kinds of superconducting wires. While the MgB_2 armature windings being cooled to 20 K via liquid hydrogen, the REBCO field windings are cooled below 50 K via thermal conduction using gas helium (He), as shown in Figure 2. The gas He is filled in cryostat. The torque tube, which makes temperature gradient, is inserted between the two SC windings to keep them at low temperature. The cryostat of this FSCM is made of FRP and stainless steel (SUS). Especially, as shown in Figure 2, the MgB_2 armature winding part should be kept at 20 K to prevent heat penetration, therefore the FRPs (orange part) are employed, and the evacuated spaces are secured between the FRPs and the SUS cryostat walls. As for the electromagnetic shield, the authors selected two materials, SUS and copper (Cu), for this study; since the specific density of SUS, namely 7360 kg/m^3, is lighter than that of Cu, 8960 kg/m^3, both the lighter weight shield and the better mechanical characteristics, including resistance to centrifugal force, can be expected. However, the electric resistivity of SUS is lower than that of Cu; the tradeoff relationships should be discussed.

Table 2 shows the design specifications of the proposing FSCM. As mentioned in Table 1, the 5.5 MW×8 FSCMs have the capacity to generate 44 MW take-off power. The number of poles is set to 4 poles. The motor rotation speed, 5,000 rpm, come from the value for the low-pressure turbine shaft speed of turbofan engines; the FSCMs are to be connected to this part. The mechanical gap between the field windings and the armature windings is 20 mm and the electromagnetic shield having the thickness of 5 mm is inserted into this space. Also, mechanical gap between the MgB_2 armature windings and the back yoke is 50 mm; this space is for FRP and SUS cryostat wall and evacuated space for thermal insulation.

Table 3 shows the specifications of two superconducting wires [5, 6]. In this paper, already commercialized SC wires are used for the design; the REBCO wires, 4 mm × almost 0.1 mm, for field coils has critical current density of 250 A at 50 K, 2.0 T. On the other hand, 36-filament MgB_2 wires for armature windings are also employed; the filament diameter, twist pitch and critical current density are 63 μm and 25 mm and 1780 A/mm^2 at 20 K, 2.0 T, respectively.

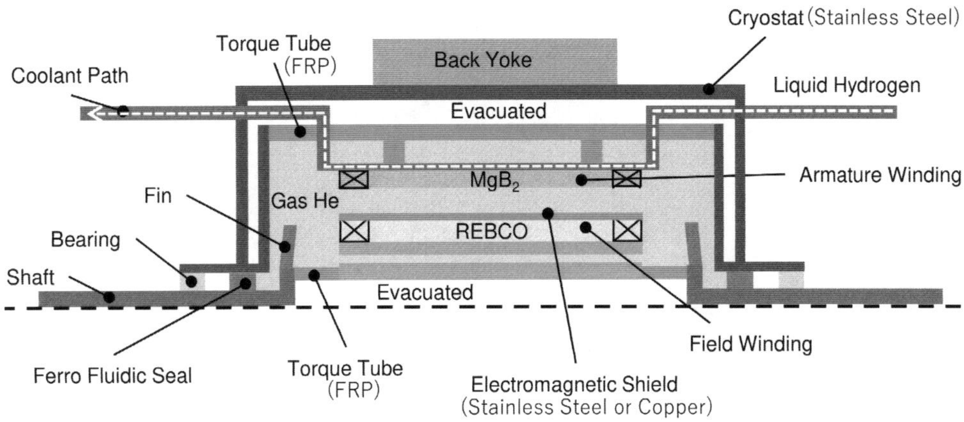

Figure 2. Conceptual illustrations of fully superconducting motors (FSCMs)

Table 2. Design specifications of a fully superconducting motor

Parameters	Values
Output power P	5.5 MW
Number of poles p	4
Motor speed N_{rot}	5000 rpm
Line voltage	1410 V
Gap among field and armature windings	20 mm
Gap among armature windings and back yoke	50 mm
Thickness of electromagnetic shield	5 mm
Armature winding operating temperature	20 K
Field winding operating temperature	≦ 50 K

Table 3. Specifications of superconducting wires [5, 6]

Parameters	Values
REBCO	
Width	4 mm
Thickness	~0.1 mm
Critical current at 50 K, 2.0 T	250 A
MgB_2	
Diameter D_o	0.84 mm
Filament diameter d_f	63 μm
Num. of filaments	36
Fill factors of superconductors λ	0.19
Twist pitch l_p	25 mm
Current density at 20 K, 2.0 T	1780 A/mm^2

4. Electromagnetic design of the FSCMs

4.1. Motor optimal design via analytical equations

Optimal design based on the values in Table 2 is conducted in this section. It is known that a motor output is expressed by the following parameters:

$$P = \frac{\pi^2}{\sqrt{2}} k_w B_{max} A_s D^2 L_{eff} \frac{N_{rot}}{60} \quad (1)$$

In the above equation, the symbols, namely k_w (= 0.95), B_{max} [T], A_s [A/m], D [m] and L_{eff} [m], stand for the winding coefficient, the magnetic flux density amplitude at armature, the electric load, the average armature diameter and the motor effective length, respectively. As shown in the previous work, the authors used the solver "Evolutionary" mounted on Microsoft Excel® for this optimal design [7]. For obtaining the highest output density design, the optimal conditions were like the followings:

$$B_{max} \leqq 0.8 \text{ T} \quad (2)$$
$$350 \text{ kA/m} \leqq As \leqq 430 \text{ kA/m} \quad (3)$$
$$280 \text{ mm} \leqq D \leqq 400 \text{ mm} \quad (4)$$
$$D \leqq L_{eff} \quad (5)$$
$$N_{rot} = 5000 \text{ rpm.} \quad (6)$$

The conditions were defined on the basis of our previous works [8, 9]. The maximum magnetic flux density amplitude, B_{max}, was proportional to the first and second power of the hysteresis loss and the coupling loss, respectively; therefore, the value was set to less than 0.8 T, while As to the values between 350 kA/m and 430 kA/m. The armature diameter range in the condition (4) came from the relationships between the motor diameter and the fan blade diameter, considering the speed, N_{rot} [7]. And as for the condition (5), the effective length, L_{eff}, should be longer than D because the proposed motor was a high-speed machine over 5000 rpm.

As a matter of convenience, we calculated the weight margin for the stator by means of adding all the active material weights and decreasing the weight of iron by 50% from them, while one for the rotor adding their active material weights.

Figure 3 shows the optimal design results for the FSCMs; both lines, namely solid lines for SUS and dotted lines for Cu, have the peak values at D = 300 mm, since the ratio between D and L_{eff} is almost 1 : 1 until D = 300 mm; and thereafter L_{eff} is longer than D, resulting in the volume increase of the FSCMs.

The output density of the designed FSCMs achieved 18.3 kW/kg in the case of using the SUS shield when A_s = 430 kA/m. And all the output densities using Cu shield, dotted lines, were lower than those of SUS shields, due to the difference of their specific densities. Table 4 shows the designed values of the FSCMs, under the conditions of A_s = 430 kA/m and D = 300 mm. Comparison of the weights of electromagnetic (EM) shields revealed the Cu shield was almost 2 kg heavier than the SUS shield; the difference in weight between the Cu and SUS shields resulted in the decrease by 0.2 kW/kg in output density. The loss characteristics of the two cases are to be analyzed via FEM in the next session.

4.2. Loss analysis using FEM

Based on the results in the previous section, the Finite Element Method analysis (FEM) using JMAG-Designer® was conducted to estimate motor losses. Figure 4 shows the two-dimensional analysis model; there are 5 types of losses, namely iron loss, stator SC AC loss, EM shield eddy current loss, rotor SUS eddy current loss and rotor SC AC loss. We assume that the hysteresis loss, $W_{h,}$ and the coupling loss, $W_{c,}$ are dominant in the stator SC AC losses. Especially, the equations of the SC AC losses at the rotor and the stator are as shown in the equations of (7) – (11) [11, 12]. After the simulation of the magnetic flux density distributions using the JMAG-Designer®, the obtained values about the armature windings and the field windings were substituted in the equations (7) – (11) to calculate AC losses. The iron loss at the back yoke and the eddy current loss at the EM shield were calculated directly in the JMAG-Designer®.

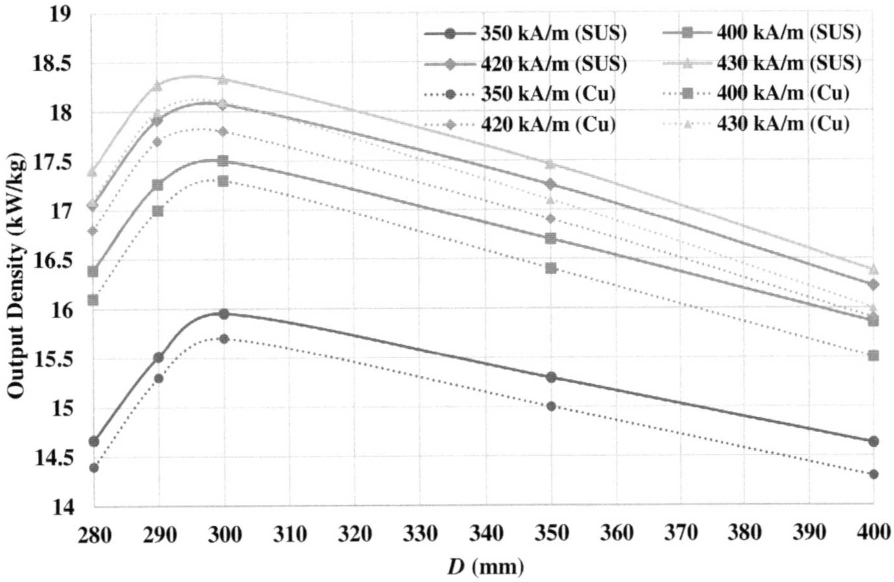

Figure 3. Optimization results of output density of the FSCMs

Table 4. Designed values of the FSCM, A_s = 430 kA/m, D = 300 mm

Parameters	Values
Motor diameter	439.4 mm
Effective length	439.2 mm
B_{max}	0.58 T
Stator weight	
Back Yoke	84.2 kg
MgB_2 armature winding	5.2 kg
Stator weight margin	47.8 kg
Rotor weight	
REBCO field winding	8.9 kg
Shaft	10.2 kg
Electromagnetic Shield	16.9 kg (SUS) / 19.1 kg (Cu)
Torque tube	0.8 kg
Exciter [8]	14.0 kg
Rotor weight margin	50.9 kg (SUS) / 53.0 kg (Cu)
Total weight	301.1 kg (SUS) / 304.4 kg (Cu)
Output density	18.3 kW/kg (SUS) / 18.1 kW/kg (Cu)

$$W_h = \frac{16}{3\pi} f \lambda B_m J_{cf} d_f \Omega(\eta) \tag{7}$$

$$W_c = \frac{1}{2\rho_e} \left(f B_m l_p\right)^2 (1+\eta)^2 \tag{8}$$

$$W_{FC} = \frac{\mu_0}{\pi} f {I_c}^2 F\left(\frac{\Delta B}{B^*}\right) \tag{9}$$

$$F(x) = 2\ln[\cosh x] - x \tanh x \tag{10}$$

$$B^* = \frac{\mu_0 I_c}{2a\pi} \tag{11}$$

where, f [Hz], B_m [T], J_{cf} [A/m^2], ρ_e[Ω m], I_c [A], ΔB [T] and $2a$ [m] are the frequency, the magnetic flux density which includes with field windings and self-field by armature windings, the critical current density under B_m, the electrical resistivity, the critical current, the fluctuating magnetic flux density and the REBCO wire width, respectively. The electric resistivities of SUS and Cu components at 50 K were 5.0×10^{-7} Ω m and 1.2×10^{-9} Ω m, respectively. Also, $\Omega(\eta)$ is a function representing the circularity of rotation in a rotating magnetic field. In this paper, the rotating magnetic field was approximated as a circle and therefore, η and $\Omega(\eta)$ were 1 and $\pi/4$, respectively. Figure 5 and Table 5 show the magnetic flux density distribution in the rotor where rotating magnetic field is applied from the armature windings under no field current, and the fluctuating magnetic-flux density in the field windings, respectively. The order of the fluctuating magnetic flux density in the field windings using the SUS shield was almost ten times higher than that of the Cu shield, due to the difference of the electric resistivity at 50 K shown in the above sentence.

The rotor AC losses were calculated and summarized with other losses in Figure 6. Total efficiency of the two designs achieved 98.6% with the Cu shield and 98.5% with the SUS shield, respectively. The stator AC losses were dominant, and their values were over 60 kW, in both cases; it was a challenging value for removing the big heat load via conduction cooling. However, if the authors employ over 100-filament MgB_2 wires with smaller diameter like D_o = 0.32 mm, l_p = 5 mm [12], the stator AC losses should be below 2.0 kW. On the other hand, the rotor AC losses in the two EM shields were 1.1 kW with the SUS shield and 0.01 kW with the Cu shield, respectively. We assessed both the rotor SUS losses, less than 1%, were negligible. It is clear that the Cu shield is better choice in terms of electric characteristics; however, the trade-off relationship between mechanical and electrical characteristics should be discussed continuously.

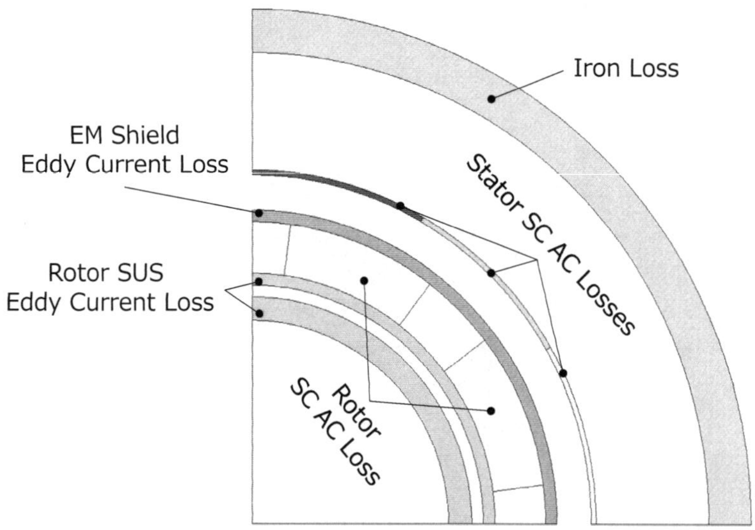

Figure 4. FEM analysis model for loss estimation of the FSCMs

Journal of Physics: Conference Series **2545** (2023) 012027 doi:10.1088/1742-6596/2545/1/012027

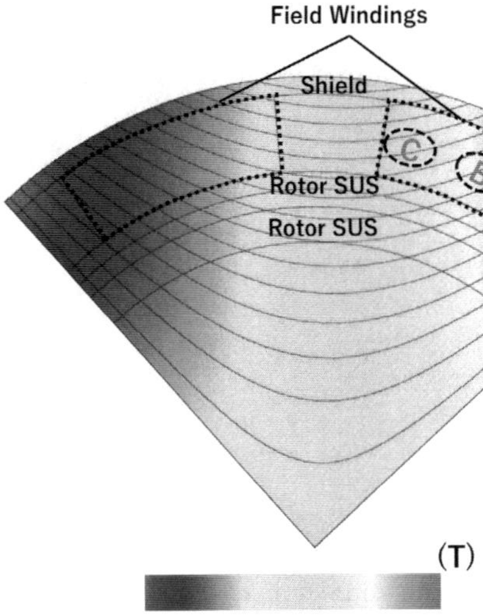

Figure 5. Magnetic flux density distribution at the rotor side applying rotating magnetic field under the condition of no field current

Table 5. Fluctuating magnetic flux density, ΔB at field windings

	A	B	C
SUS shield	21.0 mT	12.5 mT	9.0 mT
Cu shield	1.5 mT	1.1 mT	0.8 mT

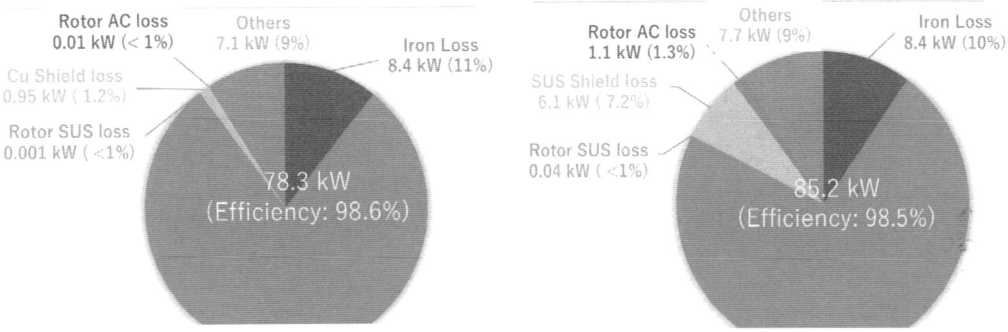

(a) Cu shield (b) SUS shield

Figure 6. Loss estimation results of the FSCMs

5. Conclusions

The authors have proposed the fully superconducting machines employing the dilute-gas rotor and the liquid hydrogen-stator cooling structure and conducted analytical calculation and the FEM analysis. The output density of the FSCM using the SUS shield via analytical calculation is 18.3 kW/kg. According to the estimation for the losses via FEM, over 98.5% was obtained in case of the electromagnetic shield made of SUS and Cu. Since Cu is effective material for electromagnetic shield, we should assess its structural design carefully, while discussing the way to increase its output density and_efficiency, simultaneously.

In the next phase, the 3-dimensional analysis and the physical coupled analysis will be conducted for discussing the cooling method, the mechanical design and so on.

References

[1] Terao Y, Ishida Y, Ohsaki H and Oyori H 2019 *SAE International Journal of Advances and Current Practices in Mobility*, **2** pp. 828-837

[2] Boll M, Corduan M, Biser S, Filipenko M, Pham Q H, Schlachter S, Rostek P and Noe M 2020 *Supercond. Sci. Technol.* **33** 4 044014

[3] Kato M, Terao Y and Ohsaki H 2021 *Journal of Physics: Conference Series*, **1857** 1 012016

[4] Okai K, Taguchi H, Terao Y, Ishida Y and Ohsaki H 2020 Joint Propulsion Conference, AIAA Propulsion and Energy Forum

[5] SuperPower Inc. Available: https://www.superpower-inc.com/specification.aspx

[6] Hypertech Research Inc. *Hyper Tech MgB_2 and Nb_3Sn Brochure Rev 10*

[7] Terao Y, Ishida Y and Ohsaki H 2021 *Journal of Physics: Conference Series*, **1857** 1 012017

[8] Terao Y, Seta A, Ohsaki H, Oyori H and Morioka N 2019 *IEEE Trans. Appl. Supercond.*, **29** 5 5202305

[9] Terao Y, Kong W, Ohsaki H, Oyori H and Morioka N 2018 *IEEE Trans. Appl. Supercond.*, **28** 4 5208005

[10] Kalsi S S, Hamilton K A and Badcock R A 2018 Joint Propulsion Conference, AIAA Propulsion and Energy Forum

[11] Brandt E H and Indenbom M 1993 *Physical review B*, **48** 17 12893

[12] Balachandran T, Lee D, Salk N and Haran K S 2020 *IOP Conf. Series: Materials Science and Engineering* **756** 012030

Thermal Characteristic Analysis of Fully Superconducting Motors Employing Dilute Gas Rotor Cooling Structure

Kazuki Akasaka[1], Yutaka Terao[1], Hiroyuki Ohsaki[1], Keiichi Okai[2] and Hideyuki Taguchi[2]

[1] Graduate School of Frontier Sciences, The University of Tokyo, 5-1-5 Kashiwanoha, Kashiwa, 277-8561, Japan

[2] Environmentally Sustainable Engine Team, Aviation Environmental Sustainability Innovation Hub, Aviation Technology Directorate, Japan Aerospace Exploration Agency, 7-44-1 Jindaiji-Higashi, Chofu, Tokyo 182-8522, Japan

Email: ohsaki@k.u-tokyo.ac.jp

Abstract. We have proposed a rotor cooling structure using dilute gas for a fully superconducting motor for aircraft application. Numerical analysis was performed to investigate the feasibility of the proposed cooling method and structure, which can make a fully superconducting motor lighter and more compact than a partially superconducting motor. The helium gas pressure in the gap, the rotor rotational speed, and the rotor loss are key parameters of the cooling, and thermal analysis was performed to clarify the dependence of the field winding temperature on these parameters. In the analysis, the motor model has the maximum output of 5.5 MW and the maximum speed of 5000 rpm. The superconducting armature windings are cooled down to 20 K by conduction cooling with liquid hydrogen, and the superconducting field winding is cooled down to about 50 K indirectly using helium gas in the gap. Numerical analysis results indicate that the helium gas pressure should be increased when the rotor loss is larger, but if the pressure is increased too much, the temperature of the field winding becomes higher due to wind loss. When the rotor loss is several hundred watts, the helium gas pressure should be in the range of 10^4 to 10^5 Pa.

1. Introduction

Electrification of aircraft propulsion systems is one of the key solutions for realizing Net Zero of CO_2 emission in the aviation sector by 2050, which was given by ATAG (Air Transport Action Group). The use of electric motors and power converters in aircraft propulsion systems is an effective way to improve system efficiency. However, the weight of the electrical equipment used there should be lighter than that of state-of-the-art motors and power converters. There is an estimate that more than 16 kW/kg is needed [1].

Superconducting technology has the potential to meet or excess the requirements. The current density of superconducting (SC) wire cooled below 77 K that would be used in rotating machines are tens or hundreds of times higher than that of copper wire. Also, the iron teeth in the stator can be eliminated thanks to the high magnetic field generated by the superconducting windings; therefore, the weight of the superconducting motors can be lighter than that of conventional motors.

There are two kinds of SC rotating machines: partially SC motors (PSCMs) and fully SC motors (FSCMs). While the PSCMs have SC field windings and copper armature windings, the FSCMs use

SC wires for both field and armature windings. In the FSCMs, the armature and field windings are installed in the same cryostat and the magnetic gap between the field and armature windings can be reduced in comparison with that of the PSCMs. Therefore, the power density of the FSCMs can be higher than that of the PSCMs.

The authors have been studying the FSCMs using MgB_2 armature windings and REBCO field windings cooled at 20 K with liquid hydrogen [2][3]. MgB_2 SC wires have a multifilament structure, which has the potential to reduce AC losses [4]. On the other hand, REBCO wires have better J_C - B characteristics at 20 K or higher. However, this FSCM structure needs two coolant paths for SC field and armature windings and results in a complicated cooling structure [5]. Therefore, the authors have proposed and studied the rotor cooling using dilute gas in the airgap, while the stator side is still cooled at 20 K with liquid hydrogen. Figure 1 shows a conceptual diagram of the motor cooling structure. The REBCO field windings are operated around 50 K by the cooling based on heat conduction and convective heat transfer via dilute gas. This structure can simplify the rotor cooling structure. Thermal analysis using finite element method was carried out to investigate the effect of the dilute gas cooling for SC field windings. By giving electromagnetic losses as thermal load to the rotor and changing parameters such as the helium gas pressure, the rotational speed, and so on, the cooling characteristics of this method is discussed.

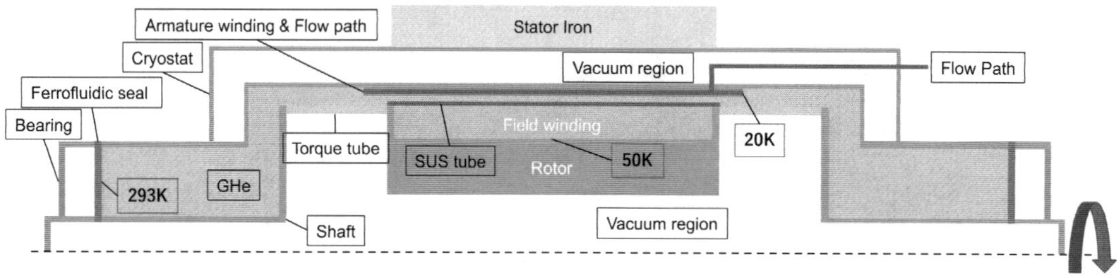

Figure 1. Conceptual diagram of the proposed cooling structure

2. Numerical Analysis

2.1. Method

Steady-state thermal conduction analysis was performed using COMSOL Multiphysics®, a finite element analysis software program. The governing equation of the analysis is as follows:

$$-\nabla \cdot k(T)\nabla T = Q \tag{1}$$

where T is the temperature, k is the thermal conductivity, and Q is the heat generation. The thermal conductivity of the dilute gas in the stator-rotor gap depends on the relationship between the mean free path l [m] of the gas and the gap length d [m], which is the distance between the heat transfer surfaces, and is expressed by the following equation [6].

$$k(T) = \begin{cases} \dfrac{9\gamma - 5}{4}\eta(T)C_V & (l \ll d\text{: viscous flow}) \\ \dfrac{1}{2}\dfrac{\gamma + 1}{\gamma - 1}ap\sqrt{\dfrac{k_b}{2\pi mT}} & (l > d\text{: } molecular\ flow) \end{cases} \tag{2}$$

$$\eta(T) = 2.669 \times 10^{-6} \times \frac{\sqrt{MT}}{\sigma^2 \Omega_v} \tag{3}$$

where γ is the ratio of specific heat, η is the viscosity coefficient, C_V is the specific heat at constant volume, a is the accommodation coefficient, p is the pressure, k_b is the Boltzmann constant, m is the mass of the gas, M is the molecular weight, σ is the collision diameter, and Ω_v is the collision integral. To reflect the effect of dilute gas convection due to the rotation of the rotor, the effective thermal conductivity k_e, related with Nusselt number Nu, is used.

$$k_e = Nu \cdot k \tag{4}$$

The Nusselt number and the dimensionless numbers used to find the Nusselt number were obtained by applying the following empirical equation for convective heat transfer in a concentric rotating double cylinder [7].

$$Nu = \begin{cases} 2 & (Ta < Ta_c) \\ 0.42(TaPr)^{0.25} & (Ta_c < Ta < 10^8) \\ 0.092(TaPr)^{\frac{1}{3}} & (10^8 < Ta < 5 \times 10^{12}) \end{cases} \tag{5}$$

$$Ta = \frac{\omega^2 R_1}{\nu^2} \left(\frac{D_h}{2}\right)^3 \tag{6}$$

$$Pr = \frac{\nu}{\alpha} \tag{7}$$

where Ta is the Taylor number, Pr is the Prandtl number, Ta_c is the critical Taylor number ($Ta_c \approx 1700$), ω is the rotational velocity, R_1 is the rotor radius, ν is the kinematic viscosity, D_h is the hydraulic diameter, and α is the thermal diffusivity.

In addition, the rotation of the rotor generates wind loss, resulting in heat generation on the rotor surface. The wind loss generated in a rotating double cylinder was considered with reference to [8], and the heat source was given to the rotor surface using the following equation.

$$Q_{WC} = C_{mc} \rho \omega^3 r_1^2 r_2^2 h \tag{8}$$

where C_{mc} is the coefficient of friction moment, r_1 is the rotor radius, r_2 is the inner radius of the cryostat, and h is the rotor length. The coefficient C_{mc} is the dimensionless number determined by the rotational Reynolds number R_ω as shown below:

$$C_{mc} = 0.036 R_\omega^{-0.194} \tag{9}$$

The rotational Reynolds number R_ω is defined by Equation (10).

$$R_\omega = \frac{(r_2^2 - r_1^2)\omega}{2\nu} \tag{10}$$

2.2. *Specifications of FSCMs*

In this study, the analysis was performed for FSCMs with a power output of 5.5 MW. The main specifications of the FSCMs are listed in Table 1 and the materials used in the model in Table 2. Material properties of the winding areas in the thermal analysis were approximated by those of copper. Figure 2 shows the geometry model used for the thermal analysis.

Table 1. Specifications of the FSCM

Output power	5.5 MW
Rotational speed	5000 rpm
Number of poles	4
Rotor radius	128.9 mm
Rotor – stator gap	20 mm
SUS tube thickness	1 mm

Table 2. Materials used in the FSCM model

Components	Materials
Rotor, Shaft, Cryostat, SUS tube	SUS
Superconducting field and armature windings	Copper
Torque tube, Armature winding support	FR4

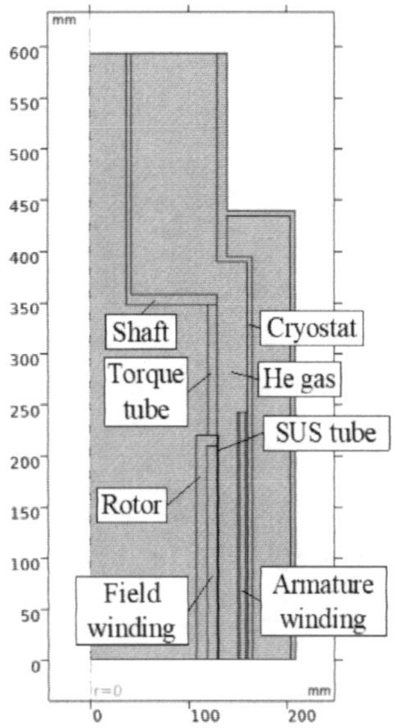

Figure 2. Analysis model

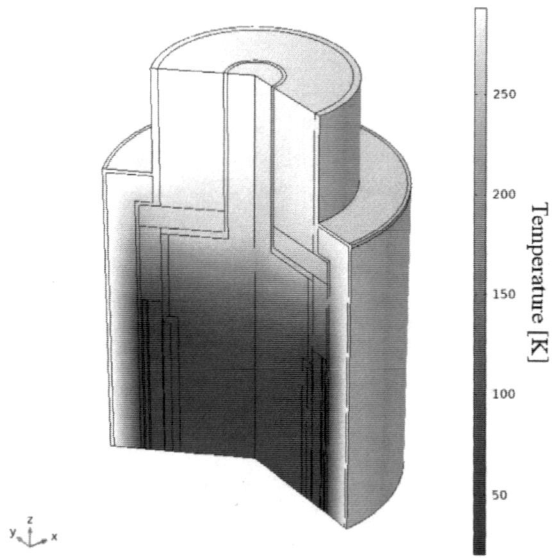

Figure 3. Temperature distribution at 10^4 Pa and 5000 rpm

2.3. *Results and discussion*

Thermal analysis was performed to investigate the dependence of field winding temperature on the helium gas pressure and the rotor loss. The eddy current rotor loss was given to the SUS tube as a heat source. Figure 3 shows an example of the analysis result, giving the temperature distribution in the motor at the helium gas pressure of 10^4 Pa, the rotational speed of 5000 rpm, and no heat loss.

The dependence of the field winding temperatures on the helium gas pressure were obtained for the cases of no loss, 10 W, 100 W, and 1000 W in the rotor. The results are shown in figure 4. For the rotor loss of 0 W and 10 W, the field winding temperature is lowest when the pressure is 10^4 Pa, and for the rotor loss of 100 W, the temperature is lowest when the pressure is 10^5 Pa. This is considered to be due to the increase in the Nusselt number, as shown in figure 5. As the pressure of the helium gas increased, the density of the helium gas also changed, which increased both the Taylor number and the Nusselt number, accelerating the cooling effect of convection.

On the other hand, if the helium gas pressure is too high, the temperature of the field winding rises due to wind loss.

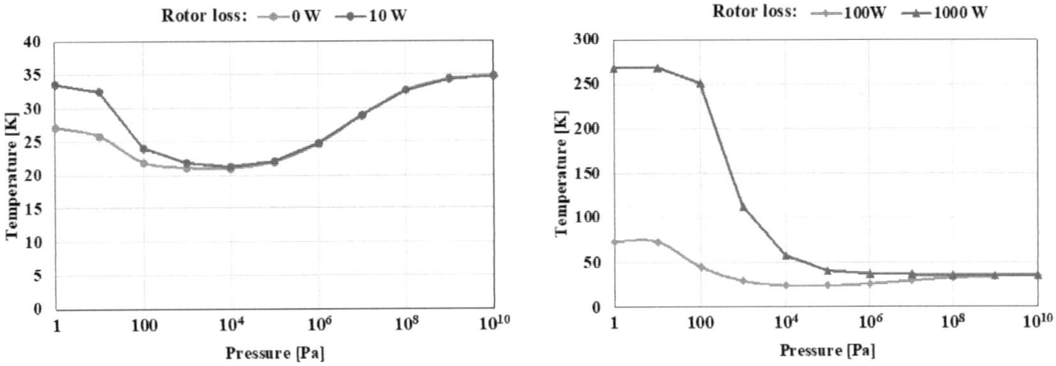

(a) (b)

Figure 4. Dependence of field winding temperature on He gas pressure and rotor loss

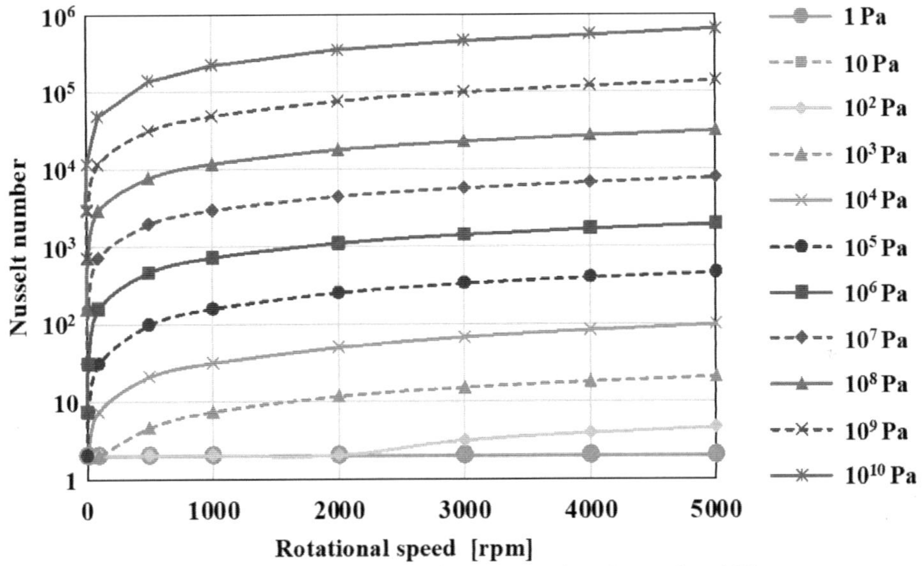

Figure 5. Dependence of Nusselt number on rotational speed and He gas pressure

Dependence of the field winding temperature on the rotational speed and the helium gas pressure was also analyzed for the cases of no loss and 100 W loss in the rotor. The results are shown in figure 6 and figure 7. When the helium gas pressure is 1 Pa, the temperature is constant regardless of the rotational speed, but as the pressure is increased, it is confirmed that the rotation promotes cooling. When the helium gas pressure is higher than 10^4 Pa, the field winding temperature is sufficiently cooled even at low rotational speeds. On the other hand, when the helium gas pressure is set to 10^6 or higher, the field winding temperature rises due to the increase in wind loss caused by the increase in rotational speed.

Based on these results, a helium gas pressure of about 10^4 to 10^5 is appropriate for rotor losses up to several hundred watts.

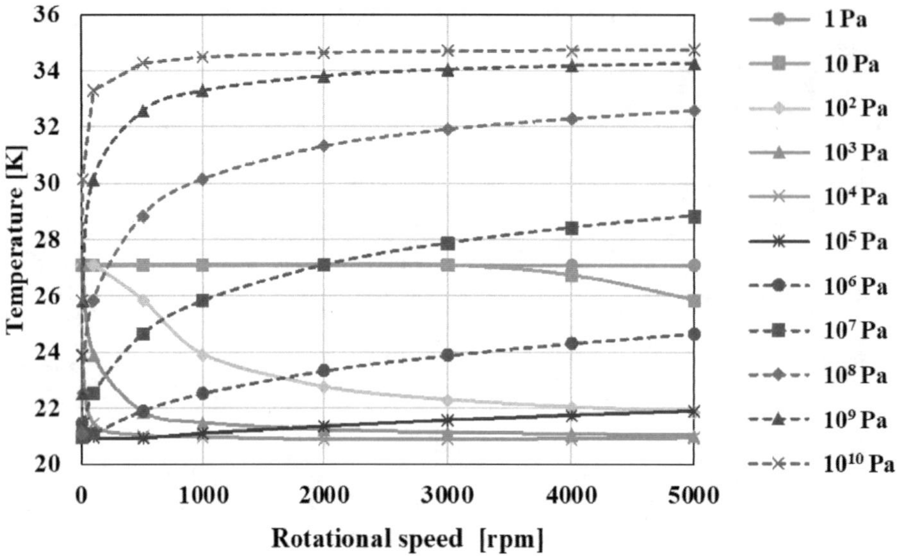

Figure 6. Dependence of field winding temperature on rotational speed and He gas pressure: No rotor loss case

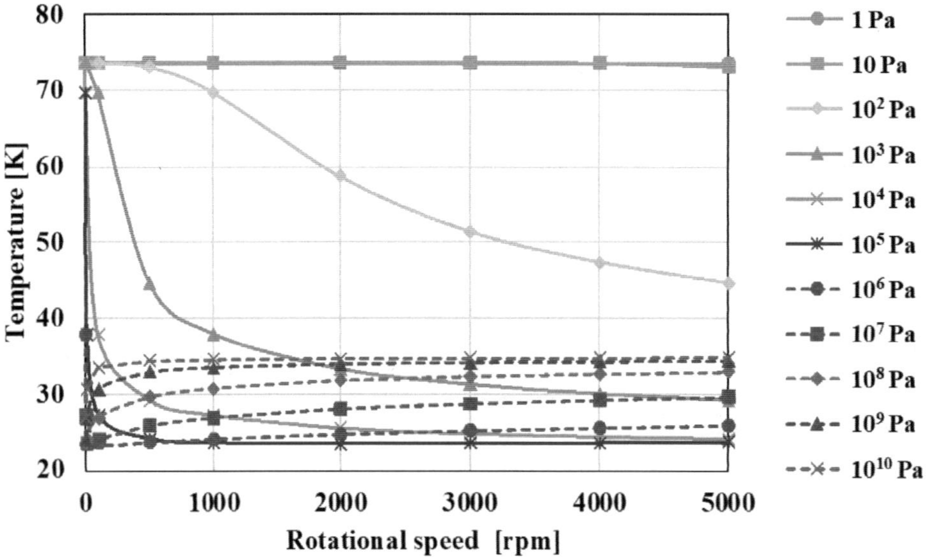

Figure 7. Dependence of field winding temperature on rotational speed and He gas pressure: 100 W rotor loss case

3. Conclusion

In this study, the numerical analysis was performed to investigate the feasibility of rotor cooling using dilute gas for fully superconducting motors. If the rotor losses are large, the helium gas pressure for cooling should be increased, but if it is too high, the temperature of the field winding rises due to wind loss. The helium gas pressure should be in the range from 10^4 to 10^5 Pa for a rotor loss of a few hundred watts.

For a practical design of the motor using the proposed cooling method, further electromagnetic and thermal design of the motor is needed to reduce the rotor losses and to achieve more efficient rotor cooling.

References

[1] Łukasik B 2017 *ASME Turbo Expo 2017: Turbomachinery Technical Conf. and Exposition GT2017* **1** GT2017-63834
[2] Terao Y, Ishida Y, Ohsaki H and Oyori H 2020 *SAE Int. Journal of Advances and Current Practices in Mobility* **2** pp 828-37
[3] Kato M, Terao Y, and Ohsaki H 2021 *Journal of Physics: Conf. Series* **1857** pp 012016
[4] Terao Y, Kong W, Ohsaki H, Oyori H and Morioka N 2018 *IEEE Transactions on Applied Superconductivity* **28** 5208005
[5] Terao Y, Seta A, Ohsaki H, Oyori H and Morioka N 2019 IEEE Transactions on Applied Superconductivity **29** 5202305
[6] ULVAC 2002 *Vacuum Handbook* (Tokyo: Ohmsha)(Japanese)
[7] Fénot M, Bertin Y, Dorignac E and Lalizel G 2011 *Int. Journal of Thermal Sciences* **50** pp 1138-55
[8] Yamada N, Fujimori T and Wakashima S 2012 *Transactions of the JSME (B)* **78** pp 1095-1106

Journal of Physics: Conference Series **2545** (2023) 012029 doi:10.1088/1742-6596/2545/1/012029

Influence of low electric field parts of electric field–current density curves of superconductors on their ac loss characteristics

Y Sogabe and N Amemiya*

Department of Electrical Engineering, Kyoto University, Kyoto 615-8510, Japan

E-mail: *amemiya.naoyuki.6a@kyoto-u.ac.jp

Abstract. We performed numerical electromagnetic field analyses to investigate the influence of the low electric field parts of the electric field (E)–current density (J) curves of superconductors on their ac loss characteristics. A simple one-dimensional model across the width of a coated conductor was used for the analyses. Whereas we fixed the n value of the high E part of E–J curves of superconductors, we varied the n values of low E parts. J distributions as well as E distributions were calculated in coated conductors carrying ac currents and/or exposed to ac magnetic fields. We examined the influence of the n values at low E part of E–J curves on ac loss characteristics and discussed the electric field regions that are relevant to ac loss generation. We also discussed the applicability of the results to copper-plated multifilament coated conductors.

1. Introduction

When a superconductor is carrying ac current and/or is exposed to ac magnetic fields, a non-ignorable electric field could may appear in the superconductor due to the electromagnetic induction. This consideration raises the following questions.

1) What is the dominant magnitude of electric field in superconductors operated at such ac conditions?
2) Which part of an E–J curve is relevant to ac loss generation?

Regarding to the second question, one of our recent concerns is E–J curves of low gradients in low E regions, which we observe in copper-plated multifilament coated conductors by transport method frequently as shown in figure 1. In the figure, I means the transport current. This may be due to the existence of local defects, where critical current is locally low, in the superconductor filaments and the electric field generated by transverse currents that bypass these defects and flow through the copper layer as shown in figure 2 [1]. Various studies have been conducted on the characteristics of coated conductors with local defects [1]–[7], but none have investigated the influence of local defects in copper-plated multifilament coated conductors under ac conditions on their electromagnetic characteristics. It is necessary to investigate the influence of E–J curves of low gradients in low E regions on electromagnetic characteristics of the coated conductors in ac conditions.

For prediction of critical currents and ac losses in coils wound with coated conductors, numerical electromagnetic field analyses are important tools. At relatively high frequency, for example 50 Hz, much higher electric fields compared with the criterion electric field E_0 (generally 100 μV/m) for

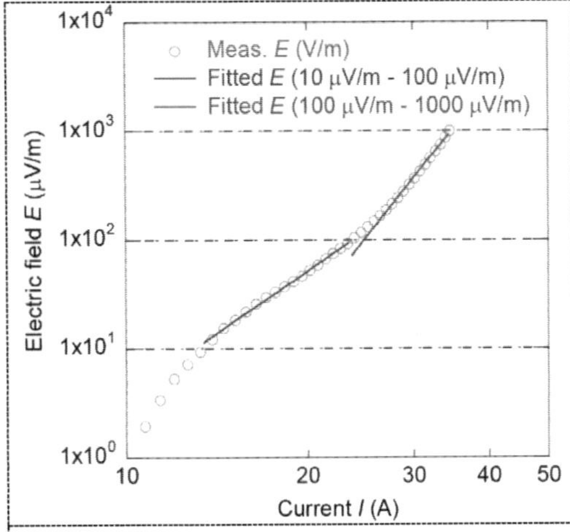

Figure 1. Typical dc measurement result of E–I characteristics of a copper-plated multifilament coated conductor by using transport method.

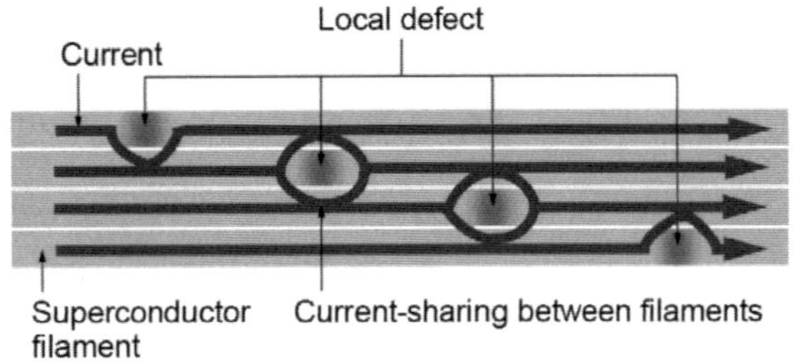

Figure 2. Schema of current path in a copper-plated multifilament coated conductor with local defects carrying transport current [4].

definition of critical current I_c, may be induced in the coated conductors. The formulation of E–J characteristics of superconductors, which is used in electromagnetic field analyses, is generally determined from dc transport measurement results of E–I characteristic in the electric field region of 10 μV/m–1,000 μV/m. Namely, the electric field region in a relatively high frequency ac conditions may not match it used to formulate the E–J characteristics. Since this mismatch adversely affects the accuracy of the analyses, it is necessary to quantitatively evaluate the electric field distribution induced in the coated conductors in ac environment.

We evaluated electric fields and ac losses in coated conductors carrying ac currents and/or exposed to ac magnetic fields by simple one-dimensional electromagnetic field analyses. The analysed coated conductors had different E–J curves in which we varied the n-value of low E parts, but fixed the n-values of the high E parts. In section 2, we demonstrate the numerical electromagnetic field analysis method briefly and show the analysis conditions. In section 3, the analysis results are presented, such as J and E distribution in coated conductors, time variation of E and ac losses, and dependence of the E–J curves in low E regions on E and ac losses. Also we discuss the validity of approximation of influence of local defects with E–J curves of low gradient in low E regions in this session. In session 4, we conclude.

2. Numerical electromagnetic field analysis model and conditions

2.1. *Numerical electromagnetic field analysis model [8]*

We analysed a single coated conductor by using one-dimensional analysis model. In the model, we used current vector potential T-formulation, which was derived from Faraday's law, Biot-Savart's law, Ohm's law, and definition of T. We applied a thin-strip approximation and assumed that the analysed coated conductor is infinitely long. In this situation, we need only consider the electromagnetic field distribution in the width direction of superconductor layer, and the equation to be solved in the analysis is expressed as follows:

$$-\frac{\partial}{\partial y}\frac{1}{\sigma}\frac{\partial T}{\partial y}+\frac{\mu_0 t_s}{2\pi}\frac{\partial}{\partial t}\int\frac{1}{r(y,y')}\cdot\frac{\partial T'}{\partial y'}\mathrm{d}y'=0 \qquad (1)$$

Here, T and T' are normal components of current vector potentials at field point and source point, respectively. y and y' are local coordinates of field point and source point, respectively. r is distance from source point to field point, and t_s is the thickness of superconductor layer. σ is the equivalent conductivity of the superconductor layer, which is calculated based on the E–J characteristics.

2.2. Analysed coated conductors and analysis conditions

We analysed single infinite-long monofilament coated conductors. The specifications of the analysed coated conductors are listed in table 1. We used the power-law model as the E–J characteristics of the coated conductors, and we did not consider the magnetic field dependence of critical current density J_c and n-value in the analyses to simplify the model. We analysed three different coated conductors with different E–J characteristics, in which we varied the n-value of low E parts, but fixed the n-values of the high E parts. The combinations of n-values are listed in table 2.

Table 1. Specifications of analysed coated conductors

Analysed object	Single infinite-long coated conductor
Tape width	4 mm
Tape thickness	0.2 mm
Thickness of superconductor layer t_s	2 μm
Critical current density J_c	2.5×10^{10} A/m^2
Critical current I_c	200 A
Electric field criterion E_0 for J_c definition	100 μV/m

Table 2. n-value patterns used in numerical analyses

n-value patterns	Low E region ($E < E_0$)	High E region ($E \geq E_0$)
Pattern 1	20	20
Pattern 2	10	20
Pattern 3	5	20

The time-evolutions of transport current $I(t)$ and external magnetic field $B(t)$ are described as follows:

$$I(t) = I_t \sin(2\pi f t), \quad B(t) = B_{ext} \sin(2\pi f t) \tag{2}$$

The combinations of amplitude of transport currents I_t and amplitude of external magnetic fields B_{ext} in the analyses were as follows: (I_t, B_{ext}) = (100 A, 0 T), (0 A, 0.1 T), and (100 A, 0.1 T). The frequencies f of currents and external magnetic fields are 50 Hz, and they are in-phase.

3. Analysis results

3.1. Time evolution of distributions of current density, electric field, and ac loss in a coated conductor

The calculated current density J in the coated conductor with n-value pattern 1 were shown in figure 3. When (I_t, B_{ext}) was (100 A, 0 T) (current-only condition), J at the lateral centre of the coated conductor was changing because the non-linear E–J curve of the coated conductor was considered in the model. When external magnetic field was applied, magnitude of J was gradually decreased from t = 10 ms to t = 15 ms due to finite resistivity in the coated conductor and reduction of dB_{ext} / dt. In contrast, J distribution changed drastically from t = 15 ms to t = 20 ms because of reversal of dB_{ext} / dt at t = 15 ms.

The calculated E in the coated conductor with n-value pattern 1 were shown in figure 4. In figure 4, the magnitude of E were plotted on the logarithmic axis. Under current-only condition, E was

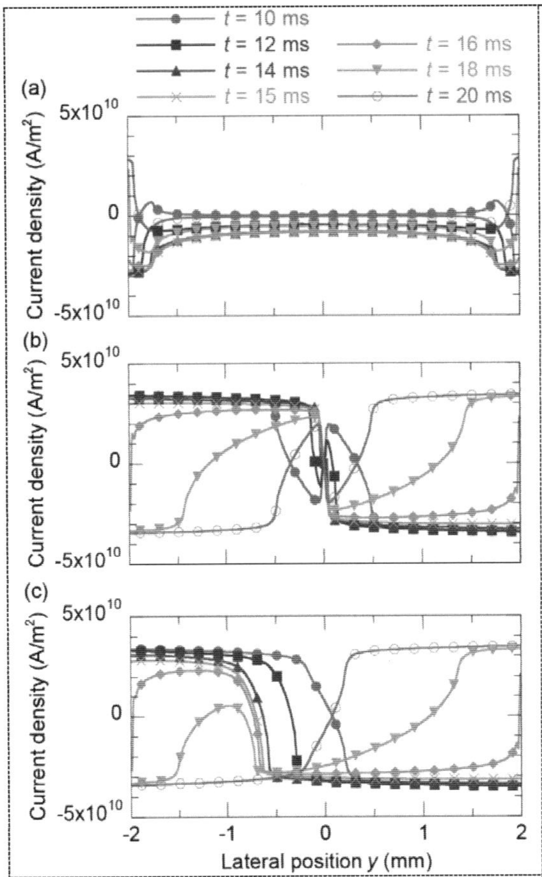

Figure 3. Current density J distribution in the analysed coated conductor with n-value pattern 1; (a) when (I_t, B_{ext}) = (100 A, 0 T), (b) when (I_t, B_{ext}) = (0 A, 0.1 T), and (c) when (I_t, B_{ext}) = (100 A, 0.1 T).

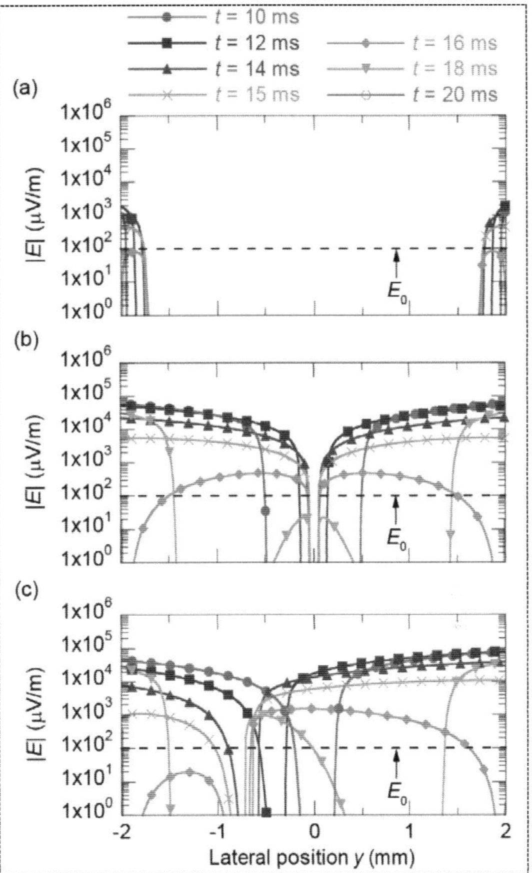

Figure 4. Electric field $|E|$ distribution in the analysed coated conductor with n-value pattern 1; (a) when (I_t, B_{ext}) = (100 A, 0 T), (b) when (I_t, B_{ext}) = (0 A, 0.1 T), and (c) when (I_t, B_{ext}) = (100 A, 0.1 T).

concentrated at the edge of the coated conductor. On the other hand, when an external magnetic field was applied to the coated conductor, E was induced near the lateral centre of the coated conductor. This was because the self-field generated by the transport current was much smaller than the applied external magnetic field. Focusing on the magnitude of E at the edge of the coated conductor, it exceeded E_0 under all conditions. Especially when an external magnetic field is applied, magnitude of E is exceeding 1×10^4 μV/m.

The time variation of $|E|$ and ac loss per length in the coated conductor were shown in figure 5. Here, the analysed conductor was with n-value pattern 1 and was carrying ac current under ac magnetic field ((I_t, B_{ext}) = (100 A, 0.1 T)). E at y = 2 mm in figure 4 was plotted in the figure. These results showed that the induced large E in the coated conductor dominated the time variation of ac loss, and it is in the order of 10^4 μV/m.

3.2. Dependence of electric fields and ac losses on n-value in lower electric field region

Figure 6 showed the dependences of E and ac losses on n-values in low E region. Here, magnitude of E at the edge of coated conductors (y = 2 mm) at t = 20 ms were plotted in figure 6(a). We did not see any dependences of E and ac losses on n-value in low E region. This was a natural consequence of the result

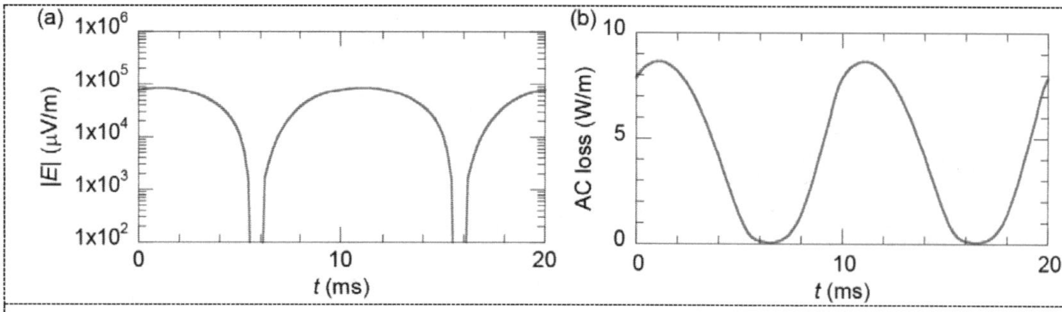

Figure 5. (a) Time variation of magnitude of E at y = 2 mm and (b) time variation of ac loss per length. The analysed coated conductor was with n-value pattern 1 and was carrying current under external magnetic field.

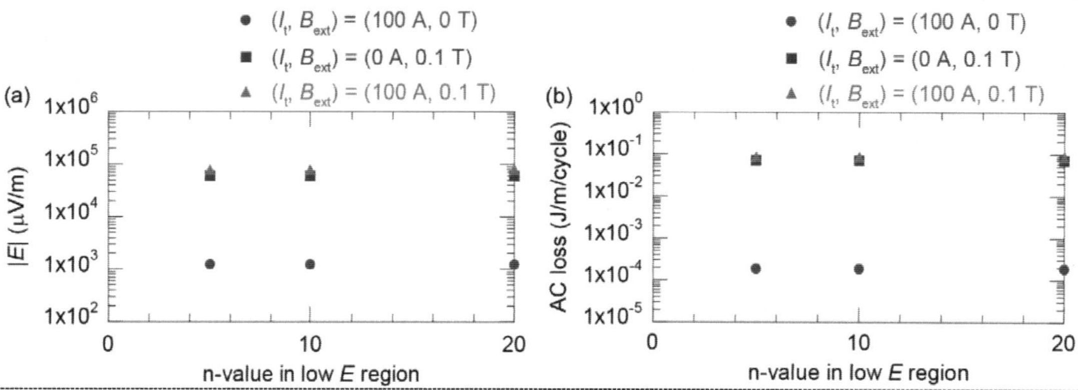

Figure 6. n-value in low E region dependence of (a) magnitude of E and (b) ac losses per cycle per length. E were at the position of y = 2 mm, and at t = 20 ms.

that E induced by the time variation of the transport current and external magnetic field were much higher than E_0 as shown in figure 4. This result suggests that low n-value in low E region of coated conductors will not be problem from the view point of ac losses in relatively high frequency environment.

3.3. Discussion on applicability of the results to copper-plated multifilament coated conductors with local defects

When evaluating the influence of local defects in copper-plated multifilament coated conductors on the electromagnetic behaviour of the coated conductors, it is necessary to discuss whether it is appropriate to approximate the influence of local defects by the E–J curves of low gradients in low E regions.

When copper-plated multifilament coated conductors carry ac currents without external magnetic fields, transport currents should flow through the copper layers to bypass the local defects as in the E–I characteristic measurements with dc transport currents as shown in figure 2. In this situation, the copper-plated multifilament coated conductors with the local defects can be approximated with monofilament coated conductors with low n-value in low electric field region.

When copper-plated multifilament coated conductors exposed to external ac magnetic fields without transport currents, the shielding current path induced in the coated conductors is very different from the distribution when carrying dc currents. If the filaments are decoupled, i.e., if the frequency of the external magnetic field is sufficiently lower than the characteristic frequency of the coupling current, the current distribution in the copper-plated multifilament coated conductor is as shown in figure 7(a). In this case, no coupling currents are induced and the induced shielding currents are confined in each

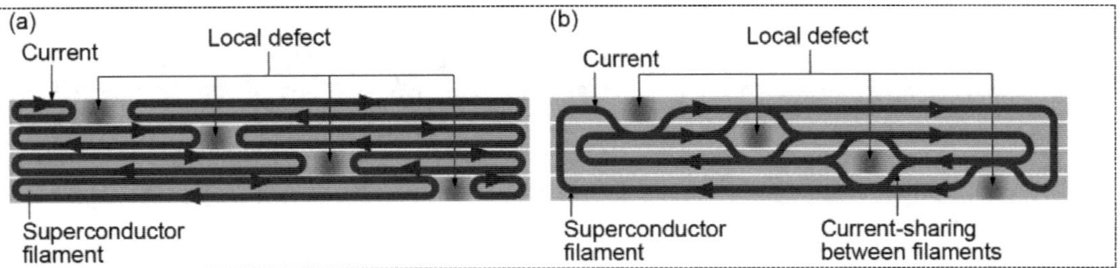

Figure 7. Schemas of current distribution in a copper-plated multifilament coated conductor with local defects under ac external magnetic field B_{ext}; (a) when frequency of B_{ext} is much lower than characteristic frequency of coupling currents, and (b) when frequency of B_{ext} is much higher than characteristic frequency of coupling currents

filament. However, the shielding currents are blocked by local defects and are divided in longitudinal direction. Here, since the transverse resistance between filaments does not affect the electromagnetic characteristics of the coated conductor, such as electric field and ac loss, it is not reasonable to approximate the copper-plated multifilament coated conductor with local defects as a monofilament coated conductor with low n-value in low electric field region.

Conversely, if the filaments are coupled, i.e., if the frequency of the external magnetic field is sufficiently higher than the characteristic frequency of the coupling current, the current distribution in the coated conductor is as shown in figure 3(b). In this case, the coupling currents that bypass the local defects are induced, and the transverse resistance between filaments affects the electromagnetic characteristics of the coated conductor. Therefore, it is appropriate to approximate the influence of local defects as E–J curves of low gradients in low E regions.

This paper focuses on the relatively high frequency range of 50 Hz, which is much higher frequency than characteristic frequency of coupling currents in a straight copper-plated multifilament coated conductor [9]. Therefore, we can reasonably apply the results of analyses to the copper-plated multifilament coated conductors.

4. Conclusion

In coated conductors carrying ac currents and/or exposed to ac magnetic fields, high electric fields appear due to the electromagnetic induction. With calculated conditions (frequency = 50 Hz, transport current = 100 A, applied magnetic field = 0.1 T), very high electric fields, whose order of magnitudes were 10^4 μV/m, appeared in coated conductors. As the consequence, high E parts of E–J curves are relevant to ac loss generation, and low E parts of E–J curves are not relevant to ac loss generation. Note that the dominating current densities could be rather higher than the critical current density, if n value is small. In such a situation, analytical value of ac loss using critical current density could contain large error [10].

The conclusion shown above is reasonably applicable not only to monofilament coated conductors but also to copper-plated multifilament coated conductors although we used a simple model, in which we did not consider their composite structure.

Acknowledgments
This work was supported in part by JST-Mirai Program Grant Number JPMJMI19E1, Japan, and in part by JSPS KAKENHI Grant Number JP22K14238.

References

[1] Amemiya N, Zhao Y, Luo X, Xu G, Li Y and Sogabe Y 2022 Current-sharing between filaments and voltage–current characteristics of copper-plated multifilament coated conductors *IEEE Trans. Appl. Supercond.* **32** 5900809
[2] Polak M, Barnes P N, Mozola P and Levin G A 2009 Critical current in YBCO coated conductors

in the presence of a macroscopic defect *IEEE Trans. Appl. Supercond.* **19** 2921–4

[3] Mbaruku A L, Trociewitz U P, Wang X and Schwartz J 2007 Relationships between conductor damage, quenching and electromechanical behavior in YBCO coated conductors *IEEE Trans. Appl. Supercond.* **17** 3044–9

[4] Luo X, Zhao Y, Sogabe Y, Sakamoto H, Yamano S and Amemiya N 2022 Thermal runaway of conduction-cooled monofilament and multifilament coated conductors *IEEE Trans. Appl. Supercond.* **32** 6600609

[5] Liu W, Yong H and Zhou Y 2016 Numerical analysis of quench in coated conductors with defects *AIP advances* **6** 095023

[6] Kwon C, Young J L, James R G, Levin G A, Haugan T J and Barnes P N 2007 Local current transport and current sharing between filaments in striated coated conductors with artificial defects *IEEE Trans. Appl. Supercond.* **17** 3191–4

[7] Haberkorn N, Kim J, Suárez S, Lee J H and Moon S H 2015 Influence of random point defects introduced by proton irradiation on the flux creep rates and magnetic field dependence of the critical current density J_c of co-evaporated $GdBa_2Cu_3O_{7-\delta}$ coated conductors *Supercond. Sci. Technol.* **28** 125007

[8] Amemiya N, Murasawa S, Banno N and Miyamoto K 1998 Numerical modelings of superconducting wires for ac loss calculations *Physica C* **310** 16–29

[9] Amemiya N, Tominaga N, Toyomoto R, Nishimoto T, Sogabe Y, Yamano S and Sakamoto H 2018 Coupling time constants of striated and copper-plated coated conductors and the potential of striation to reduce shielding-current-induced fields in pancake coils *Supercond. Sci. Technol.* **31** 025007

[10] Amemiya N, Miyamoto K, Banno N and Tsukamoto S 1997 Numerical analysis of ac losses in high Tc superconductors based on *E-j* characteristics represented with *n*-value *IEEE Trans. Appl. Supercond.* **7** 2110–3

Development of Exciter System for Power Supply driven HTS-MRI

Yasuyuki Shirai[1], Yuto Uchida[1], Hideaki Miura[2], Taisuke Hattori[2], Tetsuya Matsuda[2]

[1]Kyoto University, Yoshida-honmachi, Sakyo-ku, Kyoto 606-8501, Kyoto Japan
[2]Mitsubishi Electric Corporation Advanced Technology R&D Center, Amagasaki, Hyogo 661-8661, Japan

Abstract. Exciter systems is examined for helium free high-temperature superconducting (HTS)-MRI magnet. We have developed an active magnetic field stability control by use of a highly stabilized power supply instead of the persistent current operation. The magnetic field stability was measured and evaluated using a HTS-MRI test magnet under the power supply driving mode of a dropper-type and of a switching-type direct-current exciter system. A high stable direct current power supply is normally designed using the dropper-type power supply which is suitable for constant current operation of less current ripple. However, it is of large power loss and is not suitable for charging/discharging operation. On the other hand, the switching-type power supply is suitable for variable current control operation and of less power loss. However, it is of large current ripple in the constant current operation which would degrade the magnetic field stability in HTS-MRI imaging operation. The availability of the power supply system was discussed in MRI imaging of a mouse using an experimental HTS-MRI system. HTS-MRI imaging of mouse (~50 mm).

1. Introduction

The development of helium free high-temperature superconducting (HTS)-MRI magnet is expected. An HTS-MRI magnet made of HTC tape superconductors has difficulties in the persistent current exciting mode for producing a highly stabilized magnetic field for MRI imaging, due to a screening current in REBCO tape [1], [2]. Furthermore, development is necessary in non-resistive joint and persistent current switch of REBCO MRI magnet [3].

One solution for these problems is an active magnetic field stability control by use of a power supply instead of the persistent current operation. We have developed the power supply system consists of a exciter power supply and a small trimming current supply which compensates the magnetic field deviation due to the screening current. The MRI magnetic field feedback/feedforward control test was carried out, that is, the overshooting and the current trimming control, using a HTS-MRI test magnet [4], [5].

In this paper, we measured and evaluated the magnetic field stability, using a HTS-MRI test magnet in the power supply driving mode using a dropper-type and of a switching-type direct-current exciter system. A high stable direct current power supply is normally designed using the dropper-type power supply which is suitable for the constant current operation of less current ripple. However, it is of large power loss and is not suitable for charging/discharging operation. On the other hand, the switching-type power supply is suitable for the variable current control operation and of less power loss. However, it is

Journal of Physics: Conference Series **2545** (2023) 012030 doi:10.1088/1742-6596/2545/1/012030

of large current ripple in the constant current operation which would degrades the magnetic field stability in HTS-MRI imaging operation.

A resistor of 1.2 Ω was set between the magnet terminal in parallel to the MRI magnet in order to improve the magnetic field stability in the constant current mode for the clear MRI imaging, at the same time to alleviate the current instability in the charging mode with the dropper-type power supply and to reduce the magnet current ripple due to the voltage ripple in the constant current mode with the switching-type one.

The availability of the power supply systems were discussed in MRI imaging of a mouse using an experimental HTS-MRI system. The MRI image of the mouse was obtained under the power supply driven operation of the dropper-type power supply and the switching-type power supply.

2. Power supply for HTS-MRI magnet

2.1. Experimental HTS-MRI magnet

Our research group is developing an HTS-MRI magnet using REBCO wire [6]-[10]. The HTS-MRI magnet was originally designed for 3T with 148A [9]. The degradation of the critical current of the magnet was observed (down to 20A) in the charging test [10]. The HTS-MRI magnet was re-arranged for the imaging test with 0.27 T@20A, 102 H (Figure 1). In this study, magnetic field stability and MRI imaging tests under power excitation were performed using this magnet at a current of 20 A and a center field of 0.27 T. The results showed that the magnet was stable under the excitation of the power supply, and that the magnet was stable under the excitation of the power supply.

To obtain clean MRI images, magnetic field fluctuations must be kept below 1ppm, which corresponds to coil current fluctuations on the order of μA. To achieve this, conventional low-temperature superconducting MRI magnets operate with permanent current as a superconducting short circuit. To achieve the same operation in a power-supply-driven high-temperature superconducting MRI magnet, a dropper-type constant-current power supply with a small output current ripple is considered suitable. However, when magnets with large inductance are excited or demagnetized, control tends to become unstable. On the other hand, switching power supplies, which are voltage sources, have good current controllability, but output voltage and current ripple cannot be small. ,

Figure 1. Experimental HTS-MRI magnet. (102 H, 0.27 T@20A).

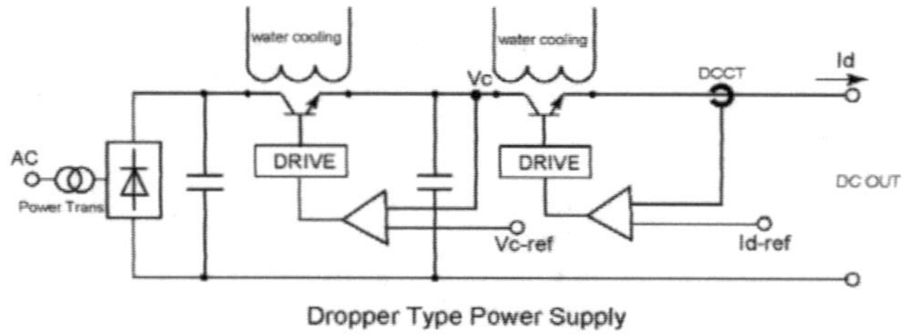

Figure 2. Circuit diagram of a dropper type DC power supply.

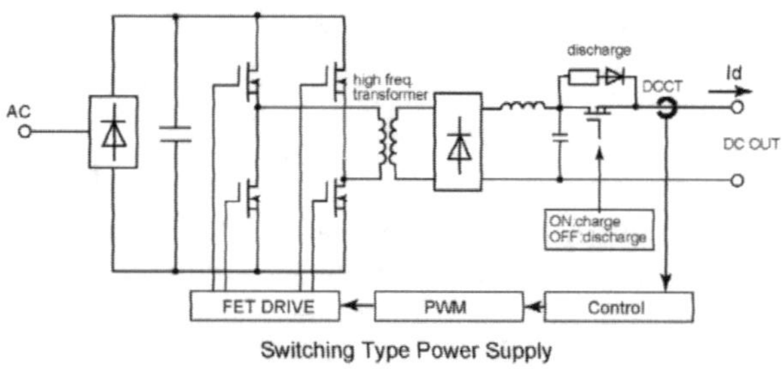

Figure 3. Circuit diagram of a switching type DC power supply.

2.2. Dropper-type direct-current stabilized power unit
Figure 2 shows a circuit diagram of a dropper-type power supply. In order to obtain a stable constant current source, the output current is detected by the DCCT and compared with the current command value Id-ref to control the base current of parallel-connected BJT(Bipolar Junction Transistor) through the feedback control system. The collector voltage is variable with respect to the magnitude of the specified current, and this voltage is also controlled by BJT dropper from the diode rectified DC voltage, thus reducing switching noise as much as possible. The BJTs with constant current flow need to be cooled by water, resulting in large operating losses.

2.3. Switching power supply unit
Figure 3 shows a circuit diagram of a switching-type power supply. The switching power supply converts the diode-rectified DC voltage into AC voltage of variable voltage and frequency according to the desired DC output current by PWM control, which is isolated by a high-frequency transformer, diode-rectified again, and supplies DC current to the load through an LC filter. Since this power supply cannot generate negative DC voltage, the current is reduced by a discharge resistor through the discharge circuit shown in the figure when the magnet is discharged.

2.3.1. Specifications of power supplies
Specifications of power supplies are listed in Table 1. The current stability of the dropper voltage is 0.1ppm/hour, whereas it is 5ppm/hour for the switching power supply. The voltage ripple of the dropper power supply is 1.0 nVp-p as a current source specification, while the switching power supply has a current ripple of 20 ppmRMS as a voltage source specification. 224 A of current flow requires water cooling with 5644 W of power dissipation for the dropper power supply, while 900 W can be air-cooled for the switching power supply.

Table 1. Specifications of power supplies.

	Dropper	Switching
Output current (A)	500	500
Output voltage (V)	±20	±20
Current stability (ppm/hour)	0.1	5
Voltage ripple (nVp-p)	1.0	-
Current ripple (ppmRMS)	-	20
Power loss :@224A (W)	5644	900
Cooling type	water	air

3. Excitation test of HTS-MRI magnet

3.1. Overshoot excitation
Figure 4 shows the results of an operation in which the HTS-MRI magnet is excited from 0 A to 20 A with charging rate of 0.05 A/s by a dropper power supply and held at a current that produces the desired magnetic field (0.27 T). The magnet current was kept 20 A for about 3 hours. The magnet field stability at the centre of imaging area, the magnet current and the magnet terminal voltage were examined. In normal excitation operation, even after the target current is reached, the central magnetic field of the magnet space measured by the MRI probe gradually increases due to the effect of attenuation of the shielding current generated during excitation. In contrast, the operation in which the target current is once increased by 0.5 A and then discharged to 20 A is called overshoot excitation operation [2]. This discharge operation attenuates the shielding current and suppresses the magnetic field fluctuation. Following tests were conducted with overshoot excitation.

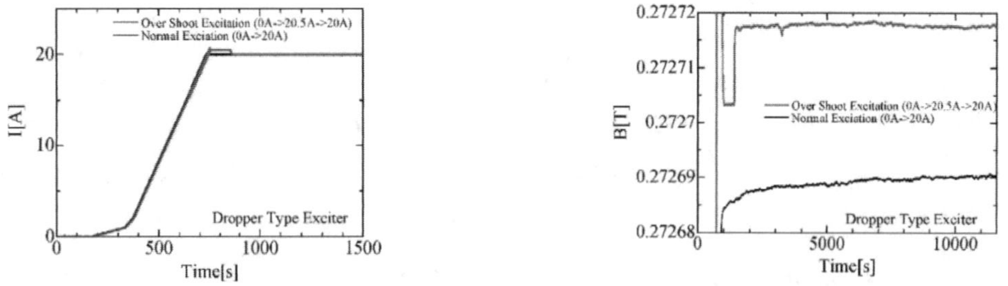

(a) Current (b) magnetic field

Figure 4. Current and magnetic field trend with normal excitation and overshoot excitation (dropper type power supply).

Figure 5 shows the current and terminal voltage of the magnet and the magnetic field at the center of the imaging area. In case of the dropper-type exciter, the terminal voltage instability was observed in the charging operation as shown in Figure 5 (a) and the magnetic field stability was evaluated as 5.5 ppm/h in 1.5 hours in the second half of the constant current mode as shown in Figure 6 (a).

In case of the switching-type exciter, the magnet current was smoothly built up without any voltage instability as shown in Figure 5 (b), and the magnetic field stability was evaluated as 6.6 ppm/h in 1.5 hours in the second half of the constant current mode as shown in Figure 6 (b), but the voltage ripple (high frequency fluctuation) was observed as shown in Figure 5 (b).

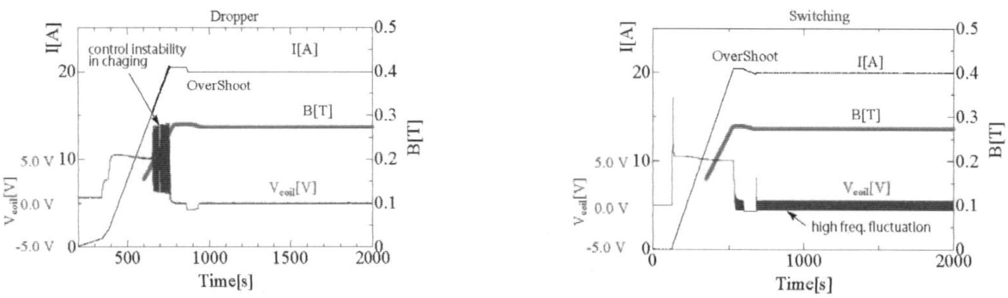

(a) dropper power supply (b) switching power supply

Figure 5. Current and terminal voltage of the magnet and the magnetic field at the center of the imaging area during the overshoot excitation by dropper type power supply (left) and switching type power supply (right).

(a) dropper power supply (b) switching power supply

Figure 6. Magnetic field for 1.5 hours (a) dropper (b) switching.

3.2. Overshoot excitation with parallel resistor

To improve the magnetic field stability in the constant current mode for the clear MRI imaging, a resistor of 1.2 Ω was set between the magnet terminals in parallel to the MRI magnet (inductance: 102 H) as shown in Figure 7. The protection resistor circuit (40mΩ resistor + Diode) can be omitted by substituting a parallel resistor.

Figure 8 shows the current and terminal voltage of the magnet and the magnetic field at the center of the imaging area during overshoot excitation with parallel resistor ((a) dropper-type, (b) switching-type). Comparing to Figure 5, the voltage instability in the charging mode with the dropper-type power supply is diminished and the voltage ripple in the constant current mode with the switching-type one is reduced.

Figure 9 shows the source and coil voltage throughout the charging, holding and discharging operation by the dropper type power supply without (a) and with (b) parallel resistor, by the switching type power supply without (c) and with (d) resistor. Each enlarged view shows the voltages between 5000 to 5006 s. The components of variation of the coil terminal voltage and supply voltage were greatly reduced by inserting the parallel resistors, especially in the case of switching type power supply. Sinusoidal fluctuation component of the coil voltage with the switching type is clearly reduced.

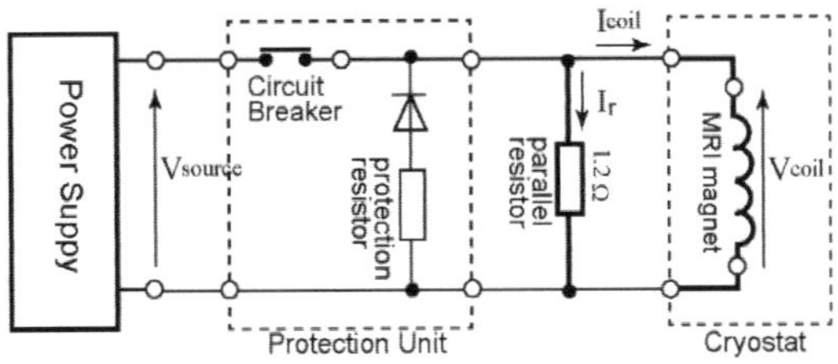

Figure 7. Circuit diagram of the excitation test with parallel resistor (1.2 Ω) to MRI magnet.

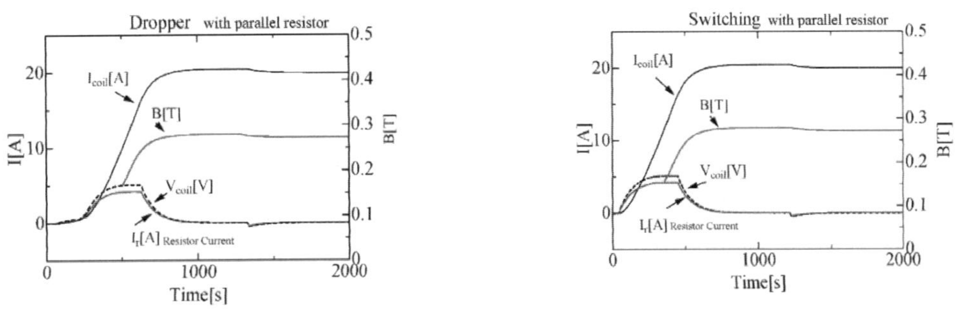

(a) dropper power supply (b) switching power supply

Figure 8. Current and terminal voltage of the magnet, the parallel resistor current and the magnetic field at the center of the imaging area during the overshoot excitation by dropper type power supply (a) and switching type power supply (b) with parallel resistor.

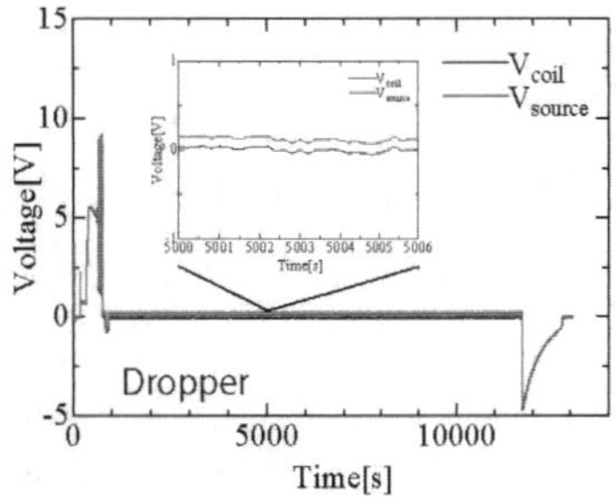

(a) dropper power supply without resistor

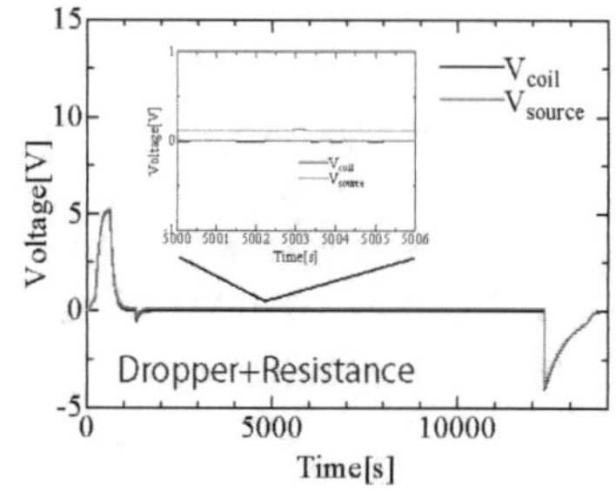

(b) dropper power supply with resistor

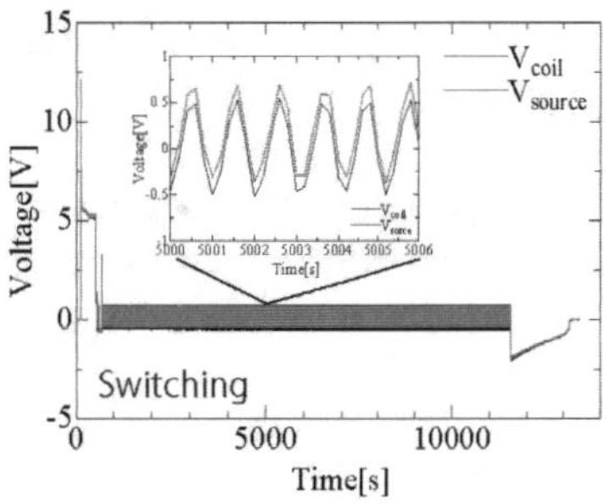

(c) switching power supply without resistor

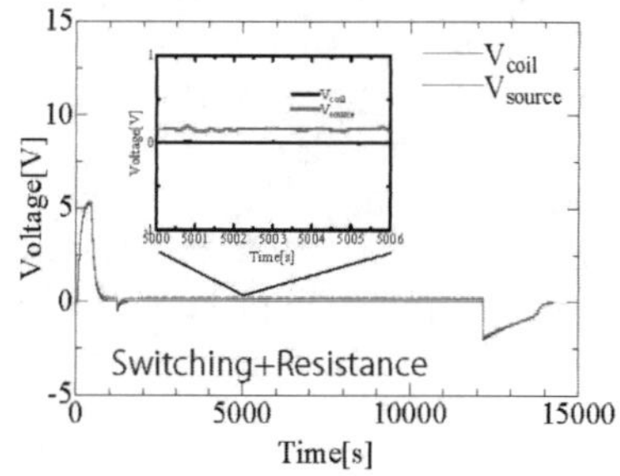

(d) switching power supply with resistor

Figure 9. Source output and coil voltage throughout the charging, holding and discharging operation by the dropper type power supply without (a) and with (b) parallel resistor, by the switching type power supply without (c) and with (d) resistor. Each enlarged view shows the voltages between 5000 to 5006 s.

3.2.1. Magnetic field stability with parallel resistor

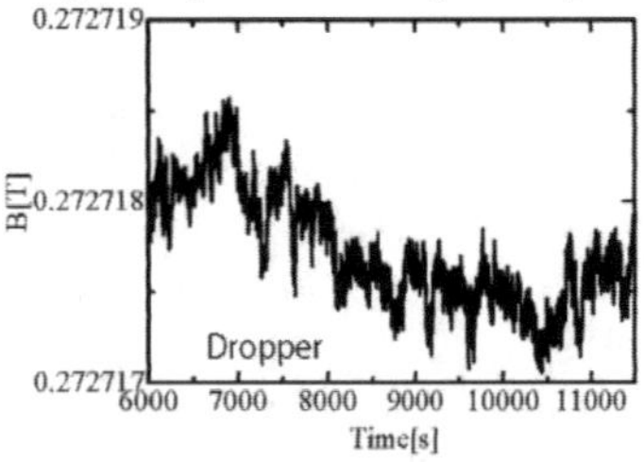

(a) dropper power supply without resistor

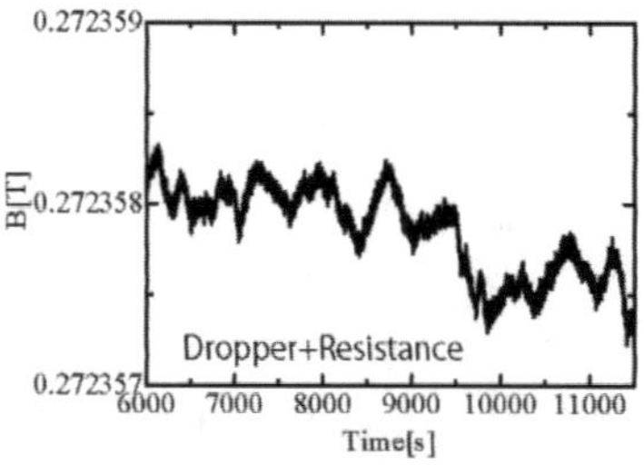

(b) dropper power supply with resistor

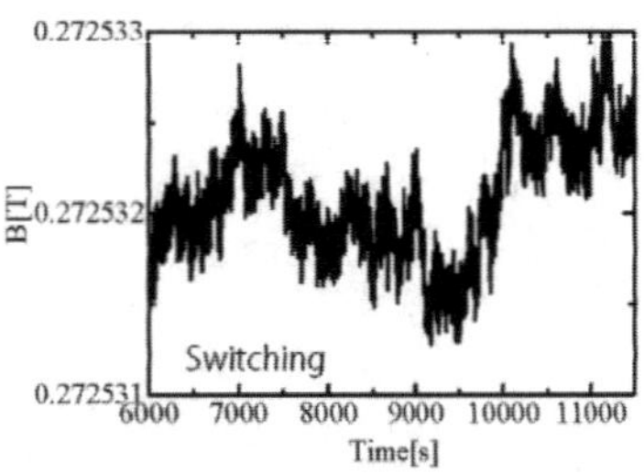

(c) switching power supply without resistor

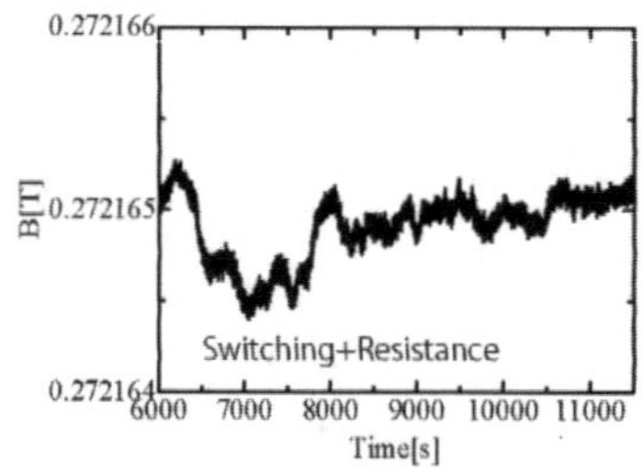

(d) switching power supply with resistor

Figure 10. Central magnetic field of a dropper power supply without (left) and with (right) a parallel resistor during the latter half of a 3-hour constant-current operation.

The upper part of Figure 10 shows the central magnetic field of a dropper power supply without (left) and with (right) a parallel resistor during the latter half of a 3-hour constant-current operation. The stability of the magnetic field is improved from 5.5 to 3.3 ppm/h by inserting a parallel resistor. The lower part of Figure 10 shows the case of a switching power supply, where the parallel resistance improves the magnetic field stability from 6.6 to 3.0 ppm/h and reduces the fast fluctuation. The fluctuation components in the magnetic field are greatly attenuated by inserting a parallel resistor, whether it is a dropper power supply or a switching one.

4. Imaging test

Imaging tests at 0.27 T were performed for the purpose of imaging with an HTS-MRI magnet under power supply drive. The excitation rate was set to 20.5 A at a sweep rate of 0.05 A/sec using either a dropper power supply or a switching power supply, and then overshoot excitation was performed to set the excitation rate to 20 A. The imaging object was a mouse about 5 cm in size as shown in Figure 11, which was set inside the RF coil and gradient field coil covered with aluminium foil. Imaging was performed using the gradient echo method, and the number of times the images were accumulated and the repetition time (TR), which is the time interval of data collection during image creation, were varied to obtain clear images despite the longer imaging time.

Figure 12 shows the images taken by the dropper power supply with 15 minutes operation without parallel resistor. Figure 13 shows images taken by a switching power supply. With the switching power supply, there were minute and periodic magnetic field fluctuations that could not be seen in the NMR probe measurement, and even the approximate shape of the mouse could not be captured. These magnetic field fluctuations were almost identical to the period of the voltage oscillation described earlier, and this voltage oscillation was thought to be the cause.

Figure 14 shows the results of a 15 minutes imaging session in which the parallel resistor of 1.2 Ω was inserted and overshoot excitation was performed again by the switching power supply. The image resolution is better than that of figure 12 taken by the dropper power supply without resistor. In the present imaging test, the minimum SN ratio of 20, which is the practical level, was obtained by both power supplies despite the test at the low magnetic field (0.27 T).

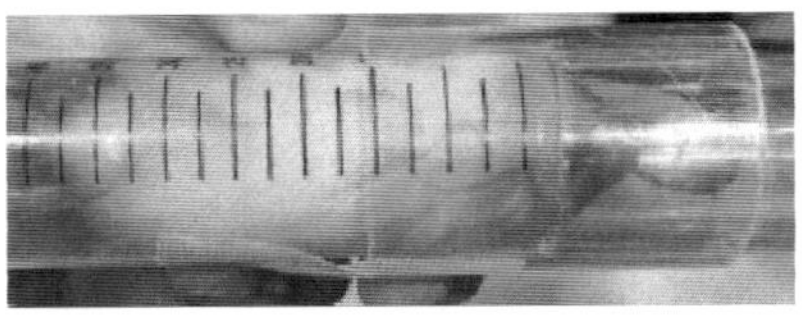

Figure 11. Imaging object (a mouse about 5 cm in size).

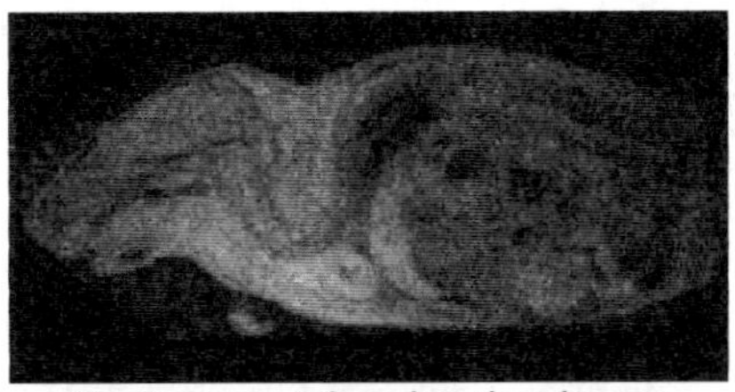

Figure 12. Images taken by the dropper power supply.

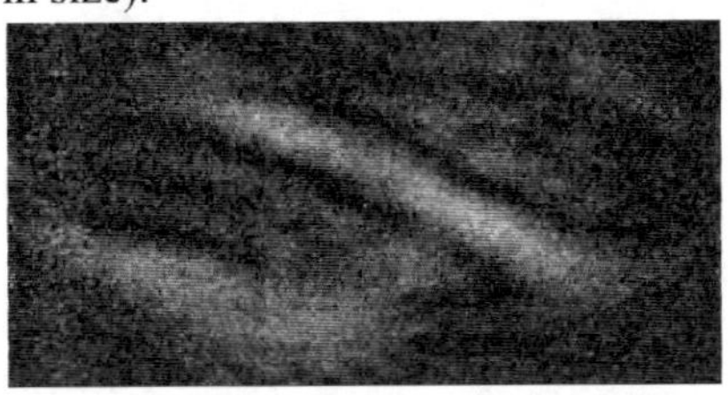

Figure 13. Images taken with a switching power supply.

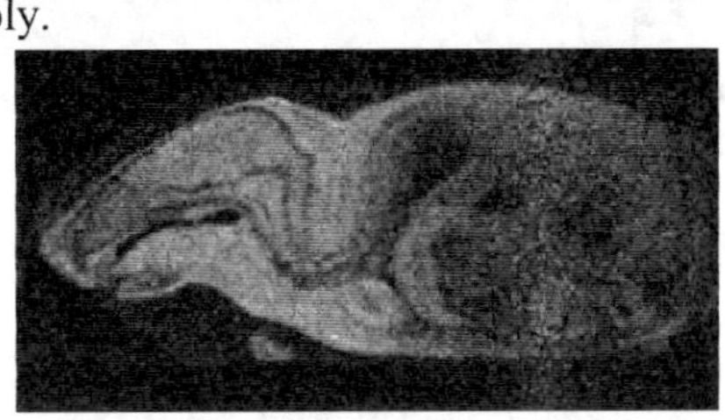

Figure 14. Images taken with a switching power supply with parallel resistor.

5. Conclusion

(1) The magnetic field stability were measured and evaluated using a HTS-MRI test magnet in the power supply driving mode using a dropper-type and of a switching-type direct-current exciter system.

(2) The HTS-MRI magnet which is fabricated by REBCO tape superconductor was charged up to 20.5A with ramp rate of 0.05 A/s and then discharged to target current 20A (0.27 T), that is the overshoot charging operation for compensating the screening current effect on the magnetic field stability.

(3) The magnet current was kept 20 A for about 3 hours and the magnet field stability at the center of imaging area, the magnet current and the magnet terminal voltage were examined. In case of the dropper-type exciter, the terminal voltage instability was observed in the charging operation and the magnetic field stability was evaluated as 5.5 ppm/h in 1.5 hours of the constant current mode. In case of the switching-type exciter, the magnet current was smoothly built up without any voltage instability, and the magnetic field stability was evaluated as 6.6 ppm/h in 1.5 hours of the constant current mode but the voltage ripple was observed.

(4) The availability of the power supply system was discussed in MRI imaging of a mouse using an experimental HTS-MRI system. The MRI image of the mouse was obtained under the power supply driven operation of the dropper-type power supply. On the other hand, the clear MRI image of the mouse was not obtained under the power supply driven operation of the switching-type power supply due to the small but repetitive fluctuation in the magnetic field due to the voltage ripple.

(5) A resistor of 1.2 Ω was set between the magnet terminal in parallel to the MRI magnet in order to improve the magnetic field stability in the constant current mode for the clear MRI imaging, at the same time to alleviate the current instability in the charging mode with the dropper-type power supply and to reduce the magnet current ripple due to the voltage ripple in the constant current mode with the switching-type one.

(6) As a result, the magnet current was smoothly built up and the magnetic field stability was improved to 3.3 ppm/h with the dropper-type power supply. The magnetic stability was improved to 3.0 ppm/h with the switching-type power supply. The MRI image of the mouse was obtained under the power supply driven operation of the switching-type power supply, since the magnetic field ripple due to the voltage ripple was reduced. The power loss of the switching-type power supply is about 1/6 of that of dropper-type one.

Acknowledgments

This work was supported by NEDO "Development of an HTS magnet system with highly stable magnetic field" as part of the "Promotion technology development for HTS practical application."

References

[1] Yanagisawa Y and Maeda H 2013 *Cryogenics* **48** 165-171

[2] Yanagisawa Y, Nakagome H, Koyama Y, Hu R, Takao T, Hamada M, Kiyoshi T, Takahashi M and Maeda H 2009 Physica C 469 1996-1999

[3] Hemmi T, Yanagi N, Bansal G, Seo K, Takahata K and Mito T 2006 *Fusion Engineering Design* **81** 2463-2466

[4] Yoshikawa M, Yonemura N, Yachida T, Imura T, Shirai Y and Yokoyama S 2016 *IEEE Trans. Appl. Supercond.* **26** 4401105

[5] Yachida T, Yoshikawa M, Shirai Y, Matsuda T and Yokoyama S 2017 *IEEE Trans. Appl. Supercond.* **27** 4702905

[6] Yokoyama S *et al.* Research and development of the very stable magnetic field HTS coil system fundamental technology for MRI 2015 *Proc. of ISS2015* SA-9-INV

[7] Yokoyama S *et al.* 2017 *IEEE Trans. Appl. Supercond.* **27** 4400604

[8] Miura H, Miyazaki J, Ide M, Miyagi D, Tsuda M and Yokoyama S 2018 *IEEE Trans. Appl. Supercond.* **28** 4401705

[9] Oya M *et al.* 2019 *IEEE Trans. Appl. Supercond.* **28** 4401205

[10] Miura H, Matsuda T, Nomura K and Inoue T 2020 *J. Phys.: Conf. Ser.* **1559** 012125

Journal of Physics: Conference Series		**2545** (2023) 012031	doi:10.1088/1742-6596/2545/1/012031

Removal of SUS304 fine particles transformed into martensite by the high gradient magnetic separation under dry condition

Chen H[1*], Yang Y[1] and Miura O[1]

[1] Department of Electrical Engineering and Computer Science, Tokyo Metropolitan University, Hachioji 192-0397, Japan

e-mail: chen-haozhou@ed.tmu.ac.jp, miurao@tmu.ac.jp

Abstract. The interfusion of impurities such as metallic wear debris has been a major issue in the manufacturing process of foods, medicines and industrial products. Such debris originates with wearing of stainless-steel pipe joints or mechanical moving parts. Since the debris shows ferromagnetic properties by undergoing the strain-induced martensitic transformation of non-magnetic SUS304 (X5CrNi18-10), the magnetic separation system is much efficient to remove such debris from raw materials. In order to study the magnetic separation properties for the martensitic transformation fine particles, the several kinds of magnetic powder with different magnetization were prepared by controlling the amount of martensitic transformation by heat-treating SUS304 powders. The magnetic separation performance was evaluated by the high gradient magnetic separation (HGMS) experiments using the multilayer unidirectional magnetic wire filter under dry condition. The experimental results and FEM particle trajectory simulations revealed that the SUS304 powders transformed into martensite were removed at high speed of 0.1 m/s by the superconducting HGMS in a relatively low magnetic field.

1. Introduction

The interfusion of impurities such as metallic wear debris has been a major issue in the manufacturing process of various industries. For example, if such debris is mixed with carbon black powders used in the production of carbon electrodes for lithium batteries, serious product safety issues such as electric sparks may occur [1]. This debris originates from the wear of stainless-steel pipe joints or mechanical moving parts, and shows ferromagnetism after undergoing the strain-induced martensitic transformation [2] [3]. In our previous works [4], we developed a new multilayer magnetic filter using rotational unidirectional magnetic wires, and successfully removed almost 100% of fine (average size 48 microns) iron particles from carbon black powders through high gradient magnetic separation (HGMS) with the developed magnetic filter without clogging the filter under dry condition. Further evaluation of the performance of this magnetic separation system is required.

In this study, the superconducting HGMS using this magnetic filtering system was applied to remove martensitic transformation SUS304 fine particles from mixed powders under dry condition. In order to investigate their magnetic separation properties, the magnetic SUS304 fine particles with different magnetization were prepared by controlling the amount of martensitic transformation in the crystals due to heat treatments. Then, using them in carbon black powder, dry magnetic separation experiments and FEM particle trajectory simulations were performed.

2. SUS304 transformed into martensite preparation method

2.1. Martensitic transformation of SUS304

Iron changes phase in the sequence $\alpha \rightarrow \beta \rightarrow \gamma \rightarrow \delta$ on heat treatment, and α phase iron is stable at room temperature. The austenitic carbon steel SUS304 obtained by high temperature quenching and cooling is the non-magnetic γ phase with the body-centred cubic lattice, whose atomic arrangement is shown in figure 1(a). On the other hand, α' martensite structure derived from α phase, has the atomic arrangement containing carbon as a solid solution shown in figure 1(b), which has the body-centred cubic lattice and ferromagnetism [5].

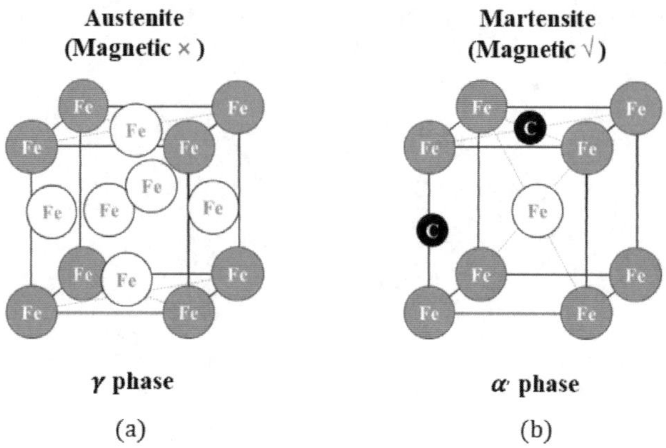

Figure 1. (a) Face-centered cubic lattice (γ phase). (b) Body-centered cubic lattice (α′ phase).

In the process of forming the martensitic transformation, γ phase steel is heated at high temperatures to firmly dissolve the carbide, and then quenched to partially convert the carbide to martensite. On the other hand, carbon steel in γ phase can also be altered to martensite when large forces are applied, which is the main reason for the mixing of these ferromagnetic debris in production.

Figure 2 shows the microstructures in commercially available SUS304 steel sheets obtained and depicted through the application of Lorentz TEM. The examination reveals the presence of strain-induced martensite, characterized by a combination of γ phase and α' phase. [6].

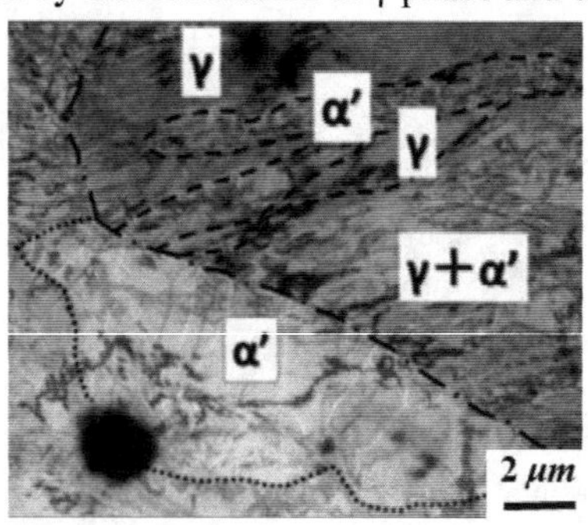

Figure 2. Lorentz TEM images obtained from the fractured specimen by strain-induced, broken lines show the grain boundaries between phases. [6].

On the other hand, heat treatment also induces martensitic transformation on SUS304, which can change the magnetic properties [5]. In order to better quantify the experiments, we used heat treatment method to simulate the martensitic transformation on SUS304 instead strain for high-gradient magnetic separation studies. The resulting changes in density and volume are negligible and do not significantly affect the results of magnetic separation.

2.2. Martensitic transformation of SUS304 by heat treatment

In order to assess the capability of removing minuscule metal debris, we used commercially available γ phase SUS304 powders with average grain size of 40 microns to transform martensite α' phase of SUS304. An appropriate amount of γ phase SUS304 powder was vacuum sealed in a quartz tube. Then the heat treatment was carried out for 2 hours in an electric furnace with a set temperature from 300 to 900 °C. In the process, γ phase structure of SUS304 powder transformed into martensite α' phase evenly and little by little as time passes. After the heat treatment, the quartz tube was quenched and cooled in cold water.

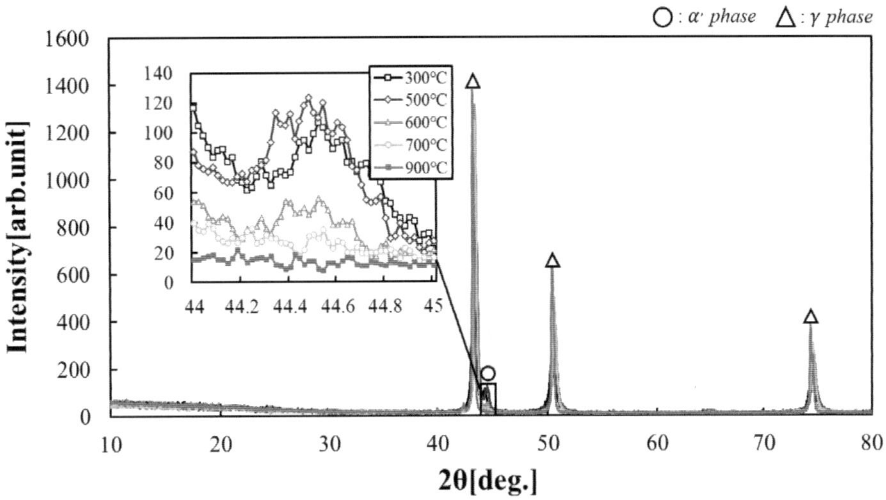

Figure 3. XRD patterns of SUS304 powder heated at different temperatures. We confirmed that the content of the α' phase differs after heat treatment at different temperatures by XRD measurements.

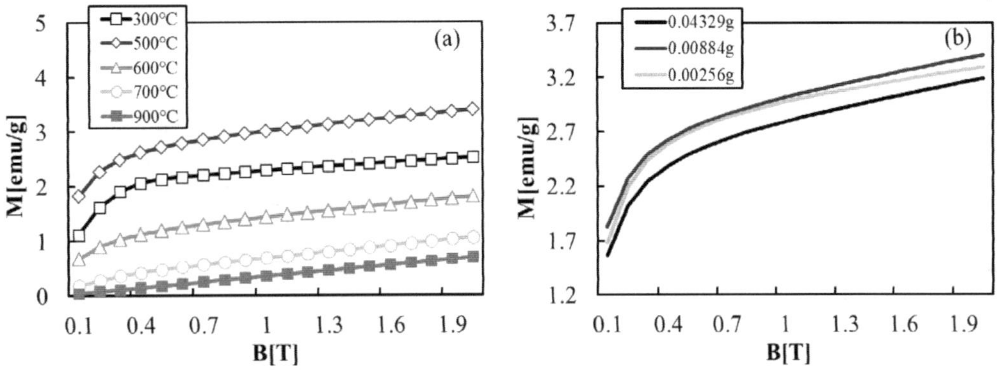

Figure 4. (a) Magnetization curves of SUS304 powders after heat treatment at different temperatures. (b) The determination of the magnetization of different masses of heat-treated SUS304 powder at 500°C.

Figure 3 shows the XRD patterns of SUS304 powders heated at different temperatures. When the heat treatment temperature is 300°C, SUS304 contains more martensite. And when the temperature

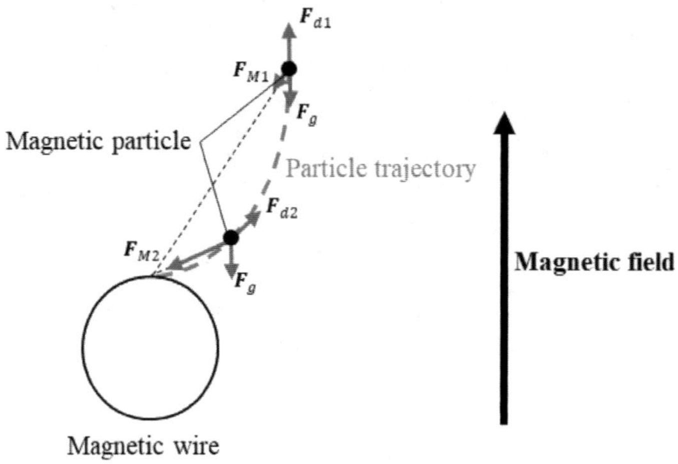

Figure 5. Forces applied to a magnetic particle and the particle trajectory in high gradient magnetic separation system [4].

reaches 500°C, the martensite content is the highest. As the temperature continues to increase gradually, the martensite content is low and almost no martensite at 900°C. This is because as the temperature increases, the γ phase of SUS304 undergoes a heat-induced transformation into an intermediate state, the α' phase of martensite, before returning to the γ phase after exceeding the critical temperature [7]. Since the critical temperature varies for different types of steel, we also measured the magnetization of the SUS304 powder heated at different temperatures.

Figure 4(a) shows the magnetization curves of SUS304 powders after heat treatment at different temperatures. The magnetization of SUS304 gradually increases as the temperature rises until 500°C. As the heat treatment temperature continues to increase, the magnetization begins to decrease. When the heat treatment temperature is 900°C, the degree of magnetization is already weak. This result is consistent with the amount of α' phase in the crystal in figure 3.

Figure 4(b) shows the magnetization of different masses for heat-treated SUS304 powder at 500°C. The magnetization values are almost the same, indicating that uniform martensitic transformation in the powder occurs due to the heat treatment.

3. Principle of high gradient magnetic separation

Figure 5 shows forces applied to a magnetic fine particle in the powder when it falls vertically toward a magnetic wire placed perpendicular to the magnetic field. When a vertical magnetic field is applied to a magnetic wire, a high magnetic field gradient is generated around the magnetic wire, and the magnetic force generated can attract magnetic particles. The forces that the particles receive at that time are magnetic force, gravity, and drag force from the powder as a pseudo fluid, which are expressed by the following equations.

$$\boldsymbol{F}_m = V_p \mu_0 \boldsymbol{M}^* \mathrm{grad} \boldsymbol{H} \quad (1)$$

$$\boldsymbol{F}_d = 6\pi\eta r_p (\boldsymbol{v}_f - \boldsymbol{v}_p) \quad (2)$$

$$\boldsymbol{F}_g = (\rho_p - \rho_f) V_p g \quad (3)$$

where $\boldsymbol{F}_m$ is the magnetic force, μ_0 is the permeability of vacuum, V_p is a particle volume, $\boldsymbol{M}^*$ is an effective magnetization of particle, $\boldsymbol{H}$ is an external magnetic field, $\boldsymbol{F}_d$ is the drag force, η is a fluid viscosity, r_p is a particle radius, v_f is a fluid velocity, v_p is a particle velocity, $\boldsymbol{F}_g$ is the gravity, ρ_p is a

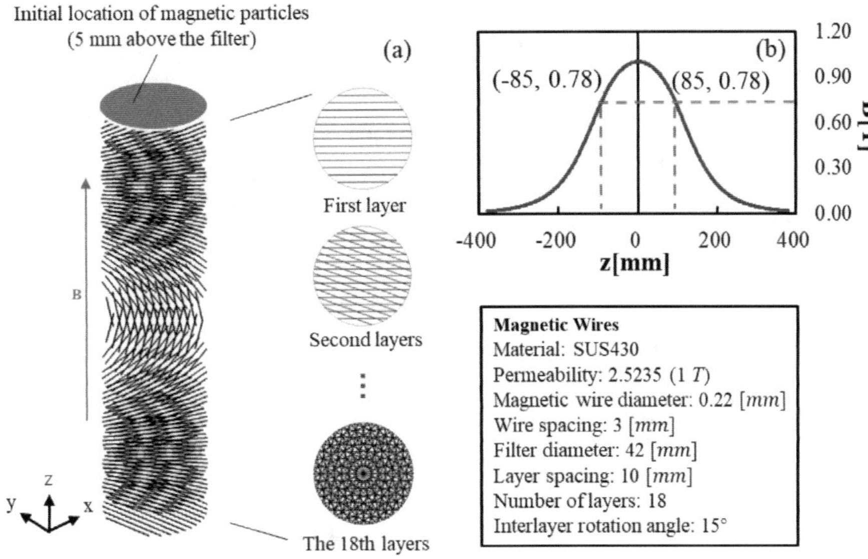

Figure 6. (a) Projection drawing of the developed multilayer Magnestain® magnetic filter with rotating unidirectional magnetic wires. (b) The magnetic field distribution in the z direction.

particle density, ρ_f is a fluid density and g is the gravitational acceleration [8]–[10]. If the magnetic force on the particles is larger than the drag force and gravity, the particles will be captured by the magnetic wire, and therefore the magnetic separation will be achieved.

4. Particle trajectory simulation experiments of SUS304 martensite formation

We carried out particle trajectory simulation by using the commercial software (COMSOL Multiphysics ver. 5.6) with the finite element method. Figure 6 shows the projection drawing of the developed multilayer Magnestain® magnetic filter with rotating unidirectional magnetic wires [4]. The diameter of the magnetic wires is 0.22 mm with the wire spacing of 3 mm. The diameter of the cylindrical multilayer filter is 42 mm with the height of 170 mm. It consists of 18 single-layer filters with a layer spacing of 10 mm and 15° rotation between layers. In the simulation experiments, the magnetic field is applied in the z direction.

The parameters used in the simulation were shown in table 1. The pseudo laminar flow consisting of carbon black powder enters the magnetic filter vertically from 5 mm above the inlet direction at an

Table 1. Parameter for simulation

Relative magnetic permeability of pseudo fluid	1
Applied magnetic field	0.3 T; 0.5 T; 1 T
Relative magnetic permeability of magnetic wire	3.47 (0.3 T) 3.39 (0.5 T) 2.52 (1 T)
Viscosity of pseudo fluid	0.89 mPa•s
Bulk density of pseudo fluid	0.13 g/cm^3
Initial velocity of pseudo fluid	0.1 m/s
Density of magnetic particle	7.9 g/cm^3
Diameter of magnetic particle	40 μm
Number of magnetic particles	600
Initial velocity of magnetic particle	0.1 m/s
Diameter of magnetic wires	0.22 mm

initial velocity of 0.1 m/s. The pseudo laminar initial velocity was set to 0.1 m/s, referring to the velocity of carbon black in the actual production. Since the viscosity of carbon black as a pseudo fluid is considered to be smaller than the water's value of 0.89 mPa·s, the actual capture width is expected to increase. In our previous works, we investigated the feasibility of using pseudo laminar flow to simulate the flow of carbon black powder in magnetic separation [4]. SUS304 martensite particles flow down flow at the same initial velocity.

Figure 7 shows the particle capture simulation results. It can be seen that for heat-treated SUS304 particles at 500°C, the filter captures more than 96% of them under the experimental magnetic field. For heat-treated SUS304 particles at 700°C, the capture rate (CR) of the filter was significantly lower than the former, which is due to the lower magnetization of heat-treated SUS304 particles at 700°C, as shown in figure 4(a), the magnetic force applied to the particle is lower according to equation 1.

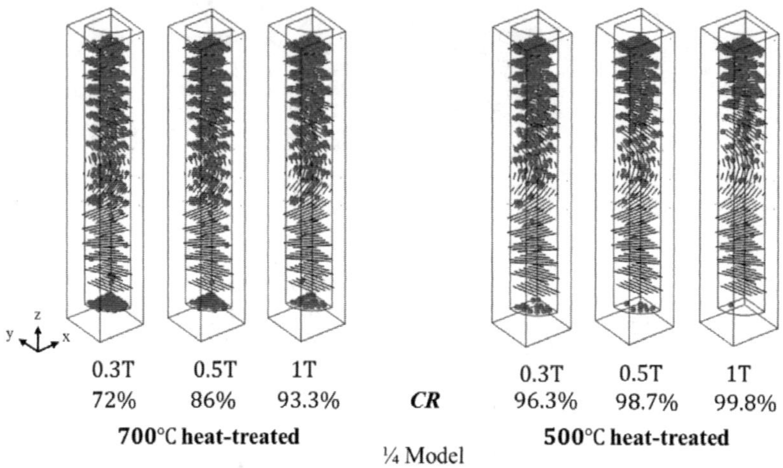

Figure 7. The heat-treated SUS304 particles capture simulation results by the multilayer filter, CR = capture rate.

5. Magnetic separation experiments of SUS304 martensite

5.1. Experimental Method

The magnetic filter was set into the superconducting electromagnet shown in figure 8. The magnetic separation experiments of SUS304 particles with an average particle size of 40 μm in carbon black powder with an average particle size of 36 nm were carried out. As a pre-treatment, the carbon black was heat-treated in an electric furnace for 2 hours while maintaining the temperature at 130°C. 700°C heat-treated SUS304 particles were selected due to the substantial contrast in capture rate as shown in figure 7. For the experimental procedure, a powder mixture was prepared by combining 0.017 g of the treated SUS304 powder and 5 g of carbon black powder (SUS304 powder volume fraction 0.56‱). The mixed powder was placed in the canister 433mm above the magnetic filter. Then the valve was opened and the mixture fell down through the magnetic filter into the collection bottle.

After the magnetic separation, it was observed that there was no clogging of the magnetic filter by the carbon powder. In order to evaluate the magnetic particles captured by the filter, both the mixed powder remaining in the collection bottle and inside the filter were stirred in water, and the magnetic particles sunk at the bottom were collected with a 1 T permanent magnet. After drying, the particles were measured with an electronic balance.

5.2. Experimental Results

Figure 9 shows the results of capture rate of SUS304 particles after heat treatment at 700°C at 0.3 T, 0.5 T and 1 T. Table 2 shows the capture values. It can be seen that the experimental results were close to the simulation experiments. The simulation experiment has a high reliability.

Table 2. Experimental results of SUS304 particles after heat treatment at 700°C

	Applied magnetic field		
	0.3 T	0.5 T	1 T
Total mass (*g*)	0.017	0.017	0.017
Residual mass (*g*)	0.0119	0.0141	0.017
Capture rate	70%	83%	100%

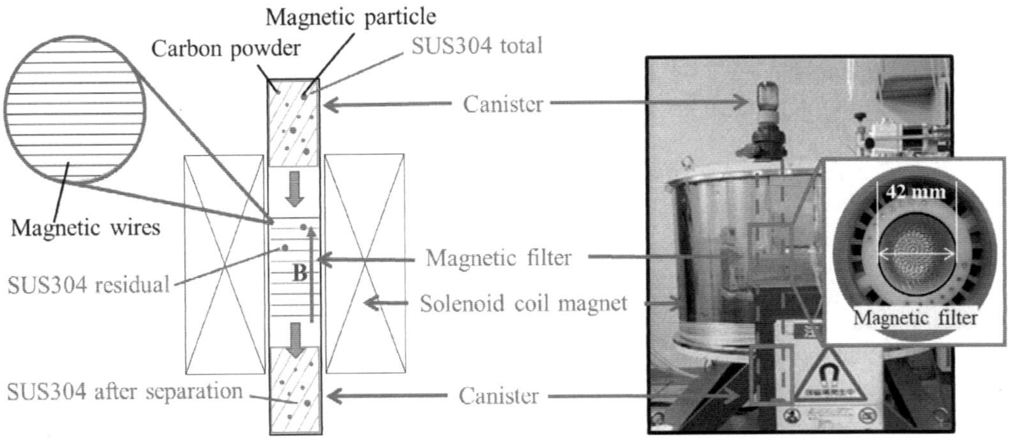

Figure 8. Equipment set up for the HGMS.

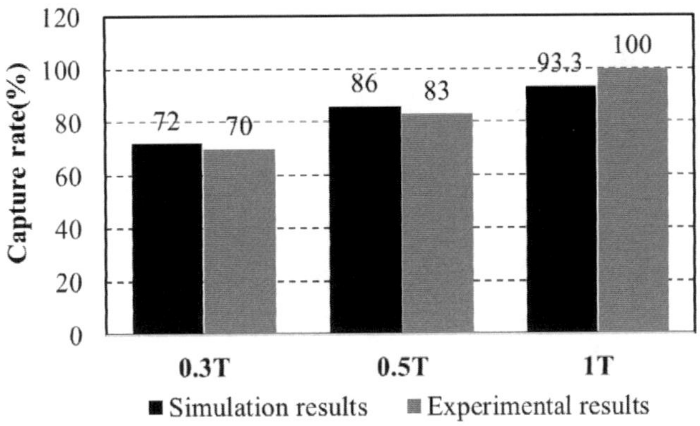

Figure 9. Simulation experiments and magnetic separation experimental results of 700°C heat-treated SUS304 under 0.3 T, 0.5 T, 1 T.

6. Conclusion

The removal of SUS304 martensite foreign matter from carbon black powder was evaluated by the high gradient magnetic separation (HGMS) using the multilayer unidirectional magnetic wire filter under dry condition. By controlling the heat treatment temperature, we successfully prepared SUS304 particles with different magnetization. We used the designed multilayer filter to capture martensitic transformed SUS304 fine particles under dry condition. We succeeded in removing almost 100% of 700°C heat-treated SUS304 particles with the size of 40 micron in carbon black powder in low magnetic fields under 1 T. The simulation results almost identical agree with the experiments ones. This reveals that the magnetic separation system is very effective in removing strain-induced SUS304 martensite particle in the powder under dry conditions. Small differences between the simulations and the experiments are believed to come from differences in fluid velocity and viscosity of carbon black powder, as well as eddy current effects of SUS304 particles. These will be considered in future studies.

References

[1] David G M and Bill M 2001 Recycling the lithium battery *Ind. Chem. Libr.* **10** 263-93

[2] Hutchings I M 1993 Mechanisms of wear in powder technology: A review *Powder Technol.* **76** 3-13

[3] Nakai Y, Mishima F, Akiyama Y and Nishijima S 2010 Development of high gradient magnetic separation system under dry condition *Physica. C. Supercond.* **470** 1812-17

[4] Chen H and Miura O 2022 Removal of magnetic fine particles from carbon black powder by high gradient magnetic separation under dry condition *IEEE Trans. Appl. Supercond.* **32** 3700605

[5] Nishiyama Z 1978 *Martensitic Transformation* (New York: Academic Press) pp 2–4

[6] Hatano M, Kubota Y, Shobu T and Mori S 2016 Presence of ε-martensite as an intermediate phase during the strain-induced transformation of SUS304 stainless steel *Philos. Mag. Lett.* **96** 220-7

[7] Jordi O 1987 Thermally induced martensitic transformations: theoretical analysis of a complete calorimetric run *Thermochim Acta.* **121** 397-412

[8] Watson J H P 1973 Magnetic filtration *J. Appl. Phys.* **44** 4209-13

[9] Anzai T, Matsuura Y, Sugawara T and Miura O 2016 Removal of humic acid in water by rice hull magnetic activated carbon and magnetic separation *IEEE Trans. Appl. Supercond.* **26** 3700104

[10] Sugawara T, Matsuura Y, Anzai T and Miura O 2016 Removal of ammonia nitrogen from water by magnetic zeolite and high gradient magnetic separation *IEEE Trans. Appl. Supercond.* **26** 3701004

Fundamental Study of the Removal of Microplastic Fibers Using Swirling Flow and Magnetic Field

Satoshi Fujii, Yoko Akiyama, Yuichiro Manabe, Fuminobu Sato

Graduate School of Engineering Osaka University, 2-1 Yamadaoka, Suita, Osaka 565-0871, Japan

s-fujii@qb.see.eng.osaka-u.ac.jp

Abstract. Microplastic fibers (MPFs) derived from laundry wastewater and fishing nets, etc. have been found in the environment around the world, raising concerns about their ecological impact. Particularly for laundry wastewater, some of the MPFs are not removed by sewage plants, and are discharged into the environment. We have studied the removal method of MPFs by magnetic separation in sewage plants, but there were some problems for practical applications such as the desorption of magnetite from MPFs and blockage of magnetic filters. To solve the problems, we proposed a novel magnetic separation system combining swirling flow and opposing magnets. In this study, we investigated whether MPFs can be concentrated by particle trajectory simulations and lab-scale experiments. The simulations and experiments showed similar trends that MPFs can be concentrated by the proposed magnetic separation system. The magnetic separation performance will be improved by using a higher magnetic field of opposed superconducting bulk magnets.

1. Introduction

The impact of microplastics leaking into the environment on the ecosystem is one of the serious issues. Most of the microplastic fibers (MPFs) are generated by washing synthetic fiber clothing [1]. MPFs pass through sewage plants and are effluent into the environment [2]. There is concern that MPFs may affect the ecosystem by adsorbing toxic substances and their uptake by living organisms [3, 4]. Therefore, it is necessary to remove MPFs in the sewage plant where they are discharged.

We have been investigating the “magnetic seeding method” (to magnetize diamagnetic or paramagnetic materials) by attaching magnetite particles to MPFs and then removing it by applying an external magnetic field. Magnetic separation methods are mainly divided into two methods. One is high-gradient magnetic separation (HGMS), in which an external magnetic field is applied to magnetic filters, such as ferromagnetic mesh filters or iron spheres, installed in a magnetic field. The other is open-gradient magnetic separation (OGMS), which uses the magnetic gradient of the magnet itself without placing the magnetic materials in a magnetic field. However, when HGMS is applied to the removal of MPFs, the magnetic force acting on the magnetite attached to the MPFs exceeds the adhesion force between the MPFs and magnetite, causing the magnetite to desorb from MPFs. In addition, the magnetic filter needs to be cleaned at fixed intervals because the filters become obstructed [5]. On the other hand, OGMS has the problem that it is not suitable for mass processing due to its small capture capacity [6]. To solve the problems, we proposed a novel magnetic separation system for the continuous concentration of MPFs by combining the centrifugal force from the swirling flow and magnetic force the magnetic field and evaluated the possibility of magnetic separation of MPFs in this system.

The effect of the swirling flow in this study is basically aimed at the effect of centrifugal force, and our idea is to assist the force by outward magnetic force. However, in order to move the magnetically-seeded MPFs particles as far outward as possible in the magnetic force vessel, it is necessary to lengthen the time the particles remain in the vessel. It can be said that the swirling flow also plays a role in lengthening the residence time, which is required to move the particles to the wall surface by applying the magnetic force outward. In this study, particle trajectory simulations and magnetic separation experiments of MPFs were conducted to evaluate the feasibility of separation of MPFs using swirling flow and magnetic field by our proposed magnetic separation system.

2. Theory of magnetic separation

In magnetic separation using the magnetic seeding method, ferromagnetic particles are attached or bound to the separation target particles. This allows the separation target particles from the dispersant by applying a magnetic field. The forces acting between the separation target particles and the ferromagnetic particles are electrostatic and van der Waals forces, and the sum of these forces is $\mathbf{F}_\mathrm{A}$. The magnetic force acting on the ferromagnetic particle is $\mathbf{F}_\mathrm{m}$. In the magnetic seeding method, these forces must always satisfy equation (1), $\mathbf{F}_\mathrm{A}$ is larger than $\mathbf{F}_\mathrm{m}$, to remove the separation target particles.

During magnetic separation, magnetic force $\mathbf{F}_\mathrm{m}$, drag force $\mathbf{F}_\mathrm{d}$, gravity $\mathbf{F}_\mathrm{g}$, and buoyancy force $\mathbf{F}_\mathrm{b}$ expressed by equations (2), (3), (4), and (5) respectively, act on the ferromagnetic particles attached to the separation target particles. Here, V_m is the volume of the ferromagnetic particle [m^3], $\mathbf{M}$ is the magnetization of the ferromagnetic particle [A/m], $\mathbf{B}$ is magnetic flux density [T], r is the radius of separation target particle as a sphere considering orientation [m], μ is the viscosity of the fluid [Pa·s], $\mathbf{U}$ is the relative velocity between the fluid and separation target particle [m/s], ρ is the density of separation target particle [kg/m^3], ρ_f is the density of the fluid [kg/m^3], V is the volume of separation target particle [m^3], and $\mathbf{g}$ is the gravitational acceleration [m/s^2].

$$|\mathbf{F}_\mathrm{A}| > |\mathbf{F}_\mathrm{m}| \tag{1}$$

$$\mathbf{F}_\mathrm{m} = V_\mathrm{m}(\mathbf{M} \cdot \nabla)\mathbf{B} \tag{2}$$

$$\mathbf{F}_\mathrm{d} = 6\pi r \mu \mathbf{U} \tag{3}$$

$$\mathbf{F}_\mathrm{g} = (\rho V + \rho_\mathrm{m} V_\mathrm{m})\mathbf{g} \tag{4}$$

$$\mathbf{F}_\mathrm{b} = \rho_\mathrm{f}(V + V_\mathrm{m})\mathbf{g} \tag{5}$$

Here, equation (3) corresponds to a spherical separation target, although in the actual system, the MPs have a cylindrical shape. Since no analytical derivation has been made for the torque due to the drag force acting on the cylindrical particles, the calculations in this study are based on the approximation of a rotating ellipsoid, which is close to a cylindrical particle. For this calculation, we referred to the work of Gustavsson K et.al. [7]. We adopted this calculation method because our previous studies have shown that the effect of torque has a considerable influence on the calculation of particle trajectories and separation ratios for spheres and rotating ellipsoids. For simplicity, equation (3) is used here with the radius of the sphere converted to account for the particle orientation against the flow. Here, MPFs of 400 μm length and 10 μm diameter were used as the separation target particles because MPFs of 100-800 μm length are generated while washing [8], and magnetite particles with primary particle size 0.5 μm were used as ferromagnetic particles.

3. Particle trajectory simulations

3.1. Analysis method

First, particle trajectory simulations were conducted to verify whether the magnetic separation system proposed in this study, which combines a swirling flow and a magnetic field, is capable of separating MPFs. Figure 1 shows the magnetic separation system analyzed.

The following is an overview of this system. Simulated wastewater containing magnetite-attached MPFs flows from an L-shaped Inlet at the top of a cylinder, 92 mm in diameter and 100 mm in height, to generate a swirling flow in the channel. From the outside of the middle of the channel, a magnetic field is applied using a quadrupole-opposed magnet with a maximum flux density of 0.4 T in the channel to guide the magnetized MPFs to the outside of the cylinder. Outlet 1 and Outlet 2 were installed on the outside and inside of the lower part of the cylinder, respectively, and the effluent from Outlet 1 was defined as highly concentrated water and Outlet 2 as treated clean water. The magnetic flux density and flow velocity distributions in the magnetic separation system were analyzed using the finite element analysis software ANSYS2021R1 (ANSYS. Inc.). Figure 2 shows the analyzed flow velocity and magnetic field distributions, and Table 2 shows the ratio of the mass flow rate of the fluid passing through Outlet 1 and Outlet 2. Considering that the sum of the forces expressed in equations (2) to (5) acts on magnetite-attached MPFs, the equations of motion for translational and rotational motion of the particles were solved in time evolution by the Runge-Kutta method using the fluid velocity and magnetic field distributions. The particles were assumed to collide perfectly elastically on the wall, and when the particle velocity became less than 0.1 mm/s, the particle was regarded as adhered to the wall surface.

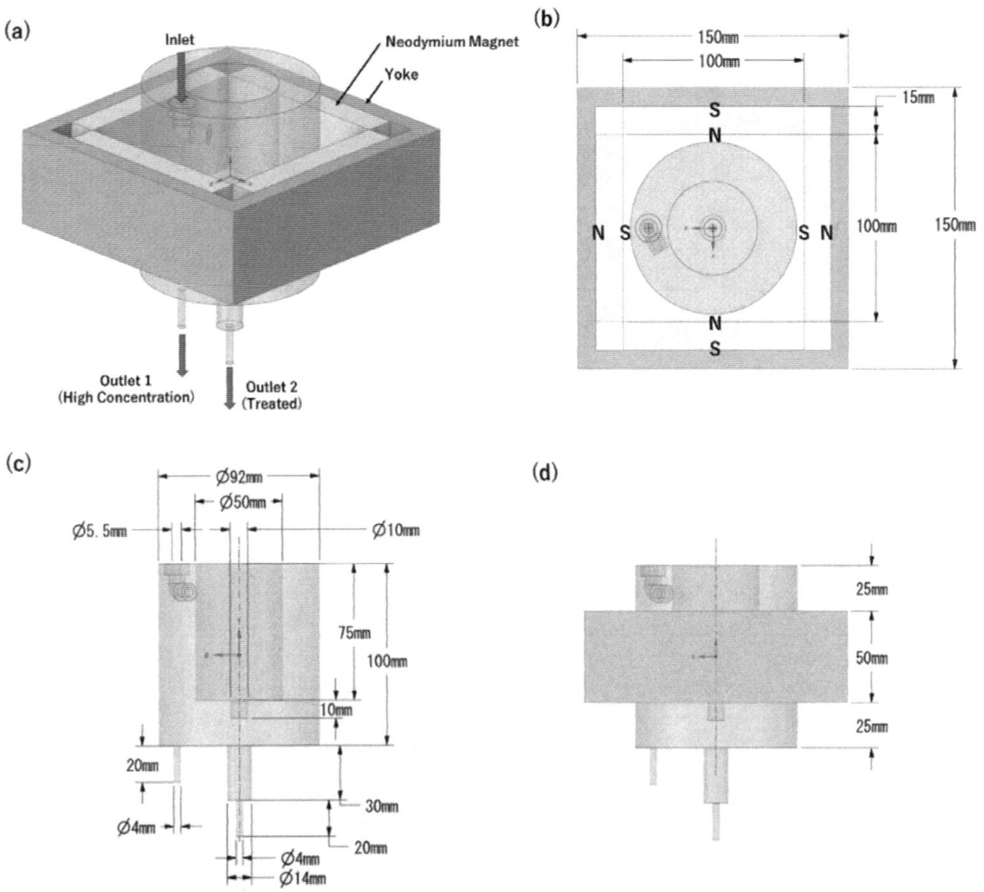

Figure 1. A model of magnetic separation system used for particle trajectory simulations: (a) Overview, (b) Top view (c) Front view (without magnet), and (d) Front View (with magnet).

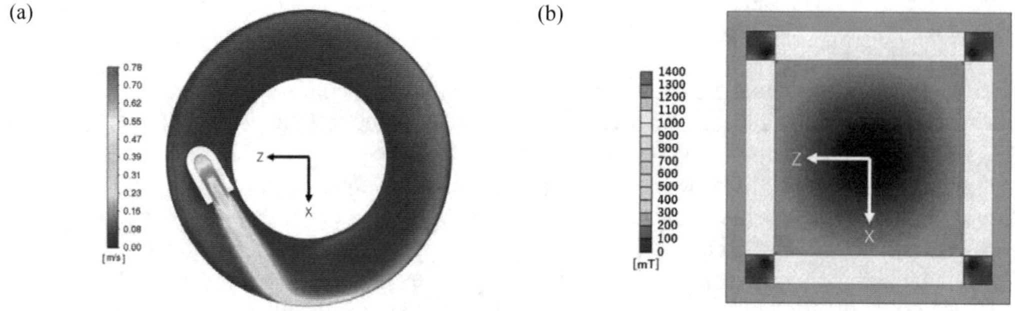

Figure 2. (a)Analysed velocity distribution (y=34.5 mm, height of L-shaped inlet center) (b)magnetic flux density distribution (y=0 mm, height of magnet center).

Table 1. Calculation conditions for particle trajectory simulations.

Inlet flow velocity	0.5 m/s
Maximum flux density in channel	0.4 T
Length of MPFs	400 μm
Diameter of MPF	10 μm
Primary particle size of magnetite	0.5 μm
Number of magnetite attached per MPF	1700
Number of MPFs analyzed	88

Table 1 shows the conditions for particle trajectory simulations. Here, the number of magnetite particles attached per MPF was estimated by optical microscopic observation and image analysis software (Fiji-Image J, ver. 1.53c[9]) shown in Figure 3. For the calculation of the number of magnetite, average of the projected area of each 50 magnetite-attached MPFs was used. Simulated MPFs wastewater was used for the estimation of the number of magnetite particles in Figure 3, in which 50 mg of polyester (φ10 μm×400 μm, Kyoto Pile Sen-i Kogyo & Co., Ltd., Japan) and 50 mg of magnetite(primary particle size 0.5 μm, sample B, Mitsui Mining & Smelting Co., Ltd., Japan) were added to 100 mL of distilled water and agitated at 400 rpm for 5 min by agitator A (SM-102, AS ONE Corporation, Japan). The initial positions of the analyzed MPFs were placed at 0.5 mm intervals in a grid pattern on the cross-section of the Inlet, and a total of 88 MPFs flowed into the separation system.

3.2. Analysis results and discussions

Fluid flow analysis showed an average velocity of 56 mm/s in the channel, in case of inlet flow velocity was 0.5 m/s. The percentage of the number of MPFs and typical examples of the trajectory of MPFs for each with and without the magnet are shown in Figures 4 and 5, respectively. It was predicted that the application of magnetic field could increase the number of MPFs adhering to the wall and in high-concentration water (Outlet 1), while reducing the number of MPFs in the treated water (Outlet 2) by about 60%. If MPFs adhered to the wall can be easily removed by cleaning with regularly increased flow velocity, 84% of MPFs may be removed in a magnetic separation system with a swirling flow.

The particle trajectory shown in Figure 5 indicates that, depending on the conditions, the particles rotate approximately 2-3 times in the flow channel, and the trajectory length is calculated to be approximately 600-900 mm. Roughly calculated from the average flow velocity of 56mm/s, the residence time of one particle in the flow channel can be estimated to be approximately 10 to 20 seconds. Since the height of the magnet is half the height of the flow path, the time required for the application of a magnetic field is about half of that time. This can be one of the reason of lowering the separation rate of MPFs.

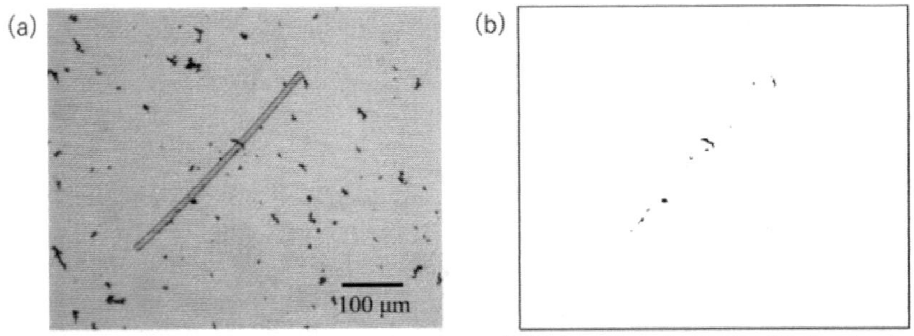

Figure 3. (a) Original optical microscope image and (b) binarized image of adsorbed magnetite particles used to estimate the number of magnetite attached per MPF.

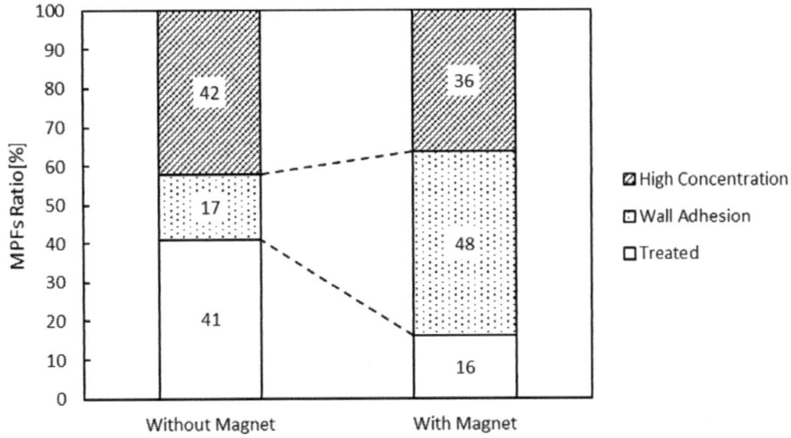

Figure 4. Number ratio of MPFs obtained from particle trajectory simulations (Treated indicates MPFs that passed through Outlet 2, Wall Adhesion indicates MPFs adhered to the wall, High Concentration indicates MPFs that passed through Outlet 1).

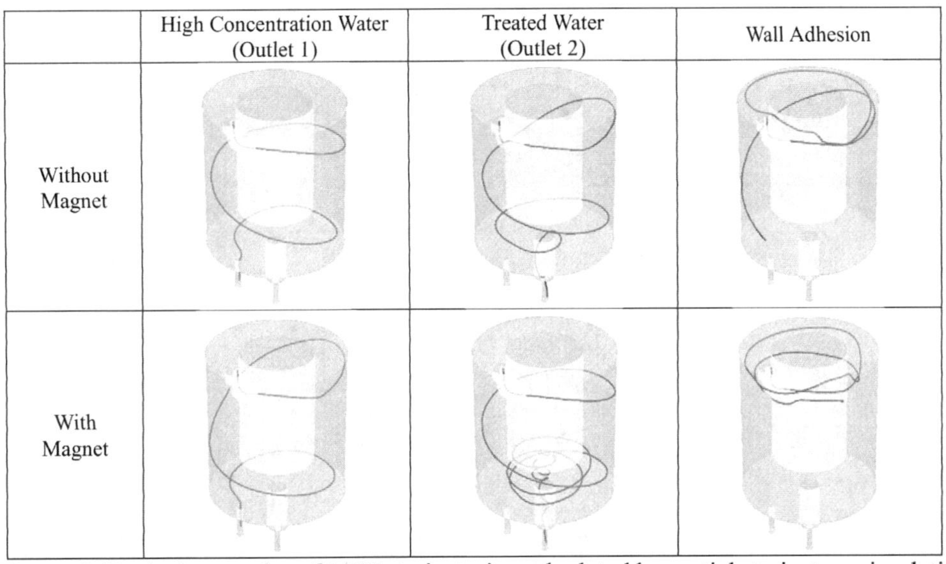

Figure 5. Typical examples of MPFs trajectories calculated by particle trajectory simulations.

4. Magnetic separation experiment

4.1. Experimental method

4.1.1. Preparation of simulated wastewater

100 mg of magnetite (primary particle size 0.5 μm, sample B, MITSUI MINING & SMELTING Co., LTD., Japan) was added to 100 mL of distilled water. Magnetite particles were dispersed by ultrasonic irradiation for 10 minutes using an ultrasonic cleaner (CPX1800-J, Yamato Scientific Co., Ltd., Japan) while agitating at 300 rpm using agitator A (SM-102, AS ONE Corporation, Japan). Next, 100 mg of polyester (10 μm in diameter and 400 μm in length, Kyoto Pile Sen-i Kogyo & Co., LTD., Japan) as model MPFs was added to 9 kg of distilled water. The magnetite suspension agitated by the previous method is added to water containing MPFs, and then distilled water was added until the total weight of distilled water reaches 10 kg. The mixture was agitated at 100 rpm for 30 minutes using agitator B (ZZ-1000, Tokyo Rikakikai Co., LTD., Japan). This was used as the simulated wastewater.

4.1.2. Magnetic separation

A magnetic separation system was constructed as shown in Figure 6, and the flow channel was filled with distilled water. The simulated wastewater was flowed from the inlet using a roller pump (RP-1000, TOKYO RIKAKIKAI Co., LTD., Japan) set at 172 rpm (inlet mean velocity 0.5 m/s, volume flow rate 9.78 mL/s). The simulated wastewater flowed while being agitated at 100 rpm using agitator B. The water that flowed out of Outlet 1 was considered as MPFs concentrated water, the water that flowed out of the Outlet 2 was considered as treated water, the water that remained in the channel after separation was considered as wall-adhesion water, and the water containing MPFs that remained in the container without flowing into the magnetic separation system was considered as untreated simulated wastewater.

(a)

(b)

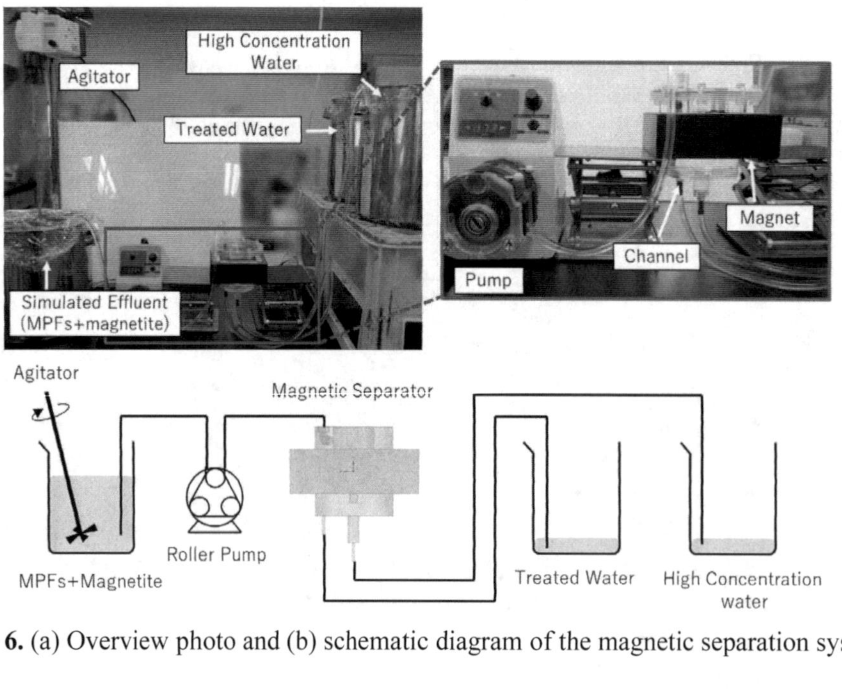

Figure 6. (a) Overview photo and (b) schematic diagram of the magnetic separation system.

4.1.3. Measurement of the concentration of MPFs in the collected water

Concentrated, treated, wall adhesion, and untreated water were each filtered by suction using a membrane filter (pore size 1.2 μm, RAWP04700, Merck Millipore Ltd.). Magnetite was dissolved by adding 5 mL of concentrated hydrochloric acid (35.0-37.0%, Sigma-Aldrich Japan, Japan) to the solid phase on the membrane, consisting of MPFs and magnetite, and allowing it to stand for 1 hour. At this moment, the MPFs in the solid phase were dispersed into the aqueous phase. Then, the MPFs were separated again from the aqueous phase using new membrane filters (pore size 1.2 μm) after dilution. The filters were dried at 70°C for 12 hours using a constant-temperature dryer (SD602, Masuda Corporation., Japan), and the weight of the filters with MPFs were separated was measured to increments of 0.1 mg. The iron ion concentration was measured on the filtrate using a Simultaneous ICP Atomic Emission Spectrometer (ICPE-9820, Shimadzu Corporation., Japan), and the MPFs weight was estimated by subtracting the weight of the membrane filter before use from the weight of the membrane filter after the second separation and drying. In this method, the mass ratios of the concentrated, treated, wall adhered, and remaining particles were calculated each for magnetite and MPFs.

4.2. Results and Discussions

Table 2 shows calculated and simulated mass ratio of fluid passing through each outlet of the concentrated and treated water by magnetic separation. The good agreement between the results of the experiment and the fluid analysis indicates that the fluid conditions of the fluid analysis are valid. The mass percentages of MPFs and magnetite obtained from the magnetic separation experiments are shown in Figures 7(a), (b), respectively. Here, the loss represents the difference between the added mass and the measured results.

The experimental results shown in Figure 7(a) show that the application of a magnetic field reduced the MPFs in the treated water by 14.2%. Considering that MPFs adhering to the wall are easily collected by temporarily increasing the inlet flow velocity, 65.0% of MPFs can be removed by the magnetic separation, as a sum of the MPFs in concentrated water and adhered to the wall.

Here, the mass ratio of magnetite and MPFs excluding loss and untreated water with and without magnet were calculated from the experimental results in Figures 7(a) and (b). Without magnet, the mass ratio of magnetite and MPFs were 47.5%, 4.9%, and 47.6% for magnetite, and 47.0%, 7.5%, and 45.5% for MPFs, respectively, in the order of concentrated water, treated water, and wall adhesion, showing a similar trend in the mass balance of magnetite and MPFs. On the other hand, with magnets, the mass ratio were 0.7%, 98.7%, and 0.6% for magnetite and 33.9%, 36.1%, and 30.0% for MPFs, in the same order above. With magnets, the mass balance of magnetite and MPFs resulted in largely different results. A magnetite particle has a smaller surface area than magnetite-attached MPFs, so the drag force acting on it is smaller, whereas the acceleration due to magnetic force is larger because of its smaller mass per particle. Therefore, it is considered that the magnetite particles are more likely to adhere to the wall when the magnetic field is applied, while the magnetite-attached MPFs are less likely to adhere to the wall. Thus, once the magnetite particle was desorbed from the MPFs, it is easily adhered to the wall.

Comparing the results of simulations and experiments, it was consistent in both results that the application of magnetic field could decrease in the amount of MPFs in the treated water and could increase in wall adhesion. This clearly indicates that impurity separation by swirling flow can be assisted by magnetic seeding and magnetic field application from outside the channel. However, there was a 15% difference in the amount of MPFs adhered to the wall between the simulations and the experiment with the magnet and a 10% difference without the magnet, with both experimental results showing considerably less adhesion to the wall. This will be due to that the magnetic field was shielded by magnetite adhering to the wall in the experiment, although the particle trajectory analysis neglects the effect of particles adhered to the wall.

This experiment showed that 65% of MPFs could be removed when a magnetic field was applied, assuming regular recovery of MPFs adhering to the wall. Membrane bioreactors (MBR) is the current method for removing microplastics, which can remove 99.9% of microplastics [10]. However, there are issues such as high energy consumption. Our system does not require a high-pressure pump, which is considered to consume less energy. The removal rate our this method is lower than MBR, but it can be further increased by using the system in multiple stages to achieve the same level of removal.

Journal of Physics: Conference Series **2545** (2023) 012032 doi:10.1088/1742-6596/2545/1/012032

Table 2. Mass ratio of fluid passing through each outlet based on fluid analysis and magnetic separation experiments. Simulated value was calculated from mass flow rate, and experimental value was calculated from mass after magnetic separation.

	Mass Ratio [%]	
	Treated (Outlet 1)	High Concentration (Outlet 2)
Simulations	50.0	50.0
Experiments	50.4	49.6

(a)

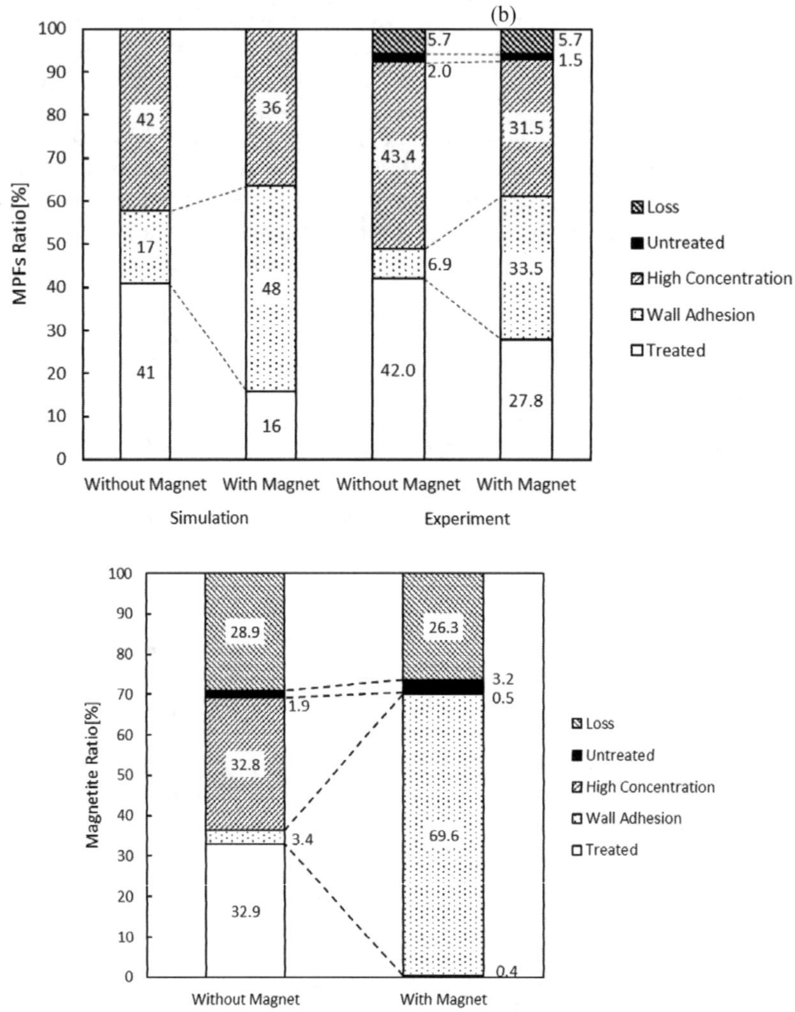

Figure 7. (a) Number ratio of MPFs obtained by calculation and mass ratio of MPFs obtained from experiments with and without magnet, and (b) mass ratio of magnetite obtained from separation experiment with and without magnet. Here, "Treated" is MPFs or magnetite that passed through Outlet 2, "Wall Adhesion" is MPFs or magnetite adhered to the wall, "High Concentration" is MPFs or magnetite that passed through Outlet 1, "Untreated" is MPFs or magnetite in the simulated wastewater remaining in the experiment, and "Loss" indicates the difference from the amount added.

The requirements for the current magnetic separation system for MPFs removal are to (i) reduce the number of MPFs in the treated water (ii) to reduce the adhesion of MPFs and magnetite to the wall, and (iii) to increase throughput. As for (i) and (ii), it is possible to improve them by optimizing and modifying the flow channel geometry and magnet arrangement, and as for (iii), it will be improved by increasing the size and the magnetic flux density of the magnet.

Permanent magnets, including Halbach magnets, have limitations on the maximum flux density and the size of the magnetic field gradient, and it is expected to be difficult to increase the size of the flow channel above a certain level. Therefore, by installing a flow channel that generates a swirling flow between opposing superconducting bulk magnets, which can apply a stronger magnetic field over a wider area, the channel radius can be further increased, which may increase the throughput largely. However, in such cases, there may be issues of high cost and the problem of too large a magnetic field or magnetic field gradient on the channel wall that promotes adhesion to the channel wall, so the magnetic field design must take this effect into account. By making improvements to these problems, we plan to design a wastewater treatment system that can be installed in textile industry plants to provide continuous treatment for wastewater containing MPFs.

5. Conclusion

In this study, magnetite was used for the magnetic seeding of MPFs and the MPFs were concentrated by a magnetic separation system using a swirling flow. Both the simulations and experimental results showed that MPFs could be concentrated by the proposed magnetic separation system, indicating that the swirling flow assisted by the magnetic field was effective in removing MPFs. The simulations and magnetic separation experiments showed similar trends, confirming the validity of the simulations. In future, we will improve the magnetic separation performance by optimizing the design of flow channel and magnet arrangement through simulations, using opposing superconducting bulk magnets to improve the throughput.

References

[1] Browne M A, Crump P, Niven S J, Teuten E, Tonkin A, Galloway T, Thompson R 2011 *Environ. Sci. Technol.* **45** 9175-9179
[2] Ziajahromi S, Neale P A, Rintoul L, Leusch F D L 2017 *Water Res.* **112** 93-99
[3] Anthony L A 2011 *Mar. Pollut. Bull.* **62** 1596-1605
[4] Murray F 2011 *Mar. Pollut. Bull.* **62** 1207-1217
[5] Mori T, Yamamoto J, Akiyama Y, Okada H, Hirota N, Matsuura H, Namba S, Sekine T, Mishima F, Nishijima S 2020 *IEEE Trans. Magn.* **56** 12
[6] Ueda R, Akiyama Y, Yuichiro M, Sato F 2022 *Prog. Supercond. Cryog.* **24** 12-18
[7] Gustavsson K, Sheikh M Z, Lopez D, Naso A, Pumir A, Mehli B 2019 *New J. Phys.* **21** 083008
[8] Hernandez E, Nowack B, Mitran D M 2017 *Sci. Technol.* **51** 7036-7046
[9] Schindelin J, Arganda-Carreras I, Frise E, Kaynig V, Longair M, Pietzsch T, Preibisch S, Rueden C, Saalfeld S, Schmid B, Tinevez J Y, White D J, Hartenstein V, Eliceiri K, Tomancak P, Cardona A. 2012 *Nat. Methods* **9** 676–682
[10] Freeman S, Booth A M, Sabbah I, Tiller R, Dierking J, Klun K, Rotter A, Ben-David E, Javidpour J, Angel D L, 2020 *J. Environ. Manage.* **266** 110642

All-optical non-contact level sensor for liquid hydrogen

M Futamura[1], T Oikawa[2], S Miura[2] and H Okamoto[2]

[1]Department of Mechanical Engineering, Akita Prefectural University, Yurihonjo 015-0055, Japan
[2]Department of Intelligent Mechatronics, Akita Prefectural University, Yurihonjo 015-0055, Japan

E-mail: futamura@akita-pu.ac.jp, okamoto@akita-pu.ac.jp

Abstract. We propose a new type of liquid hydrogen level sensor and evaluate a prototype system with liquid nitrogen. The level sensor acquires images of the liquid surface, which is illuminated by point sources of light, and estimates the liquid level by image processing. This scheme allows us to avoid immersion of the sensor body into the liquid and also eliminate the necessity for any electrical wire connected to the sensor body. The accuracy of the prototype sensor is estimated to be ≈ 4.8 mm in our experiment using a 10 liter liquid nitrogen vessel.

1. Introduction

The transition to renewable energy is underway. This will require us to deal with ever larger intermittent energy sources such as solar and wind [1]. To balance the supply and demand, we need either to expand the ability of large-distance energy transmission over the power grid [2], or to develop a large-scale, cost-efficient energy storage [3], or both. Liquid hydrogen could play significant roles in these cases, as coolant for superconducting cables [4] or as chemical energy storage. Although liquid hydrogen as electrical energy storage has relatively low round-trip efficiency, it may be generated by excess renewable energy and it may find its own use in industrial processes, in particular steel production, to decarbonize such processes. Hence accurate, convenient, and safe liquid level sensors could be a significant support technology in the era of renewable energy.

A variety of liquid hydrogen level sensors have been developed. They are based on a variety of physical principles and have been commercialized in many cases. Existing sensors include the differential pressure level gauge [5], capacitive liquid level sensor [6], and the MgB_2-based superconducting liquid level sensor [7-9]. In all these cases, however, physical components, that may include electrical wires, are immersed in the liquid hydrogen. This entails unwanted heat leakage along the sensor body and also gives rise to safety concerns. For example, the electrical wirings may be associated with a risk, or a perceived risk by the user, of potentially explosive failure when, e.g., lightning strikes the storage tank.

Here we propose a liquid level sensor that optically monitors the liquid surface from a distance. The sensor is basically a camera: It monitors the liquid level from a physically remote position, i.e. the top of the tank. Our intension is to take advantage of the increasingly powerful image processing technology available today to enable performance that has not been possible. Since nothing is immersed in the cryogen, the sensor is expected to be safe, has minimal heat leakage and is easy to maintain. On the other hand, the attainable precision with such remote sensing is expected to be an issue, which we study in the present work.

2. Operation principle and construction

Figure 1 shows the principle of operation. We illuminate the liquid surface with three point-sources of light and observe the reflected image with a camera. We then infer the position of the liquid surface from the size of the reflected image of the triangle. The image appears large when the distance L between the sensor and the liquid surface is small and *vice versa*. The use of optical fibers and a fiber bundle allows us to totally eliminate the use of electrical signal transmission. Three optical fibers are used for illumination, while the image of the reflected light is transferred through an optical fiber bundle. As mentioned above, our design avoids loss of cryogen due to heat conduction along the sensor body.

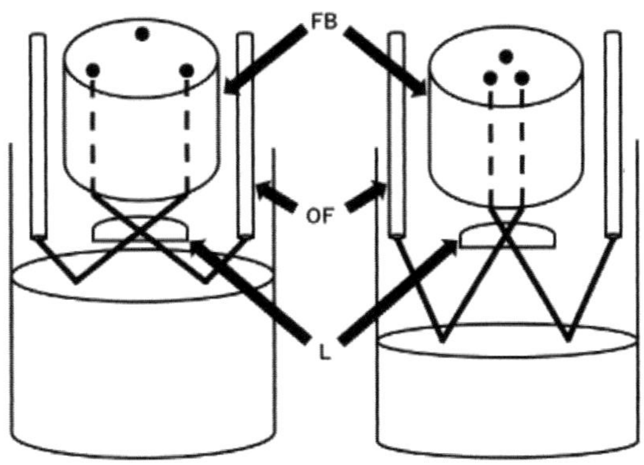

Figure 1. The principle of operation. The light emanating from the optical fibers (OF) is reflected by the cryogen surface. A lens (L) focuses the reflected image onto the bottom end of an optical fiber bundle (FB). As shown here, different cryogen liquid levels result in different sizes of the image of the triangle transmitted through the FB.

Figure 2 (a) shows the actual prototype level sensor. Three plastic fibers (Mitsubishi Chemical Co., ESKA CK60, ϕ1.5 mm) for illumination go through a large Wilson seal (Swagelok Ultra-Torr SS-4-UT-A-8) fixed on the main aluminum body, while the fiber bundle (Edmund Optics, #38-302, ϕ3.2 mm, comprising $N = 5 \times 10^4$ fibers) is fixed on a small Wilson seal (Swagelok Ultra-Torr SS-2-UT-A-6). A lens (ϕ10 mm, the focal length $f = 20$ mm, N-BK7, Edmund Optics, #63-473) is mounted in the center of an aluminum plate with the dimeter 40 mm. The tips of three plastic fibers are symmetrically placed on a 30 mm-diameter circle on the same aluminum plate, forming an equilateral triangle. Outside the cryogen vessel, we coupled each of the plastic fibers to a laser pointer (Sakura Color Products, RX-7G, the wavelength $\lambda = 532$ nm, 1 mW) in the experiment described in the next section.

Figure 2 (b) shows an optical coupler under development. The coupler connects optical sources and the plastic fibers for illumination. Each lens (ϕ10 mm, $f = 10$ mm, N-BK7, Edmund Optics, #63-471) focusses light from an LED onto the end of an optical fiber. Three LEDs, controlled by a Python program through a standard interface (NI, USB-6002), can independently be turned on/off. This capability is intended for avoiding image processing difficulties arising from overlapped images of light spots from different apexes of the triangle when the liquid surface is not still.

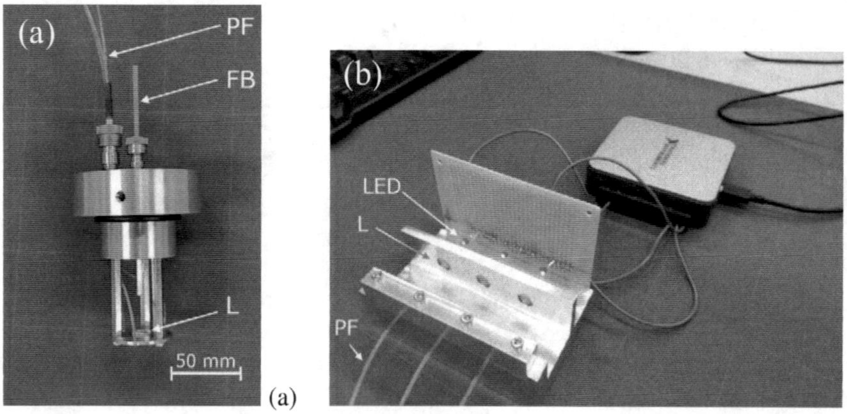

(a) (b)

Figure 2. (a) The sensor body. FB: Fiber bundle. L: Lens. PF: Plastic fibers. (b) An optical coupler under development.

The attainable measurement resolution is estimated as follows. Let the side length of the illuminating triangle, i.e., the separation between the tips of the illuminating optical fibers, be d. In the present case $d = 26$ mm. Also, let the distance between the sensor and the liquid surface be L. The angle to be measured is

$$\theta \approx d/L \tag{1}$$

and hence the change of angle $\delta\theta$ resulting from the change of the liquid level δL is

$$\delta\theta \approx d\delta L/L^2 . \tag{2}$$

In our prototype setup described below, $\theta \approx 0.1$ rad and $\delta L/L \approx 10^{-2}$ if δL is roughly $\lesssim 1$ cm. This means that the incident angle of the light onto the lens in the sensor needs to be measured with the precision $\delta\theta \approx 10^{-3}$ rad. On the other hand, the corresponding lateral shift δx of the focal spot on the end of the fiber bundle is $\delta x = f\delta\theta \approx 20$ μm, where $f = 20$ mm is the focal length of the lens. This amount of shift $\delta x \approx 20$ μm is about the diameter of each optical fiber consisting the fiber bundle. Hence, one may expect roughly a centimeter resolution if we can measure the transferred image through the optical fiber bundle at the resolution of individual optical fiber, provided that the liquid surface is still. With image processing, however, one could handle nonstationary liquid surface by e.g. averaging of multiple images. Conceivably, one could also determine the position of the light spot with the resolution smaller than the diameter of the individual fiber, by properly analyzing the light intensity emanating from individual light fiber in the fiber bundle.

3. Experimental results and discussion

We tested the prototype sensor with liquid nitrogen. Figure 3 shows an overall view of the experimental setup. The sensor body shown in Fig. 2 (a) is attached to the top of the liquid nitrogen cryogenic vessel (Cryo One, DC-10, 10 liter capacity). The camera (DITECT, HAS-U2, max. 7500 fps) on a tripod focuses on the outer end of the fiber bundle, where the image of triangular light spots shows up. While all images were taken at the rate of 60 fps in the present study, the high-speed camera could enable us to measure the level of boiling and/or sloshing liquid using advanced image processing in the future. All measurements performed in the present work has been carried out with a still liquid surface, however.

Journal of Physics: Conference Series **2545** (2023) 012033 doi:10.1088/1742-6596/2545/1/012033

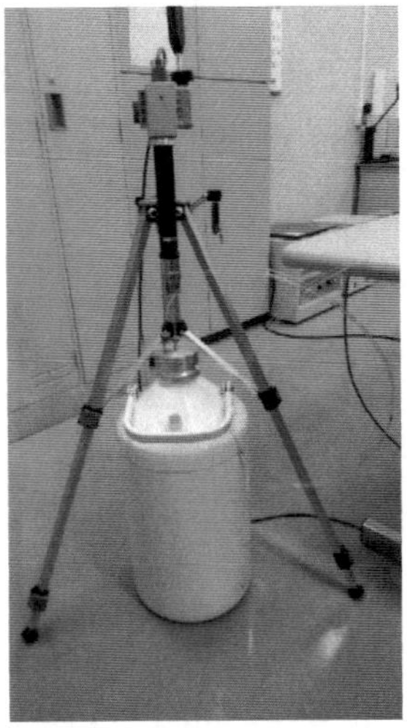

Figure 3. The measurement setup. The sensor head is mounted on top of the liquid nitrogen vessel. A camera on a tripod acquires images appearing at the end of the fiber bundle.

Figure 4 shows representative optical patterns appearing on the end of fiber bundle, as imaged by the camera. Let the physical distance between the light spots appearing on the end of fiber bundle be s. In order to evaluate our level sensor, we independently determined the position of the surface of liquid nitrogen by listening to acoustic noise generated in a tubular dipstick. We estimate the accuracy of this measurement to be few millimeters.

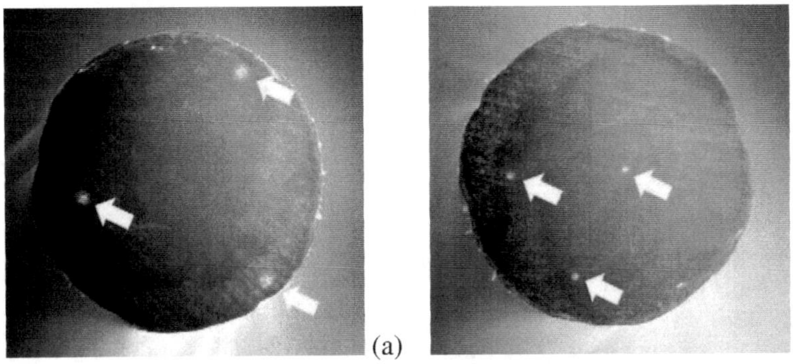

(a) (b)

Figure 4. Images appearing at the end of fiber bundle for two liquid levels. The distance L between the sensor and the liquid surface is $L = 120$ mm for panel (a) and $L = 210$ mm for panel (b).

Figure 5 shows the dependence of s^{-1} on the liquid level L. The data points form essentially a straight line passing through the origin, as they should. The dotted line represents the line of best fit obtained by the least-squares method, which is represented as

$$s^{-1} = 3.8 \times 10^{-3}\mathrm{mm}^{-2} \cdot L + 4.0 \times 10^{-3}\mathrm{mm}^{-1}. \quad (3)$$

The root mean square of the difference between the value of L, inferred from the optically determined s^{-1} using Eq. (3), and the independently determined value of L, using the hollow dipstick, is 4.8 mm. This value should represent the overall accuracy of the proposed method.

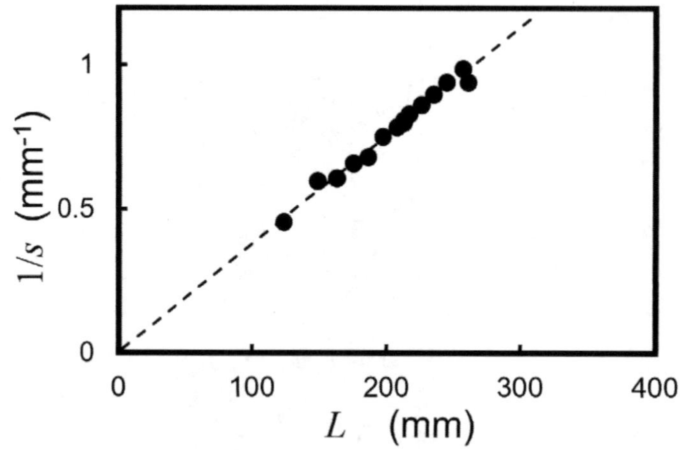

Figure 5. Dependence of s^{-1}, which is determined by the proposed optical method, on the liquid level L measured using a hollow dipstick.

Application of our method to liquid hydrogen may give rise to issues that are insignificant in the case of liquid nitrogen. One potential problem, in view of the small latent heat *per volume* of liquid hydrogen, is liquid loss due to the energy of light illumination. The power of illumination should typically be on the order of, or less than, mW and hence certainly is not large. Moreover, one could operate the light source intermittently to further reduce the heat influx. Nonetheless, averaging of multiple images would be necessary when the liquid surface is not still and the aforementioned intermittent operation would give rise to a long measurement time. Another related issue is the small refractive index of liquid hydrogen $n_{H2} = 1.11$ compared to that of liquid nitrogen $n_{N2} = 1.20$ [10-12]. The standard formula relating the reflectivity to the index of refraction

$$R = \left(\frac{n-1}{n+1}\right)^2 \tag{4}$$

for normal incidence yields $R_{H2} = 0.3\%$ and $R_{N2} = 0.8\%$ for the respective liquids. Hence we expect a larger light dose for a single measurement for liquid hydrogen, when compared to the case of liquid nitrogen.

4. Conclusions and outlook

We proposed an all-optical liquid level sensor that takes advantage of recent advances in imaging and associated data processing technology. Although the main application we have in our mind is measurement of the level of liquid hydrogen, the proposed method should be applicable to any liquid. We performed an experiment using a prototype sensor to measure the level of liquid nitrogen. The experiment revealed that the expected measurement error is 4.8 mm when the distance between the sensor and the liquid surface is in the range of 124 mm~261 mm, which is consistent with our theoretical expectation.

An immediate next task is to automate the process of obtaining the value of L from the image data. Looking ahead, it remains to be seen as to whether sufficient accuracy could be attained when the distance between the sensor and the liquid surface is much larger than what is reported here. Another challenge is accurate measurement of the liquid level when the liquid surface is not still. We expect such a situation when, for example, the tank is being filled with cryogen. Advanced imaging processing, possibly using artificial intelligence, could solve these problems.

Acknowledgments
We thank Professor Teruo Bitoh for his assistance in developing the experimental setup and Professor Hiroto Kawashima for providing us with liquid nitrogen. Professor Shin-ya Matsushita and Mr. Hiroki Dakeishi kindly offered their insights into image data processing. This research was supported in part by the grant donated by Eurus Energy Akita-Port Corporation to Akita Prefectural University.

References

[1] Brouwer A S, Broek M van den, Seebregts A and Faaij A 2014 Impacts of large-scale Intermittent Renewable Energy Sources on electricity systems, and how these can be modeled *Renew. Sust. Energ. Rev.* **33**, 443-466
[2] Ardelean M and Minnebo P 2015 *HVDC Submarine Power Cables in the World*, Joint Research Centre, European Union, Report EUR 27527 EN
[3] Breeze P 2018 *Hydrogen Energy Storage*, Chapter 8 - Power System Energy Storage Technologies, Academic Press, 69-77
[4] Grant P M, Starr C and Overbye T J 2006 *A Power Grid for the Hydrogen Economy*, Scientific American, July 2006 issue
[5] Olsen W A 1963 A Survey of Mass and Level Gauging Techniques for Liquid Hydrogen *Advances in Cryogenic Engineering*, **8**, 342-359.
[6] Matsumoto K, Sobue M, Asamoto K, Nishimura Y, Abe S and Numazawa T 2011 Capacitive level sensor for liquid hydrogen, *Cryogenics* **51**, 114-115
[7] Haberstroh Ch and Zick G 2006 A Superconductive MgB_2 Level Sensor for Liquid Hydrogen, *AIP Conference Proceedings* **823**, 679
[8] Kajikawa K et al 2008 Fundamental investigation of a superconducting level sensor for liquid hydrogen with MgB_2 wire, *Journal of Physics: Conference Series* **97**, 012140
[9] Maekawa K et al, 2017 Dynamic Level-Detecting Characteristics of External-Heating-Type MgB_2 Liquid Hydrogen Level Sensors Under Liquid Level Oscillation and Its Application to Sloshing Measurement, *IEEE Trans.Appl. Supercond.* **27**, 9000304
[10] Johns H E and Wilhelm J O 1937 The refractive indices of liquid oxygen, nitrogen, and hydrogen, *Canadian Journal of Research* **15a**, 101-108
[11] Corruccini R J 1965 Refractive index and dispersion of liquid hydrogen, *NBS Technical Note* No. 323
[12] Jacobsen R T, Steward R B, McCarty R D and Hanley H J M 1973 Thermophysical properties of nitrogen from the fusion line to 3500 R (1944 K) for pressures to 150,000 PSIA (10342×15^5 N/m^2), *NBS Technical Note* No. 648

AUTHOR INDEX